FPST-Vol. 12

Proceedings of the ASME

FLUID POWER SYSTEMS AND TECHNOLOGY DIVISION

– 2005 –

FLUID POWER CONTROL AND DIAGNOSTICS

FLUID POWER SYSTEMS AND INTEGRATION

GENERAL

presented at

2005 ASME INTERNATIONAL MECHANICAL ENGINEERING CONGRESS AND EXPOSITION
NOVEMBER 5–11, 2005
ORLANDO, FLORIDA USA

sponsored by

THE FLUID POWER SYSTEMS AND TECHNOLOGY DIVISION, ASME

A S M E

Three Park Avenue ◆ New York, N.Y. 10016

ISBN No. 0-7918-4220-7

• • WELCOME FROM THE CHAIRS • •

Welcome to the Celebration! The Congress in ASME's 125th year. And what a great place to celebrate: Orlando, sun, fun, and all the attractions. Some of the best attractions will be right here at the Congress. The Technical Program has over 600 paper and panel sessions, involving all the technical communities within ASME, with focused series of sessions on such timely topics as Design and Analysis, Energy, Advanced Materials, Ethics and Licensure, Health and Safety, MEMS and Nano Systems, Manufacturing, Sustainable Engineering and Transportation. There are also special educational programs on Project Management and Water Management. Visit the Exhibit where you can inspect state-of-the-art instrumentation, hardware, software and the latest engineering publications.

There are also some special features and events taking place at the 2005 Congress:

- ***Colonel Mike Mullane**, NASA Shuttle Astronaut and Novelist will be the **Keynote Speaker** at 5:30 pm on Monday, November 7, followed by the **Opening Reception** in the Exhibit Hall.*
- *The **Honors Assembly**, ASME's annual recognition of the stars of the engineering profession, will be held Tuesday, November 8 at 5:30 p.m. Meet and congratulate the honorees at the reception following the assembly.*
- *See the future of our profession by attending and supporting engineering students at the **Student Design Contest** and the **Student Technical Paper Competitions**. Students are invited to attend a breakfast, a luncheon and a reception, and recent graduates are encouraged to attend the **Young Engineers Forum**.*

In addition, there are Honorary Lectures, receptions, lunches and dinners sponsored by ASME divisions, as well as divisional and society business meetings where you can learn of and influence the future course of the Society. Please consult your Congress Final Program so you can maximize your participation in the multitude of opportunities for exchange of technical and professional information provided by this year's event.

Michael J. Askew
Summa Health System
Congress Committee Chair

Said Jahanmir
Mohawk Innovative Technology, Inc
2005 General Program Chair

Ming Leu
University of Missouri–Rolla
2005 Technical Program Chair

Jack Whitehead
Bechtel, Inc.
2005 Industry Program Chair

• Program Representatives and Track Managers •

Advanced Energy Systems
Program Representative: **Robert Boehm**
University of Nevada Las Vegas, Las Vegas, NV USA
boehm@me.unlv.edu

Aerospace
Program Representative: **Ajit Roy**
Air Force Research Laboratory, Dayton, OH USA
ajit.roy@wpafb.af.mil

Applied Mechanics
Program Representative: **K. Ravi-Chandar**
University of Texas at Austin, Austin, TX USA
kravi@mail.utexas.edu

Bioengineering
Program Representative: **Michael C. Murphy**
Louisiana State University, Baton Rouge, LA USA
murphy@me.lsu.edu

Board on Diversity and Outreach
Program Representative: **Mary Lynn Realff**
Georgia Institute of Technology, Atlanta, GA USA
marylynn.realff@ptfe.gatech.edu

Board on Research and Technology Development
Program Representative: **I.S. Jawahir**
University of Kentucky, Lexington, KY USA
jawahir@engr.uky.edu

Computers and Information in Engineering
Program Representative: **Chris Paredis**
Georgia Institute of Technology, Atlanta, GA USA
chris.paredis@me.gatech.edu

Design and Analysis
Track Manager: **Jack Whitehead**
Bechtel Plant Machinery Inc., Schenectady, NY USA
jrw5@att.net

Design Engineering
Program Representative: **Judy Vance**
Iowa State University, Ames, IA USA
jmvance@iastate.edu

Dynamic Systems and Control
Program Representative: **Ronald Perez**
University of Wisconsin – Milwaukee, Milwaukee, WI USA
perez@uwm.edu

Electronic and Photonics Packaging
Program Representative: **Sandeep Tonapi**
General Electric Global Research, Niskayuna, NY USA
tonapi@crd.ge.com
Program Co-Representative: **Jianmin Qu**
Georgia Tech, Atlanta, GA USA
jianmin.qu@me.gatech.edu

Energy
Track Manager: **Lawrence A. Kielasa**
DTE Energy, River Rouge, MI USA
KIELASAL@DTEENERGY.COM

Engineering Business Management
Track Manager: **Ted Aanstoos**
The University of Texas at Austin, Austin, TX USA
t.aanstoos@mail.utexas.edu
Track Co-Manager: **Steve Long**
GDS SNC Lavalin, Houston, TX USA
steve.long@gdseng.com

Environmental Engineering
Program Representative: **Mike Hightower**
Sandia National Laboratories, Albuquerque, NM USA
mmhight@sandia.gov

Ethics and Licensure
Track Manager: **Donald Westbrook**
Honeywell Aerospace, Torrance, CA USA
westbrookd2@asme.org

Fluid Power Systems and Technology
Program Representative: **Eric Barth**
Vanderbilt University, Nashville, TN USA
eric.j.barth@vanderbilt.edu

Fluids Engineering
Program Representative: **Urmila Ghia**
University of Cincinnati, Cincinnati OH
urmila.ghia@uc.edu

Health and Safety
Track Manager: **Gerald E. Miller**
Virginia Commonwealth University, Richmond, VA USA
gemiller@vcu.edu

Heat Transfer
Program Representative: **Rod Douglas**
Los Alamos National Laboratory, Los Alamos, NM USA
rwd@lanl.gov

INTEGRATION AND PACKAGING OF MICRO/NANO/ELECTRONIC SYSTEMS TRACK
Track Manager: **Gregory Kowalski**
Northeastern University, Boston, MA USA
gkowal@coe.neu.edu
Track Co-Manager: **Carolyn Fries**
Intelligent Micro Patterning, St. Petersburg, FL USA
carolynf@intelligentmp.com
Track Co-Manager: **C. Channy Wong**
Sandia National Laboratories, Albuquerque, NM USA
ccwong@sandia.gov
Track Co-Manager: **Ajay P. Malshe**
University of Arkansas, Fayetteville, AR USA
apm2@engr.uark.edu

MANUFACTURING (RESEARCH ORIENTED)
Program Representative: **Shivakumar Raman**
University of Oklahoma, Norman USA
raman@ou.edu

MANUFACTURING TRACK (INDUSTRY ORIENTED)
Track Manager: **Judith Todd**
The Pennsylvania State University, State College, PA USA
jtodd@psu.edu
Track Co-Manager: **Shivakumar Raman**
University of Oklahoma, Norman USA
raman@ou.edu

MATERIALS
Program Representative: **Mahyar S. Dadkhah**
Program Co-Representative: **Devdas Pai**
North Carolina A&T State University, Greensboro, NC USA
pai@ncat.edu

MATERIALS HANDLING ENGINEERING
Program Representative: **Bilal Maiteh**
Delphi Corporation, Troy, MI USA
bilal.y.maiteh@delphi.com

MECHANICAL ENGINEERING EDUCATION
Program Representative: **Ahmad Fakheri**
Bradley University, Peoria, IL USA
ahmad@bradley.edu

MECHANICAL ENGINEERING TECHNOLOGY DEPARTMENT HEADS
Program Representative: **Raju Dandu**
Kansas State University at Salina, Salina, KS USA
rdandu@ksu.edu
Program Co-Representative: **Ann M. Anderson**
Union College, Schenectady, NY USA
andersoa@union.edu

MICROELECTROMECHANICAL SYSTEMS
Program Representative: **Aaron Knobloch**
General Electric - Global Research, Niskayuna, NY USA
knobloch@research.ge.com
Program Co-Representative: **Taher Saif**
University of Illinois at Urbana-Champaign, Urbana, IL USA
saif@uiuc.edu

NOISE CONTROL AND ACOUSTICS
Program Representative: **Lonny L. Thompson**
Clemson University, Seneca, SC USA
lonny.thompson@ces.clemson.edu

NON-DESTRUCTIVE EVALUATION
Program Representative: **Yuris Dzenis**
University of Nebraska-Lincoln, Lincoln, NE USA
ydzenis@unl.edu
Program Co-Representative: **Michael Peterson**, Ph.D.

NUCLEAR ENGINEERING
Program Representative: **Yassin A. Hassan**
Texas A&M University, College Station, TX USA
y-hassan@tamu.edu

PLANT ENGINEERING AND MAINTENANCE
Group Representative: **Meherwan P. Boyce**
The Boyce Consultancy Group, LLC, Houston, TX USA
mpboyce@boycepower.com

POWER
Program Representative: **K. M. Bryden**
Iowa State University, Ames, IA USA
kmbryden@iastate.edu

PRESSURE VESSELS AND PIPING
Program Representative: **Luc H. Geraets**
Tractebel Electricity & Gas International, Brussels, Belgium
luc.geraets@tractebel.com
Program Co-Representative: **Young W. Kwon**
Naval Postgraduate School, Monterey, CA USA
ywkwon@nps.edu

PROCESS INDUSTRIES
Program Representative: **Samuel Sami**
Univeresity of Moncton, Moncton, NB, Canada
samis@umoncton.ca

RAIL TRANSPORTATION
Program Representative: **Roger Sims**
Sims Professional Engineers, Highland, IN USA
rds@simspe.com
Program Co-Representative: **Cameron Lonsdale**
Standard Steel, LLC, Burnham, PA USA
clonsdale@standardsteel.com

SAFETY ENGINEERING AND RISK ANALYSIS
Program Representative: **Ali Mosleh**
University of Maryland, College Park, MD USA
mosleh@eng.umd.edu

SOLAR ENERGY
Program Representative: **Jeffrey Morehouse**
University of South Carolina, Columbia, SC USA
more@engr.sc.edu

SUSTAINABLE ENGINEERING
Track Manager: **Timo Marquez Arreaza**
Consultant to ETHsustainability, Maracaibo, Zulia, Venezuela
marquezt@asme.org

TECHNOLOGY AND SOCIETY
Program Representative: **Richard Ciocci**
Penn State University, Middletown, PA USA
cioccir2@asme.org

TEXTILE ENGINEERING
Program Representative: **Stephen Bechtel**
Ohio State University, Columbus, OH USA
bechtel.3@osu.edu

TRANSPORTATION
Track Manager: **Richard Marboe**
Penn State University, State College, PA USA
rcm4@psu.edu

TRIBOLOGY
Program Representative: **Steven Schmid**
University of Notre Dame, Notre Dame, IN USA
schmid.2@nd.edu
Program Co-Representative: **Andreas A. Polycarpou**
University of Illinois at Urbana-Champaign, Urbana, IL USA
polycarp@uiuc.edu

• Engineering Research Program Committee* •

Chair
Michael J. Askew
askewm@summa-health.org

General Chair
Said Jahanmir
MiTi Heart Corporation
sjahanmir@mitiheart.com

Technical Program Chair
Ming C. Leu
University of Missouri-Rolla
mleu@umr.edu

Track Programming Advisor
Jack Whitehead
Bechtel Plant Machinery Inc.
jrw5@att.net

*formerly the Technical Program Planning Committee or International Congress Committee

CONTENTS

FLUID POWER CONTROL AND DIAGNOSTICS

IMECE2005-79498 1
Adaptive Fuzzy Gain Scheduling in Position Control of Servo Hydraulic System With Flexible Load
Hassan Yousefi and Heikki Handroos

IMECE2005-79761 9
Experimental Simulation of Piston Leakage in an Axial Piston Pump
Zeliang Li, Richard Burton, and Peter Nikiforuk

IMECE2005-80126 17
Design of an Engine Load Simulator
T. Radpukdee and P. Jirawattana

IMECE2005-80138 27
Development of Hydraulic-Electric Power Unit for Mobile Robots
Justin W. Raade, Kurt R. Amundson, and H. Kazerooni

IMECE2005-80305 35
Sliding Mode Controller and Filter Applied to a Model of an Electrohydraulic Actuator System
Shu Wang, Richard Burton, and Saeid Habibi

IMECE2005-80562 45
An Application of the Self-Excited Oscillation Method to a Hydraulic Motor Angular Velocity System
Takayoshi Ichiyanagi and Takao Nishiumi

IMECE2005-80609 51
Stability and Performance Analysis of a Metering Poppet Valve
Roger Fales

IMECE2005-80980 57
Control of Electro-Hydraulic Poppet Valves via Online Learning and Estimation
Patrick Opdenbosch and Nader Sadegh

IMECE2005-81376 63
Software Enabled Variable Displacement Pumps
Perry Y. Li, Cassie Y. Li, and Thomas R. Chase

IMECE2005-81711 73
Comparison of Hydraulic Pump Faults Diagnosis Methods: Wavelet vs. Spectral Analyses
Yingjie Gao, Qin Zhang, and Xiangdong Kong

IMECE2005-82335 79
Fluid Effects in Small Gaps: A Challenge for Sensor Application
C. Stammen

FLUID POWER SYSTEMS AND INTEGRATION

IMECE2005-79019 .... **85**
Switchmode Hydraulic Power Supply Theory
Jianwei Cao, Linyi Gu, Feng Wang, and Minxiu Qiu

IMECE2005-79329 .... **93**
Model Simplifications for Nonlinear Hydraulic Circuits
M. Alirand, N. Orand, and M. Lebrun

IMECE2005-81687 .... **99**
Selection of Operating Modes of a Multi-Functional Hydraulic Device
Amir Shenouda and Wayne J. Book

IMECE2005-82124 .... **111**
Electric Motors Coupled to Hydraulic Motors as Actuators for Hydraulic Hardware-in-the-Loop Simulation
Scott Driscoll, James D. Huggins, and Wayne J. Book

IMECE2005-82237 .... **121**
Micro-Precision Hydrostatic Actuation System
S. R. Habibi, R. Burton, and E. Sampson

IMECE2005-82350 .... **131**
Effect of Parameter Uncertainty on Reliability of Hydraulic Transmission System Simulation
A. Ellman, H. Kauranne, J. Kajaste, and M. Pietola

GENERAL

IMECE2005-79154 .... **139**
Experimental and Numerical Investigation of Centrifugal Pump Performance When Handling Viscous Fluids
M. H. Shojaee Fard, M. B. Ehghaghi, and F. A. Boyaghchi

IMECE2005-79627 .... **147**
Torque Ripple Attenuation of an Axial Piston Pump by Continuous Swash Plate Adjustment
Viral S. Mehta and Noah D. Manring

IMECE2005-80284 .... **157**
On Self-Sensing Actuators for Electrohydraulic Valves: Comparisons Between Boxcar Window Observer and Kalman Filter
Qinghui Yuan and Perry Y. Li

IMECE2005-80779 .... **165**
Model and Control for Hydraulic Excavator's Arm
Daqing Zhang, Qinghua He, Peng Hao, and Xinhai Zhang

IMECE2005-80842 .... **171**
Real-Time Simulated Pressure Relief Valves in Physical Hydraulic Circuits
Tero Eskola, Heikki Handroos, and Takao Nishiumi

IMECE2005-81743 .... **175**
Experimental Operation and Characterization of a Free Piston Compressor
Jose Riofrio and Eric J. Barth

IMECE2005-82002 .... **183**
Integrated Pressure-Flow-Temperature Sensor for Hydraulic Systems
Charlie Groepper, Tianhong Cui, Perry Li, and Kim Stelson

Author Index **193**

Proceedings of IMECE2005
2005 ASME International Mechanical Engineering Congress and Exposition
November 5-11, 2005, Orlando, Florida USA

IMECE2005-79498

ADAPTIVE FUZZY GAIN SCHEDULING IN POSITION CONTROL OF SERVO HYDRAULIC SYSTEM WITH FLEXIBLE LOAD

Hassan Yousefi
Institute of Mechatronics and Virtual Engineering
Mechanical Engineering Department
Lappeenranta University of Technology 53851,
Lappeenranta, Finland
E-mail: yousefi@lut.fi

Heikki Handroos
Institute of Mechatronics and Virtual Engineering
Mechanical Engineering Department
Lappeenranta University of Technology 53851,
Lappeenranta, Finland
E-mail: Heikki.Handroos@lut.fi

ABSTRACT

The control of hydraulic servo-systems has been the focus of intense research over the past decades. Hydraulic position servos with an asymmetrical cylinder are commonly used in industry. These kinds of systems are nonlinear in nature and generally difficult to control. Because of changing system parameters using the same gains will cause overshoot or even loss of system stability. The highly non-linear behavior of these devices makes them idea subjects for applying different types of sophisticated controllers. This paper is concerned with a second order model reference adaptive position control of a flexible load servo-hydraulic system using a fuzzy gain scheduling. In present study, to compensate the lack of damping in hydraulic system, an acceleration feedback was used. The proposed controller was verified with a common used p-controller. The results suggest that using fuzzy controller improves all performance evaluation criteria such as stability, fast response, and accurate reference model tracking in hydraulic systems.

INTRODUCTION

Electro-hydraulic servomechanisms are known for their fast dynamic response, high power-inertia ratio and control accuracy.

Many types of sophisticated controllers (adaptive, optimal, etc.) have been applied to hydraulic servo- mechanisms with varying degree of success [1]. However, these controllers have limitations. In addition, the computational complexity of implementing the algorithms has restricted their applications.

Fuzzy logic has found control applications in all facets of engineering. In the fluid power area, fuzzy logic have been used for control, identification and modeling of the system [2, 3]. The popularity of fuzzy logic can be attributed, in part, to their ability to deal with non-linear systems. Also easy tuning of such systems makes them a good candidate to control of hydraulic systems.

A conventional PID controller is usually applied in the servo-hydraulic cylinder and motor, but it is difficult to have a dynamic response as the desired specification. One must try many times to get a suitable controller from the experiment. A comparatively convenient method is to have a reference response model, which can be the tracking object of the control system. Following the model reference adaptive control theory [3, 4], an adaptive reference model is used in this study and implemented in the microcomputer to control a hydraulic variable servo-valve.

Fig. 1: Setup of the servo hydraulic system.

Gain scheduling based on the measurements of the operation conditions of the process is often a good way to compensate for variations in the process parameters or the known nonlinearities of the process.

The idea of present study motivated by the work of Liu et al [5] for which the idea of using neural network as the acceleration feedforward to compensate the lack of damping in the hydraulic system was good, however their results show that the performance of P-Controller is not satisfactory. In this study to compensate the lack of damping ratio a feedforward acceleration and fuzzy gain-scheduling controller was proposed.

The main aim of the following study is to apply the fuzzy controller as *gain scheduling* to enhance the performance of controller.

NOMENCLATURE

A state space matrix
A' state space matrix
A_1 piston area in chamber one
A_2 piston area in chamber two
B state space matrix
B' state space matrix
B_i biases (i=1,2,3)
b_0 viscous friction coefficients of cylinder
b_1 viscous friction coefficients of flexible structure
C state space matrix
C' state space matrix
c_v flow coefficient
D state space matrix
D' state space matrix
k spring constant
k_p proportional gain
L stroke
m_1 mass of rigid body
m_2 mass of flexible load
p_1 pressure in chamber one
p_2 pressure in chamber two
p_s pressure supply
p_T tank pressure
Q_1 flow rate in chamber one
Q_2 flow rate in chamber two
Q_n nominal flow
u control signal
U_a feedforward compensator
U_f Fuzzy controller effort
V_1 compressed volume of chamber one
V_2 compressed volume of chamber two
x_l load position
x_p piston position
x_r reference position
X_d desire position
Δp_n nominal pressure difference
β_e effective bulk modulus
v_0 volume of pipe between valve and cylinder port
ω_n natural frequency

SERVO HYDRAULIC SYSTEM

The hydraulic system with flexible load shown in figure 1 is comprised of a servo-valve, a hydraulic cylinder and two masses that are connected by a parallel combination of spring and damper. The schematic diagram of the system is illustrated in figure 2.

The nonlinearity of the constitutive equations as well as the sensitivity of the system's parameters to the sign of the voltage fed to the valve make the control of system too complicated [6].

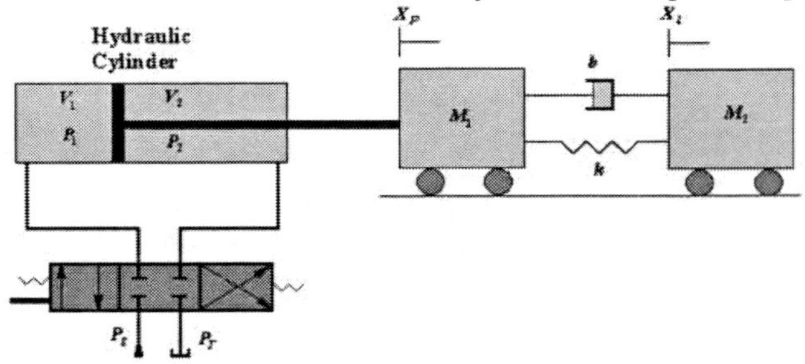

Fig. 2: Schematic diagram of system.

SYSTEM MODEL

Applying Newton's second law for each mass without consideration of coulomb friction results,

$$m_1\ddot{X}_p = -b_1\dot{X}_p - b_2(\dot{X}_p - \dot{X}_l) - k(X_p - X_l) + p_1A_1 - p_2A_2 \quad (1)$$

$$m_2\ddot{X}_l = -b_2(\dot{X}_L - \dot{X}_p) - k(X_l - X_p) \quad (2)$$

Continuity equations for the output ports of the servo valve results,

$$\begin{cases} \dot{p}_1 = \dfrac{\beta_e}{V_1}(Q_1 - A_1\dot{X}_p) \\ \dot{p}_2 = \dfrac{\beta_e}{V_2}(-Q_2 + A_2\dot{X}_p) \end{cases} \quad (3)$$

Introducing two new parameters, C_1 and C_2, defined as,

$$C_1 = \frac{V_1}{\beta_e}\ ,\ C2 = \frac{V_2}{\beta_e}$$

Equation 3 can be written in the form of,

$$\begin{cases} \dot{p}_1 = \dfrac{1}{C_1}(Q_1 - A_1\dot{X}_p) \\ \dot{p}_2 = \dfrac{1}{C_2}(-Q_2 + A_2\dot{X}_p) \end{cases} \quad (4)$$

The volumes between the valve and each side of piston are calculated as,

$$V_1 = A_1X_p + v_{01}$$

$$V_2 = A_2(L - X_p) + v_{02}$$

If the tank pressure is set to equal to zero ($p_T = 0$), the nonlinear equations of flow rate of valve can be written in the simplest form, as follow,

$$Q_1 = \begin{cases} c_v u\sqrt{p_s - p_1} & u \ge 0 \\ c_v u\sqrt{p_1} & u \prec 0 \end{cases} \tag{5}$$

$$Q_2 = \begin{cases} c_v u\sqrt{p_2} & u \ge 0 \\ c_v u\sqrt{p_s - p_2} & u \prec 0 \end{cases} \tag{6}$$

where,

$$c_v = \frac{Q_n}{u_n\sqrt{\Delta p_n}}$$

As it clear in the above equation, the system is nonlinear. Following Equations are the linearized form of 5 and 6 [7],

If $u \ge 0$

$$\begin{cases} Q_1 = K_{v1}u - K_{p1}p_1 \\ Q_2 = K_{v2}u + K_{p2}p_2 \end{cases} \tag{7}$$

$$K_{vI} = c_v\sqrt{p_s - p_{I0}} \ , K_{v2} = c_v\sqrt{p_{20}}$$

$$K_{pI} = \frac{Q_{I0}}{2(p_s - p_{I0})} \ , K_{p2} = \frac{Q_{20}}{2p_{20}}$$

The simplified equation for rate of pressures with using the aforementioned equation can be written as,

$$\begin{cases} \dot p_1 = -\frac{A_1}{C_1}\dot X_p - \frac{K_{p1}}{C_1}p_1 + \frac{K_{v1}}{C_1}u \\ \dot p_2 = \frac{A_2}{C_2}\dot X_p - \frac{K_{p2}}{C_2}p_2 + \frac{K_{v2}}{C_2}u \end{cases} \tag{8}$$

If $u \prec 0$

$$\begin{cases} Q_1 = K'_{v1}u + K'_{p1}p_1 \\ Q_2 = K'_{v2}u - K'_{p2}p_2 \end{cases} \tag{9}$$

where,

$$K'_{v1} = c_v\sqrt{p_{10}} \ , \ K'_{v2} = c_v\sqrt{p_s - p_{20}}$$

$$K'_{p1} = \frac{Q_{10}}{2p_{10}} \ , \ K'_{p2} = \frac{Q_{20}}{2(ps - p_{20})}$$

and finally, the rate of pressure for negative voltage will be,

$$\begin{cases} \dot p_1 = -\frac{A_1}{C_1}\dot X_p + \frac{K'_{p1}}{C_1}p_1 + \frac{K'_{v1}}{C_1}u \\ \dot p_2 = \frac{A_2}{C_2}\dot X_p + \frac{K'_{p2}}{C_2}p_2 - \frac{K'_{v2}}{C_2}u \end{cases} \tag{10}$$

The state space model of linear system is shown as,

$$\dot X = AX + Bu \tag{11}$$

$$y = CX + Du \tag{12}$$

The states of system defined as,

$X_1 = X_l$, $X_2 = \dot X_l$, $X_3 = X_p$, $X_4 = \dot X_p$,

$X_5 = p_1$, $X_6 = p_2$

For $u \ge 0$ from above equations,

$$\dot X_6 = \frac{A_2}{C_2}X_4 - \frac{K_{p2}}{C_2}X_6 + \frac{K_{v2}}{C_2}u$$

$$\dot X_5 = -\frac{A_1}{C_1}X_4 - \frac{K_{p1}}{C_1}X_5 + \frac{K_{v1}}{C_1}u$$

$$\dot X_4 = \frac{k}{m_1}X_1 + \frac{b_2}{m_1}X_2 - \frac{k}{m_1}X_3 - \frac{(b_1 + b_2)}{m_1}X_4 + \frac{A_1}{m_1}X_5 - \frac{A_2}{m_1}X_6$$

$$\dot X_3 = X_4$$

$$\dot X_2 = -\frac{k}{m_2}X_1 - \frac{b_2}{m_2}X_2 + \frac{k}{m_2}X_3 + \frac{b_2}{m_2}X_4$$

$$\dot X_1 = X_2$$

So, the A matrix will be,

$$A = \begin{bmatrix} 0 & 1 & 0 & 0 & 0 & 0 \\ -\frac{k}{m_2} & -\frac{b_2}{m_2} & \frac{k}{m_2} & \frac{b_2}{m_2} & 0 & 0 \\ 0 & 0 & 0 & 1 & 0 & 0 \\ \frac{k}{m_I} & \frac{b_2}{m_1} & -\frac{k}{m_1} & -\frac{b_1 + b_2}{m_1} & \frac{A_I}{m_I} & -\frac{A_2}{m_I} \\ 0 & 0 & 0 & -\frac{A_I}{C_I} & -\frac{K_{p1}}{C_1} & 0 \\ 0 & 0 & 0 & \frac{A_2}{C_2} & 0 & -\frac{K_{p2}}{C_2} \end{bmatrix} \tag{13}$$

In addition, Matrix B, C, and D could be written as follow respectively,

$$B^T = \left[0 \ 0 \ 0 \ 0 \ {K_{v1}}/{C1} \ \ {K_{v2}}/{C2}\right] \tag{14}$$

C=[1 0 0 0 0 0] (15)

D= 0 (16)

If we do the same procedure for $u \prec 0$, the following matrixes will be found,

$$A' = \begin{bmatrix} 0 & 1 & 0 & 0 & 0 & 0 \\ -\frac{k}{m_2} & -\frac{b_2}{m_2} & \frac{k}{m_2} & \frac{b_2}{m_2} & 0 & 0 \\ 0 & 0 & 0 & 1 & 0 & 0 \\ \frac{k}{m_I} & \frac{b_2}{m_1} & -\frac{k}{m_1} & -\frac{b_1 + b_2}{m_1} & \frac{A_I}{m_I} & -\frac{A_2}{m_I} \\ 0 & 0 & 0 & -\frac{A_I}{C_I} & \frac{K'_{p1}}{C_1} & 0 \\ 0 & 0 & 0 & \frac{A_2}{C_2} & 0 & \frac{K'_{p2}}{C_2} \end{bmatrix} \tag{17}$$

$$B' = \left[0 \ 0 \ 0 \ 0 \ {K'_{v1}}/{C_1} \ -{K'_{v2}}/{C2}\right] \tag{18}$$

$C' = [\,1 \;\; 0 \;\; 0 \; 0 \; 0 \;\; 0\,]$ (19)

$D' = 0$ (20)

From the Matrixes A and A' it is obvious that the dynamic of system varies depending on the sign of voltage that is fed to the servo-valve.
The setup parameters of system are shown in Table1.

$m_1 = 278\ Kg$	$L = 1\ \text{m}$
$m_2 = 110\ Kg$	$b_1 = 1000\ \text{Ns}/\text{m}$
$A_1 = 8.04 \times 10^{-4}\, m^2$	$b_2 = 500\ \text{Ns}/\text{m}$
$A_2 = 4.24 \times 10^{-4}\, m^2$	$\beta_e = 4 \times 10^{8}\,\text{Pa}$
$k = 136670\ N/m$	$P_s = 14\ \text{Mpa}$
$v_{01} = 2.13 \times 10^{-4}\, m^3$	$K_p = 5$
$c_v = 2 \times 10^{-8}\ \text{m}^3 / (\text{sVPa}^{1/2})$	$v_{02} = 1.07 \times 10^{-4}\, m^3$

Table 1: Setup Parameters.

Following, the structure of controller for positioning control of each mass is proposed.

CONTROLLER DESIGHN

Here, the aim of the controller is position tracking of a reference model. Classical approaches, like P or PD regulators for positioning of hydraulic drives, do not give satisfactory performance. For this reason, adaptive control techniques, adaptive reference model, and gain scheduling are used to improve the performance of Controller [4, 7 and 8]. The variations of parameters depending on the change in the sign of voltage fed to the valve are compensated by fuzzy gain scheduling block. The second neural network is impelled to improve the lack of damping in hydraulic systems. Following the reference model and each neural network are proposed.

REFERENCE MODEL

The desire linear, second order, reference model was selected to run parallel with the nonlinear system. The natural frequency, ω_n of this model set equal to as 7 rad/sec with the damping ratio, ζ of 0.9.

$$G_{ref}(s) = \frac{\omega_n^2}{s^2 + 2\zeta\omega_n s + \omega_n^2} \tag{21}$$

There are kinds of method to find a reference model such as ITAE or Bessel transfer functions. The Bessel transfer functions have not overshot when ω_n is equal to one, but when ω_n is greater than one the overshot appears.

The natural frequency of system was chosen in a manner that the response of system is as fast as possible. The chosen damping ratio provides the minimum overshoot of reference model.

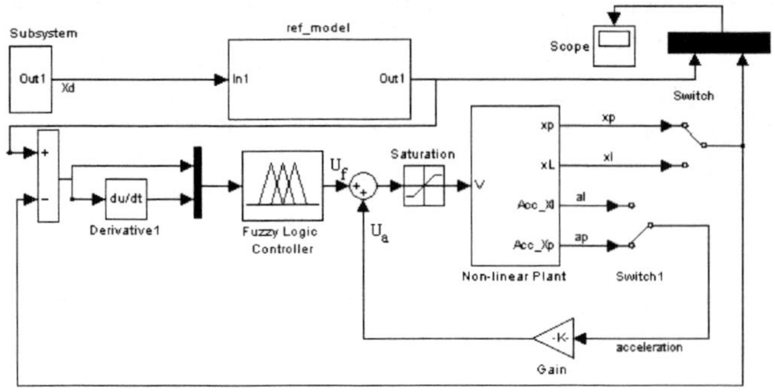

Fig. 3: Schematic diagram of controller.

The proposed controller depicted in figure 3, composed of a feedforward *acceleration* and *Fuzzy Gain Scheduling*. The feed forward gain is chosen such that the dynamic behavior of system (lack of damping) will be proper.

FUZZY CONTROLLER DESIGHN

Figure (4) shows the fuzzy logic control system. It usually consists from four parts: Fuzzification, Rule Base, Inference Engine and Defuzzification [7, 8 and 9]. The first block inside the controller is Fuzzification, which converts each piece of input data to degrees of membership by a lookup in one or several membership functions. The Fuzzification block thus matches the input data with the conditions of the rules to determine how well the condition of each rule matches that particular input instance. There is a degree of membership for each linguistic term that applies to that input variable.

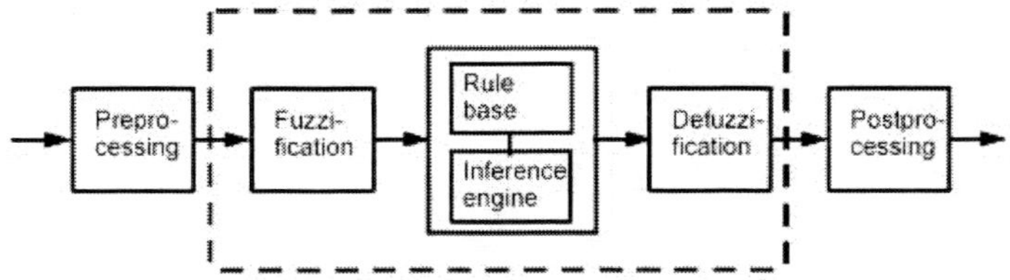

Fig. 4: Fuzzy controller system.

The Fuzzification block thus matches the input data with the conditions of the rules to determine how well the condition of each rule matches that particular input instance. There is a degree of membership for each linguistic term that applies to that input variable. The controller has two inputs and one output. The inputs are error and gradient of error. The output of fuzzy controller, U_f, will be added to the output of acceleration again, U_a, to make a proper system input. Following the membership functions are designed.

The INPUT MEMBERSHIP FUNCTION

Two inputs namely Reference position, X_r, and the position of load (piston), are considered for fuzzy controller to make the fuzzy inputs. As depicted in Figure 5, the inputs are error and gradient of error. When the input to the controller is error, the control strategy is a static mapping between input and control signal. A dynamic controller would have additional inputs, for example derivatives, integrals, or previous values of measurements backwards in time.

Fig. 5: Inputs and output of fuzzy controller.

These are created in the preprocessor thus making the controller multi-dimensional, which requires many rules and makes it more difficult to design. Here error and change in error are defined as follow,

A) ERROR

Error quantity is defined as difference between position of reference and position of load (piston). Figure 6 indicates the membership function of error.

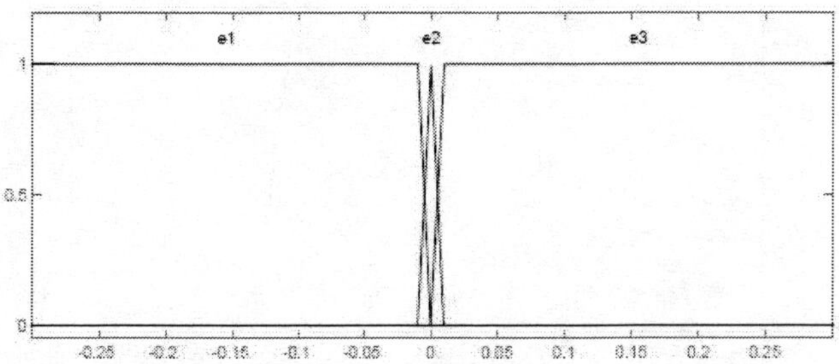

Fig. 6: Membership functions of error.

B) GRADIENT OF ERROR

Gradient of error is defined as the derivative of error. Figure 7 shows membership function of gradient error.

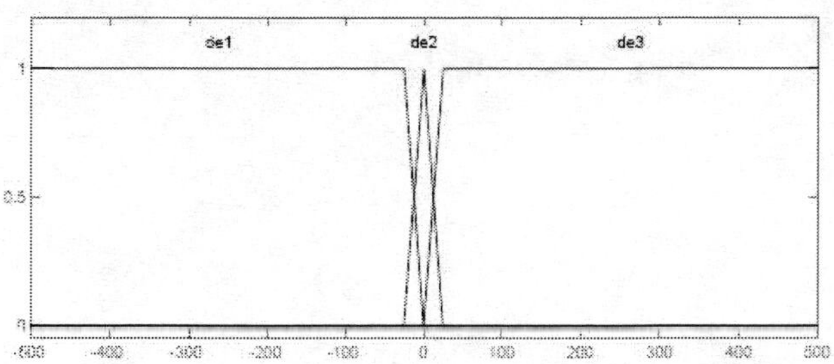

Fig. 7: Membership functions of gradient of error.

THE OUTPUT MEMBERSHIP FUNCTION

Figure (8) shows the output membership function of fuzzy controller. Here there are three output membership functions. The asymmetric form of the aforementioned diagram is due to the change of dynamic of system in extending and returning.

FUZZY RULES

Rules provide a formal way of representing directives and strategies and are often appropriate when domain knowledge results from empirical associations or experience. Rule-based systems are built upon a set of rules and use a collection of facts to make inferences.

According to the input membership functions, totally $3 \times 3 = 9$ rules are used to control the system.

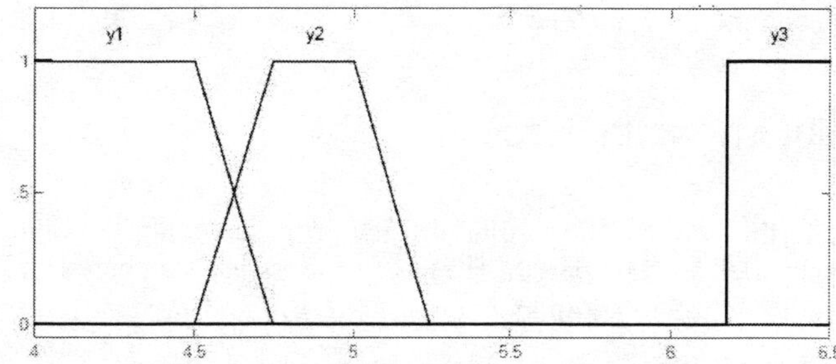

Fig. 8: Membership functions of output.

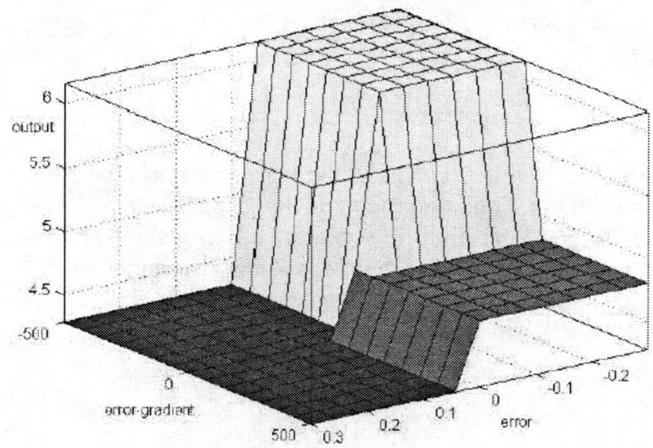

Fig.9: Control surface view.

Figure 8 shows the controlling area with respect to the selection of input membership functions and fuzzy rules.

DEFUZZIFICATION

In this study the Center Of Area (COA), Defuzzification method is used to make outputs of fuzzy controller real numbers.

Following to verify the fuzzy controller with neural compensation against a common controller, a p-controller is designed.

DESIGN OF P-CONTROLLER

To design common p-controller the linearized equations of system are used. The gain of p-controller (K_p) is determined by using Ziegler-Nichols method [4].

The schematic diagram of p-controller is shown in figure 10. The acceleration feedback is used to improve the dynamic of system, too. Depend on the position control of piston or flexible load, its related acceleration is used.

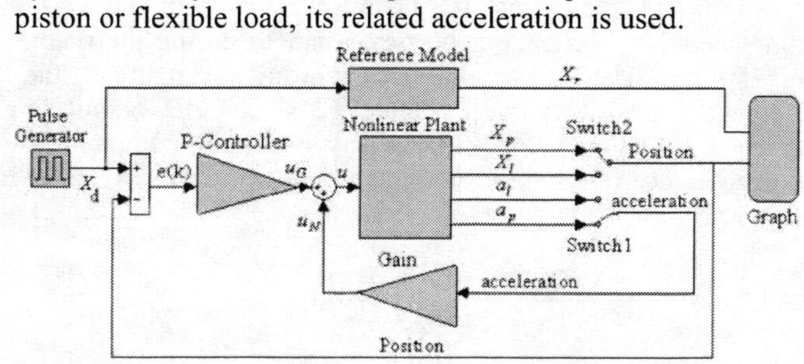

Fig. 10: Schematic diagram of P-Controller.

Following the results of using new controller is compared with common p-controller.

SIMULATION RESULTS

In this section, the results of using proposed controller and p-controller are compared. These controllers are compared in tracking of reference model for *piston* (m_1) and *flexible load* (m_2), respectively.

The desire input (X_d) are pulse inputs with the amplitude of 0-0.6 (m) and period of 4 (sec.).

PISTON TRACKING

Figure 11 is the piston tracking of the reference model using neural network with fuzzy gain scheduling. Figure 12 is its controller output that will be fed to nonlinear system.

The maximum and minimum voltages fed to the valve are limited to ± 10 Volt (limit of servo valve). As it shown in figure 12, the maximum and minimum of voltage are banded by -10 to 10 volt. A saturation element was used to avoid the controller output to be beyond of that limit.

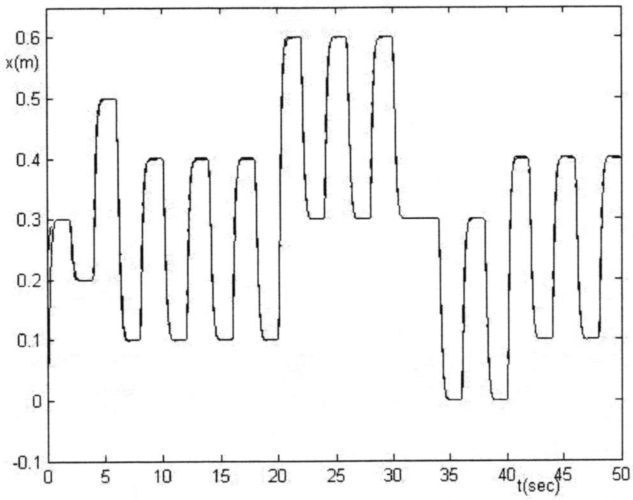

Fig. 11: Simulation result of piston mass tracking using neural network with fuzzy gain scheduling. The solid line shows the reference model and dash lines show piston load position.

Figure 11 shows that the neural network with the fuzzy gain scheduling has very good performances during the rising and falling. The aforementioned controller can track the reference model very well. Figure 12 shows the output of controller is smooth and has not any oscillation. Figure 13, zoomed figure of 11, indicates more detail of performance of proposed controller.

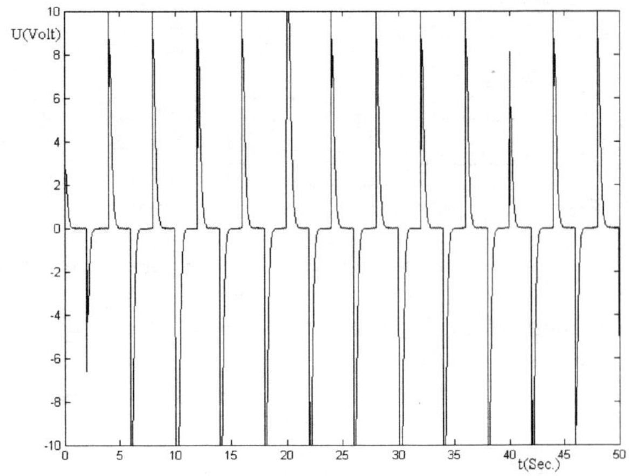

Fig. 12: The proposed Controller output.

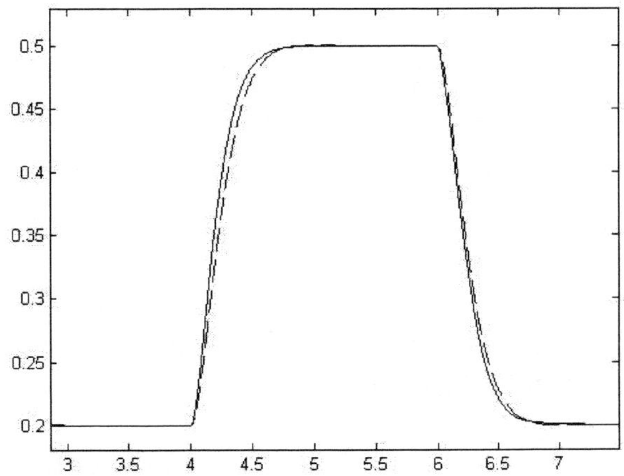

Fig. 13: Zoomed figure of 11.

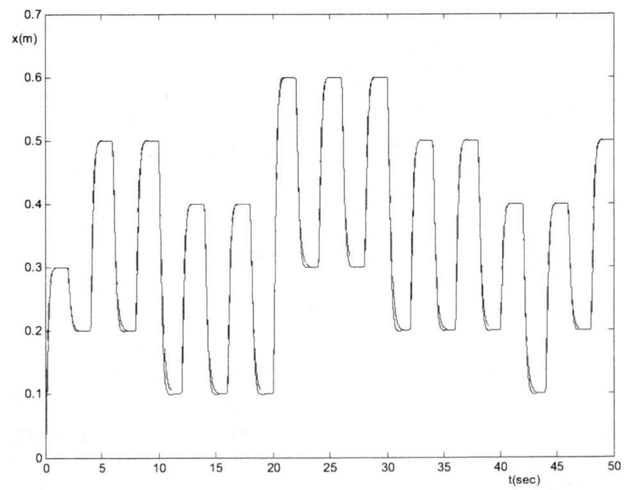

Fig. 14: Simulation result of piston tracking using P-Controller. The solid line shows the reference model and dots line show mass load position.

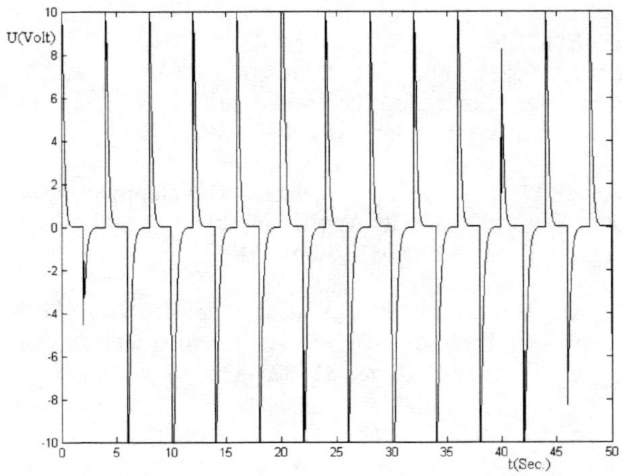

Fig. 15: Controller output using P-Controller.

It can be seen from figures 14, 15 and 16, zoomed figure of 14, that using the p-controller, the piston tracks the reference model is well during the rising, however, in the last part of fall segments the results are poor. Figure 16 provides more details of p-controller performances.

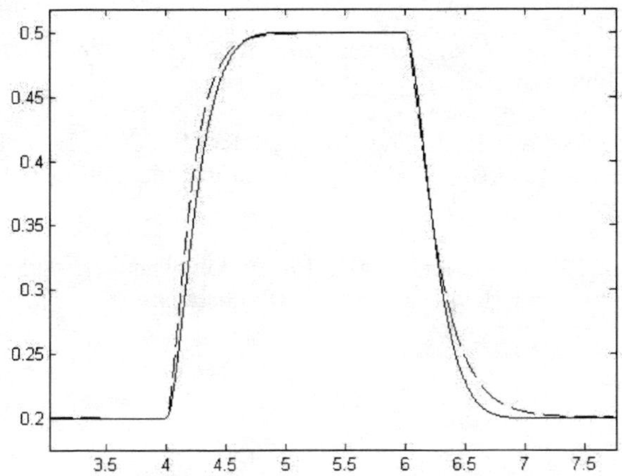

Fig. 16: Zoomed figure of 15.

FLEXIBLE LOAD TRACKING

In this section, the results of using two controllers to track of reference model for flexible load are compared. Figure 17 is tracking of the reference model by flexible load using the proposed controller.
The results are satisfactory and illustrate the capability of fuzzy controller. As it was shown in the figures 17 and 19, in rising and falling stages, the flexible load tracking is similar to that of the piston tracking.
P-Controller performances to track of reference model for flexible load are depicted in figures 20 and 21.

As it was illustrated in figure 16, the p-controller has a good tracking in rising, but due to the nonlinearities and changes in the govern equations, its behavior in the last part of falling is not good.

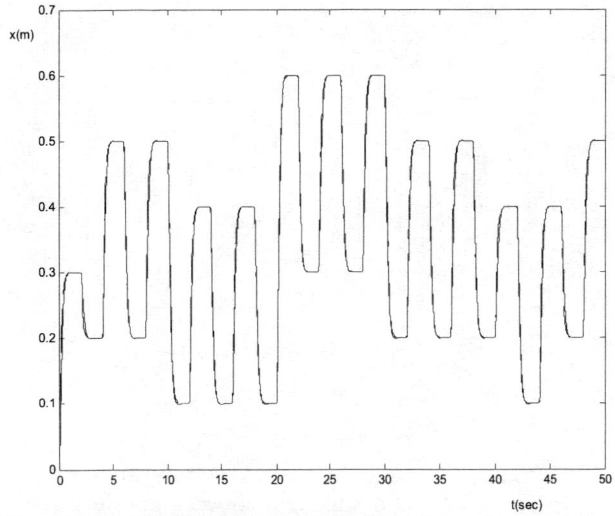

Fig. 17: Simulation result of tracking the reference model using proposed controller. The solid line shows the reference model and dash lines show flexible load position.

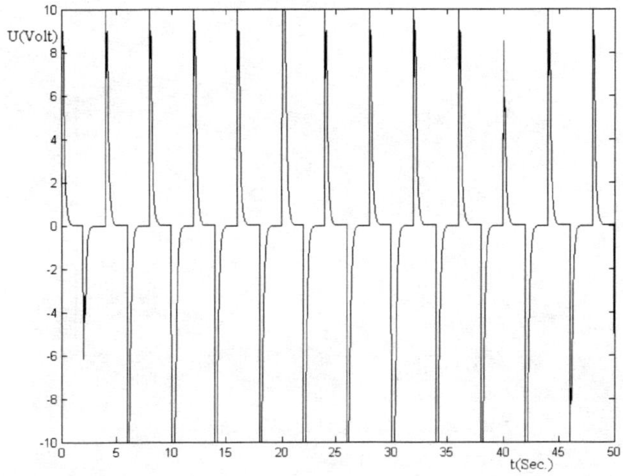

Fig. 18: Controller output using proposed controller.

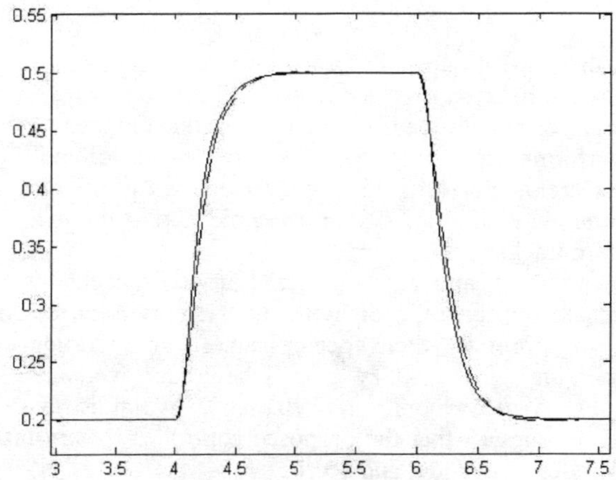

Fig. 19: Zoomed figure of 17.

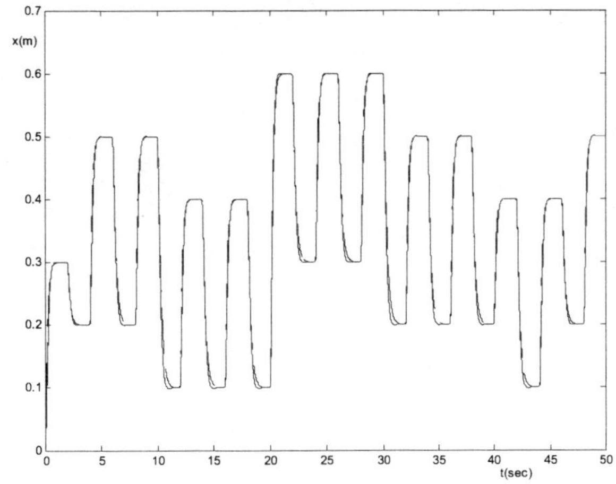

Fig. 20: Simulation result of tracking the reference model using P-Controller. The solid line shows the reference model and dash lines show Flexible load position.

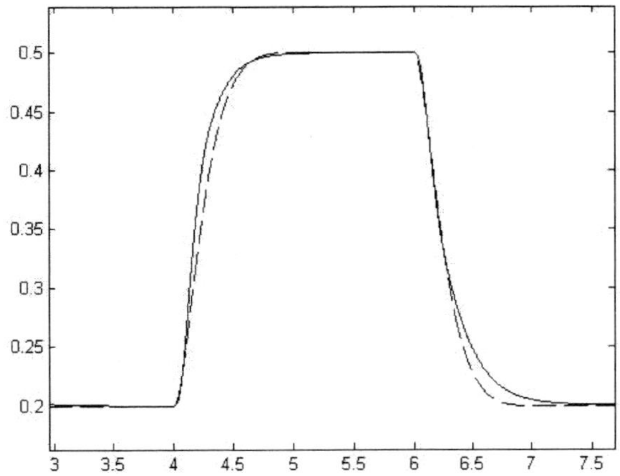

Fig. 21: Zoomed figure of 20.

CONCLUSIONS

In the present paper, the application of fuzzy controller in servo hydraulic systems was studied. The results suggest that fuzzy controller with feedforward acceleration compensator has good performance compare to the common p-controller.

The architectures of fuzzy controller are essential keys to overcome the change of system dynamic during the extension and return tracking.

In light of above, using feedforward acceleration to compensate the dynamic of hydraulic systems is an essential factor to improve their performances. In addition, the implementing of the fuzzy gain scheduling overcomes the variations of parameters for asymmetric hydraulic cylinder. The results suggest that the proposed controller is very fast to track of reference model and accrue.

The comparison between simulation results and the experimental results achieved using the set up shown in figure 1 will be illustrated in the future paper.

REFERENCES

[1] Edge, K. A. and Figueredo, K. R. A., 1987," An Adaptively Controlled Electrohydraulic Servomechanism". Proc. Instn Mech. Engrs, **201**, 175-189.

[2] Tienan Zhao, Tapio Virvalo, 1995, "Development of Fuzzy State Controller and Its Application to a Hydraulic Position Servo ".Fuzzy Sets and Systems, **70**, 213-221.

[3] Edvard Deticek, 2000, "A Fuzzy Self-learning Position Control of Hydraulic Drive" An International Journal of Cybernetics and Systems, **31**, 821-836.

[4] Gene F. Franklin, et al., 1986, Feedback Control of Dynamic Systems, Addison-Wesley.

[5] Y. Liu, T. Nishiumi and H. Handroos., 2003, "Application of a Neural Network in a Hydraulic Servo System with a Flexible load", First Inter. Nat. Conf. On Computational Methods in Fluid, power Tech., 523-533, Melbourne.

[6] Taco J. Viersma., 1980, An Analysis, Synthesis and Design of Hydraulic Servosystems and Pipelines, Elsevier.

[7] Kevin M.passino, Stephen Yukovich, 1998, Fuzzy Control, Addison-Wesley.

[8] Mohieddine Jalali and Andreas Kroll, 2004, Hydraulic Servo-Systems; Modeling, Identification and Control, Springer.

[9] H.-J. Zimmermann, 2001, Fuzzy Set Theory and Its Application, Kluwer Academic Punlishers.

Proceedings of IMECE2005
2005 ASME International Mechanical Engineering Congress and Exposition
November 5-11, 2005, Orlando, Florida USA

IMECE2005-79761

EXPERIMENTAL SIMULATION OF PISTON LEAKAGE IN AN AXIAL PISTON PUMP

Zeliang Li, Graduate Student
Richard Burton, Professor
Peter Nikiforuk, Professor Emeritus
Department of Mechanical Engineering
University of Saskatchewan, Canada
Contact Li (zel283@mail.usask.ca)

ABSTRACT

A method used to introduce "artificial leakage" into an axial piston pump to simulate leakage from a worn piston is described in this paper. A pressure control servo valve with a very high frequency response was employed to divert flow from the pump outlet in a prescribed waveform directly to tank. The purpose was to simulate piston leakage from the high pressure discharge chamber to the pump case drain chamber as the "simulated worn piston" made contact with the high pressure chamber. The system and associated control algorithms mimiced the action of a single worn piston at various degrees of wear. The experimental results indicated that the experimental system could successfully introduce artificial leakage into the pump which was consistent with a unit with a "real" worn piston. Comparisons of the pressure ripples from an actual faulty pump (with one worn piston) and the artificial faulty pump (with one simulated worn piston) are presented.

Keywords: condition monitoring, axial piston pump, artificial leakage, pressure ripple

NOMENCLATURE

A_{dn} The discharge area of P_n piston port opening to the discharge chamber in valve plate [m^2]
A_p Piston area [m^2]
A_v The discharge orifice area of the needle valve [m^2]
B Bulk modulus [Pa]
C_{d1} Flow discharge coefficient of the discharge area for piston port opening to discharge chamber
C_{d2} Discharge coefficient of needle valve orifice
d Diameter of piston [m]
h_g Gap (radial clearance) between piston and cylinder bore [m]
L Length of leakage passage [m]
m Number of pistons in the pump
n 1, 2, 3...... m
ρ Flow density [kg/m^3]
P_c Case drain chamber pressure [Pa]
P_{pn} Instantaneous pressure in P_n piston chamber [Pa]
P_s Pump output pressure (discharge chamber pressure) [Pa]
Q_{kpn} Kinematical flow delivered by P_n piston [m^3/s]
Q_{lpn} Internal leakage from P_n piston to case chamber [m^3/s]
Q_p Flow from piston chamber to discharge chamber [m^3/s]
Q_{pn} Flow from P_n piston chamber to discharge chamber [m^3/s]
Q_s Pump output flow [m^3/s]
r Radius of piston [m]
R_p Piston pitch radius on barrel [m]
P_{nd} Pressure normalized data
P_{np} Normal pressure ripple [Pa]
P_{fp} Faulty pressure ripple [Pa]
S_{pn} Stroke of P_n piston [m]
t Time [second]
V_0 Piston initial volume [m^3]
V_c Discharge chamber control volume [m^3]

Greek symbols:
α Phase delay [rad], equal to $2\pi / m$
β Angle of swash plate [rad]
μ Absolute fluid viscosity [N.sec/m^2]
ω Angular velocity of the pump drive shaft [rad/s]

INTRODUCTION

During the past few decades, a substantial amount of research has been focused towards the condition monitoring of hydraulic systems and components. In most cases, verification tests on fault detection systems require physical failures which can be very costly, unsafe and in some cases unreliable (a reasonable number of samples must be collected for statistical significance). In order to reduce the cost, and to increase safety and reliability, it is desirable that the actual components should not be damaged in introducing various types of failures.

An alternate approach is the introduction of "artificial faults" which simulate faults without damaging the components. Hindman et al [1] used "cross port calibrated orifices" to simulate internal leakage in the actuators and valves of an industrial backhoe loader. Bhojkar et al [2] used one solenoid in a solenoid proportional valve to create an external force on the spool to mimic different friction characteristics on the valve spool. Le et al [3] and Crowther et al [4] introduced artificial friction at the actuators using a connecting cylinder whose damping was adjusted by cross port calibrated orifices.

Many studies have been published during the past few decades on the topic of condition monitoring of individual hydraulic components and of complete systems. The literature reviewed to date has shown that little research has been reported which deals with introducing "artificial faults" into pumps. A possible reason for this is that, compared with valves and actuators, pumps have a much more complicated structure. For example, the variable displacement axial piston pump illustrated in Figure 1 consists of a swash plate, barrel, valve plate, control piston, return spring, drive shaft, several pistons and bearings, etc. Each part of the pump is susceptible to wear and can cause total failure if not detected early. To introduce faults directly by either running the pump to failure or by disassembling the pump to introduce a fault through part substitution, is very difficult compared to doing the same for a valve or actuator. Introducing artificial faults into the pump is equally difficult and remains a challenge for research.

Leakage is one indicator of components or system faults. There are generally three leakages paths in pumps: leakage between the valve plate and barrel, leakage between the swash plate and slippers, and leakage between the piston and cylinder bore (sleeve). Of the three, leakage between the piston and cylinder bore is most common and is directly related to the wear of the piston and hence was the focus of this study.

The objective of this research was to first model the flow and pressure characteristics of a normal pump and to assess how these characteristic waveforms change in the presence of a worn piston. The second objective was to design a test facility which would be able to mimic these changes experimentally without having to physically introduce worn pistons.

PUMP SIMULATION

A simplified pump model [5], which is presented in Appendix A, was used to simulate the pump in Matlab Simulink®. In early feasibility studies, it was ascertained that dynamic pressure ripples would be easier to measure than the flow. Therefore, the pressure is used to demonstrate the simulation and experimental results in this paper. The pressure ripple for the normal pump from the simulation result is illustrated in Figure 2. In this figure, there are exactly two pump revolutions (two cycles) with nine major ripples (total 18 ripples) in each cycle.

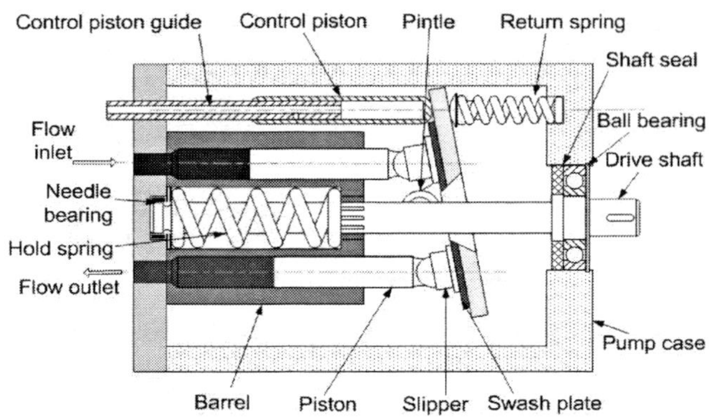

Figure 1 A variable displacement axial piston pump (Vickers PVB5)

In this research, only one piston was assumed to wear abnormally at a time. It is recognized that in most instances, wear is uniform in all pistons. However, if artificial wear can be created in one piston, then the more general case of multiple piston wear can readily be accommodated. The single piston wear is the more challenging case.

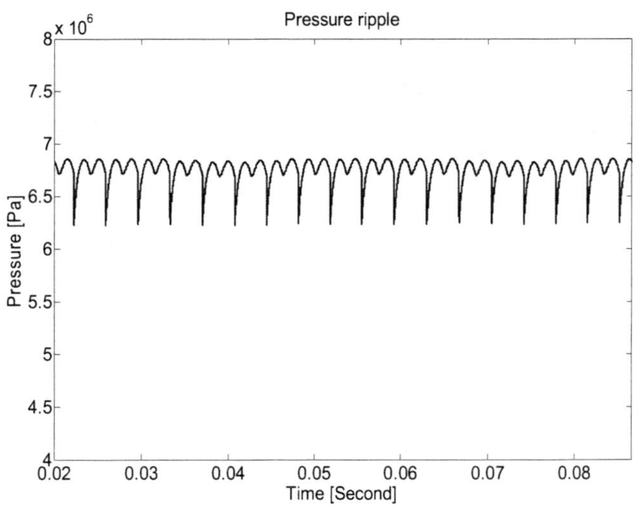

Figure 2 Pump output pressure waveform (simulated)

The simulated pressure output for one worn piston is illustrated in Figure 3. The pressure ripple waveform displayed a slight "concave dip" pattern which lasted about half a cycle. That is reasonable because in one revolution, the faulty piston delivers flow to the circuit only when it is connected with the discharge chamber. Since flow and pressure are related via the load, changes in flow will translate to changes in pressure. In this simulation, the theoretical leakages from the worn pistons with different gaps (radial clearance) were calculated from the model listed in Appendix A.

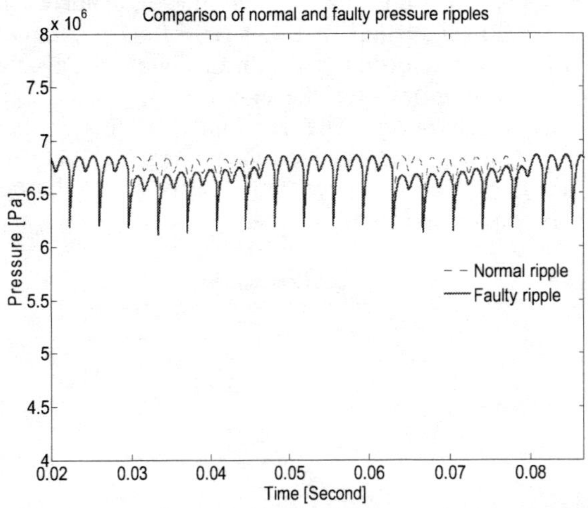

Figure 3 Comparison of normal pressure and faulty pump pressure waveform (simulated)

The magnitude of the concave dip depends on the amount the piston is worn. This is illustrated in Figure 4. It is observed that the simulated pressure ripple concave dip due to increasing wear in the piston is not proportional to the worn piston magnitude. This property is important in the development of a control strategy for the experimental fault simulator.

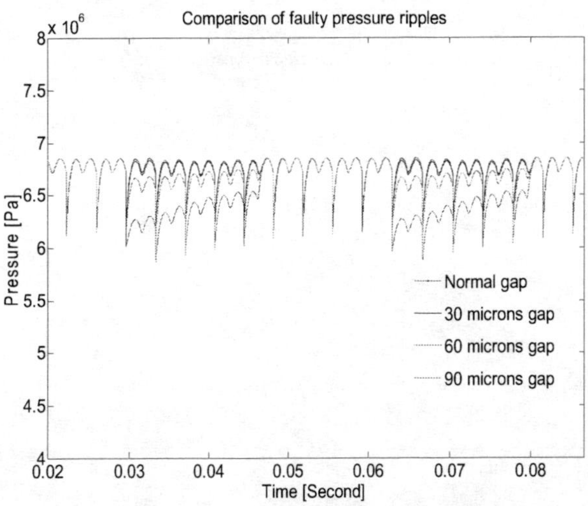

Figure 4 Comparison of pump pressure ripples with different worn pistons (simulated)

EXPERIMENTAL METHODOLOGY

To introduce artificial piston leakage, a single line and controlling valve were connected from the discharge chamber (the outlet of the pump) to the case drain or tank as illustrated in Figure 5. The valve was used to control the period and magnitude of the introduced artificial leakage. Therefore, the valve had to meet two basic requirements: a high frequency response and an adjustable flow.

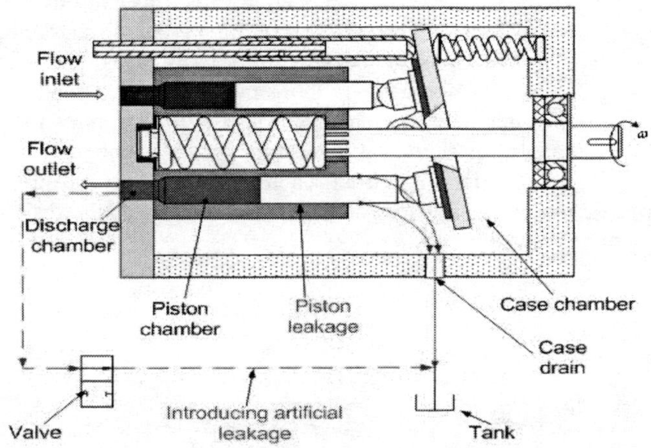

Figure 5 Method of experimentally introducing artificial piston leakage to the pump

Consider Figure 5; the line with the pressure control servo valve connects the outlet of the pump to the tank. When the servo valve is operated, some fluid is diverted from the load line to the tank according to the desired rate and amount. A pressure control servo valve usually controls pressure but since the flow and pressure are related via the orifice equation (essentially a high impedance device) flow can be regulated by changing the input to the pressure servo valve. It choice was dictated by its superior frequency response characteristics for small flow conditions. By varying both the frequency and the waveform shape of the input signal to the servo valve, the frequency and the waveform shape of the simulated leakage flow can be achieved. It is in this manner that the artificial leakage simulating one (or more if desired) worn piston was attempted.

EXPERIMENT TEST RIG

Schematic of experimental test rig

The schematic of the test system is shown in Figure 6. The test system consists of a variable displacement axial piston pump (in this study, the angle of swash plate was fixed.), a pressure control servo valve, a temperature sensor, a pressure transducer, a flow transducer (drag type), a diaphragm differential pressure transducer, three needle valves, one pick-up magnetic sensor, several amplifiers and a data acquisition system.

In Figure 6, the line with the pressure control valve (component 18) is connected to the outlet of the pump. As the value is actuated, fluid from the pump outlet line is diverted to tank via the pressure control valve.

The magnetic pick up sensor, component 9 in Figure 6, is used to trigger the electronic circuitry to open the pressure control servo valve. For each revolution, the key on the drive shaft passes once by the magnetic pick up sensor. Hence, the servo valve is opened once per revolution. It should be noted that once the appropriate piston in which the wear is simulated moves into the discharge chamber, the pressure control servo valve must open at same time. In other words, the simulated worn piston moving in the discharge chamber is synchronous to the pressure control servo valve opening. In a similar fashion, the simulated worn piston moving out of the discharge chamber must be synchronous to the pressure control servo valve closing. In this manner, the timing guarantees that the introduced artificial leakage is from a specific piston and has the same period and phase as the real worn piston leakage.

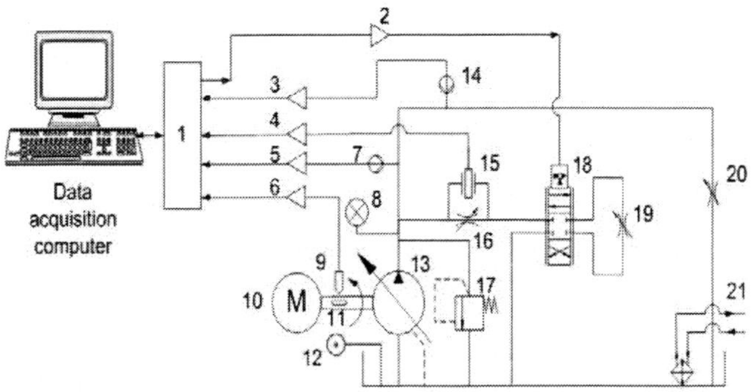

1. DAQ I/O box
2. Amplifier of pressure control servo valve
3. Amplifier of flow transducer
4. Amplifier of diaphragm transducer
5. Amplifier of pressure transducer
6. Amplifier of magnetic pick-up sensor
7. Pressure transducer 8. Pressure gauge
9. Magnetic pick-up sensor 10. Prime mover
11. Drive shaft key 12. Thermometer
13. Axial piston pump 14. Flow transducer
15. Diaphragm differential pressure transducer
16. Needle valve 17. Relief valve
18. Pressure control valve 19. Needle valve
20. Needle valve 21. Heat exchanger

Figure 6 Schematic of the experimental setup

Data acquisition system

The data for this experiment were acquired using Matlab Realtime Workshop© software to connect an I/O connector board, (component 1 in Figure 6), using the data acquisition board (NI PCI-6036E) in the PC mother board. The sampling frequency for the data collection was 2000Hz.

EXPERIMENTAL RESULTS

Artificial leakage

In order to introduce the artificial leakage in this valve, a programmed signal from the computer (via Matlab Realtime Workshop©) was input to the pressure control servo valve. The desired flow (pressure) waveform was obtained through the model developed in Appendix A; the model simulated the dynamic leakage in the pump. The measured flow rate through the servo valve for one piston is shown in Figure 7. This flow waveform was responsible for the concave dip in the ripple waveform from the pump. The amplitude of this artificial leakage could readily be adjusted via the input signal (shown in Figure 8) to the valve and hence could easily create the non-proportional property shown in Figure 4.

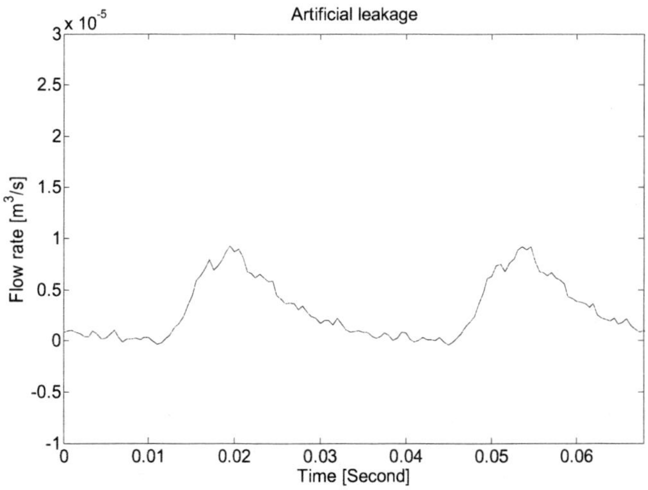

Figure 7 Introduced artificial leakage through the servo valve

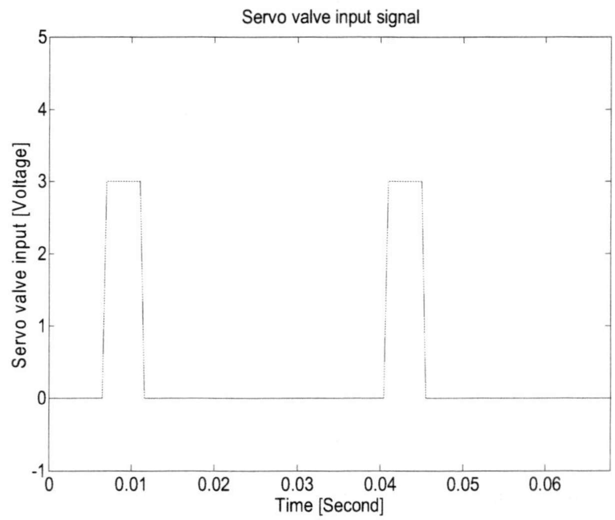

Figure 8 Servo valve input signal

Comparison of pressure ripples

Using 7MPa and 26.5 ^{0}C as a common base to compare all results, the waveforms of the pump pressure and line leakages were measured. Several pistons were precisely machined to create control clearances between the piston and its sleeve of 30 and 60 microns. The pressure ripple with one worn piston at 60 microns is shown in Figure 9. These results indicate that the measured pressure waveforms have one concave dip in each cycle and the concave form repeats every revolution as the model predicted.

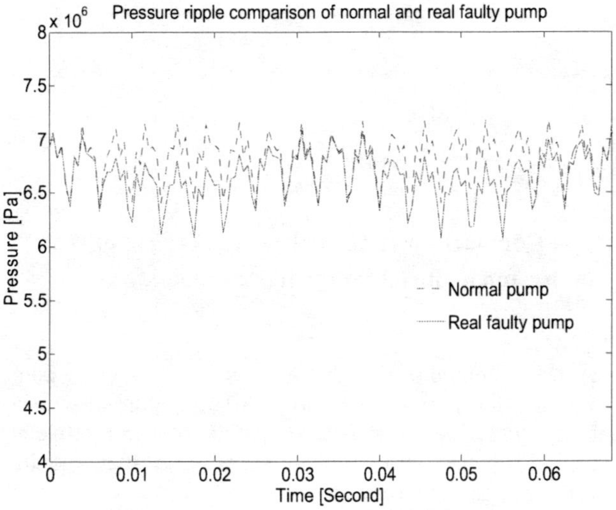

Figure 9 Comparison of the real faulty pressure and the normal pressure waveforms

In order to facilitate comparison of the measured pressure waveform with that when leakage was artificially introduced, a technique to normalize the difference between the pump with a faulty piston (actually or artificially created) and normal data was introduced. The normalization process is presented in Appendix B. The normalization data for the pressure ripple in Figure 9 is illustrated in Figure 10.

The non-normalized form of pressure ripple for the pump with the artificial fault is shown in Figure 11 and the normalized representation in Figure 12. In Figure 11, it is observed that the characteristic concave dip is present as was observed both in the model simulation and actual faulty pump results.

The normalized pressure ripples of the actual and simulated faulty pumps are superimposed in Figure 12 and in non-normalized form in Figure 13. The basic shape of the artificial fault waveform is very close to the actual waveform. However, there are some differences which need to be explained. It is observed that the slope of the artificial faulty pressure waveform was steeper than that of the real faulty pressure waveform. The reason for this was that the pressure control servo valve required some time to close due to the limitation of the pressure control servo valve, which was much longer than the period of the time to open this valve.

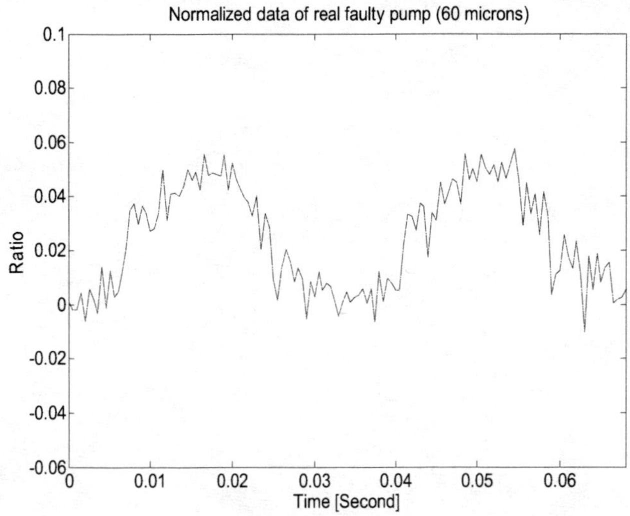

Figure 10 Normalized data (P_{nd}) of real faulty pressure ripple

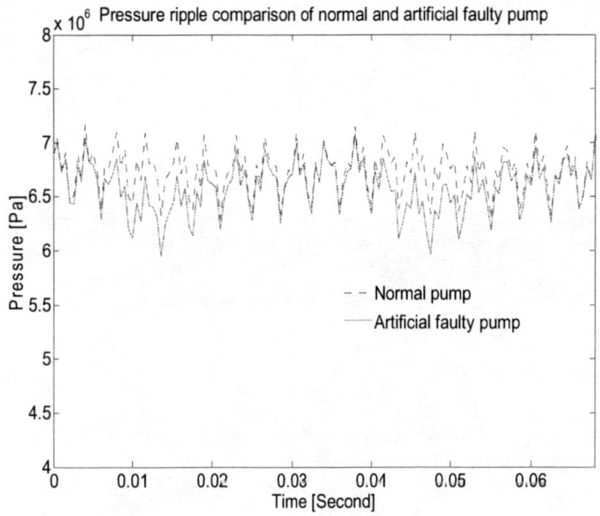

Figure 11 Comparison of the artificial faulty pressure and the normal pressure waveforms

The most important characteristic of the concave dip waveform was the synchronization of the point where the piston encountered and left the high pressure port in both the simulated and physical systems. Thus the input waveform had to be skewed slightly to ensure that the valve did indeed, completely close at the correct point. Thus, the artificial faulty pressure waveform could not be reproduced exactly as the desired waveform. This skewed form is evident in Figure 12.

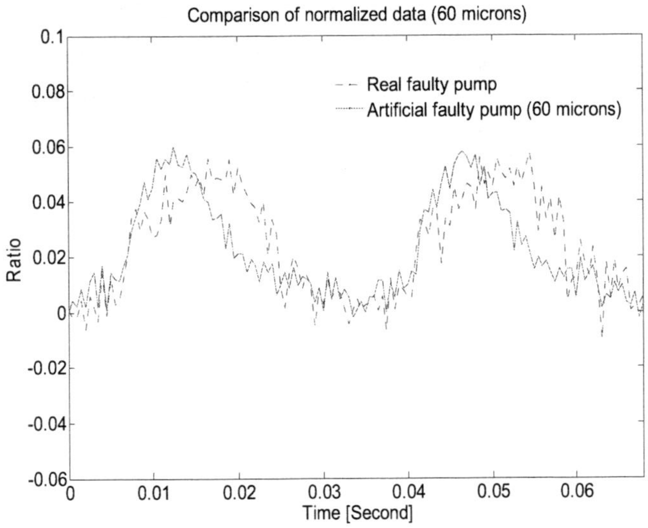

Figure 12 Comparison of normalized data (P_{nd}) of real faulty and artificial faulty pressure (60 μm)

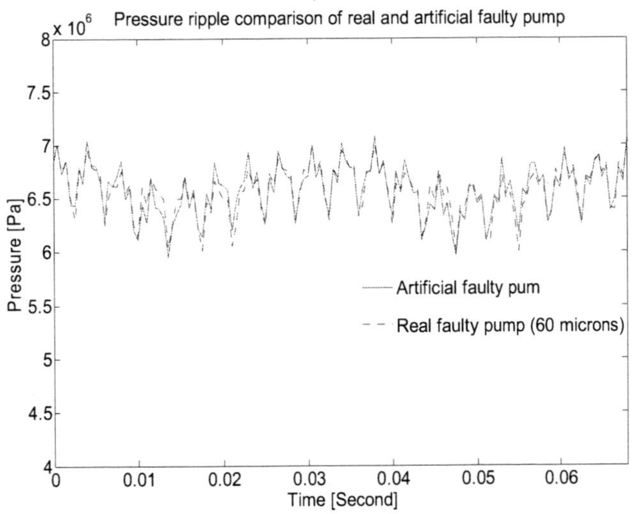

Figure 13 Comparison of pressure ripple between artificial faulty pump and real faulty pump (60 μm)

The traces do indicate that the two pressure ripples demonstrate similar trends in shape and magnitude which then implies that the artificial faulty pump displays similar characteristics to the actual faulty pump for these conditions.

It was also of interest to observe how well smaller leakage conditions could be simulated by the fault simulator. Therefore, a 30 micron worn piston was introduced into the pump. A visual comparison of the normalized data of the real fault pressure ripple with the 30 microns worn piston pump and the artificial fault pressure ripple is illustrated in Figure 14. It is noticed that the two traces matched well. Because of the smaller amplitude, the skewed input signal required for 60 microns (Figure 12) was not as evident.

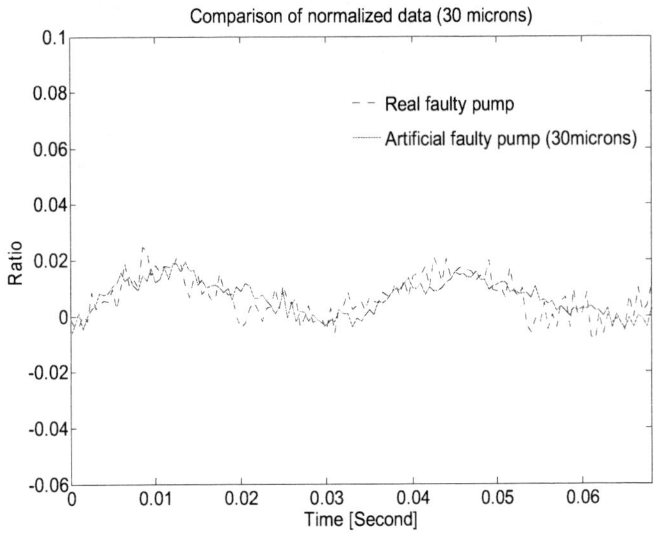

Figure 14 Comparison of normalized data (P_{nd}) of real faulty and artificial faulty pressure (30 μm)

CONCLUSIONS

A novel method of introducing an artificial leakage path into an experimental pump system to simulate piston wear in a pump has been described in this paper. The agreement between an actual worn piston pressure waveform and that of the artificial worn piston waveform is very good. It was concluded therefore that this approach was indeed very feasible for non-destructively, introducing faults into a pump. More work needs to be done in extending the system to multiple piston wear. However, it is believed this test system can now be used to evaluate condition monitoring algorithms under a variety of controlled fault conditions.

REFERENCES

1. Hindman, J. "Condition Monitoring of Valves and Actuators in a Mobile Hydraulic System Using an Artificial Neural Network and Expert Data", M. Sc. Thesis, University of Saskatchewan, 2002
2. Le, T. and Watton, J. "An Artificial Neural Network Based Approach to Fault Diagnosis and Classification of Fluid Power Systems", Proceedings Institution Mechanical Engineers Vol. 211.Part I.1997, pp.307-317
3. Crowther, W. and Edge, K. "Fault Diagnosis of a Hydraulic Actuator Circuit Using Neural Network-an Output Vector Space Classification Approach", Proceedings Institution Mechanical Engineers Vol.212.Part I. 1998, pp.57-69
4. Bhojkar, A., Burton, R., Schoenau, G. and Bitner, D. "Development of a Fault Simulator for Proportional Spool Valve", First International Conference on Computational Method in Fluid Power Technology, Melbourne Australia, 2003 (from CD version, ISBN o-9578574-2-x)
5. Ivantysyn, J. and Ivantysynova, M. "Hydrostatic Pumps and Motors", Akademia Books International Publishers & Distributor, New Delhi, India, 2001, pp.132-134

ACKNOWLEDGEMENTS

The authors would like acknowledge the financial support of the Discovery Grant from the National Science and Engineering Research Council (NSERC).

Appendix A

Pump Model Used in Simulation

$$\dot{P}_s = \frac{B}{V_c}(Q_p - Q_s) \tag{1}$$

$$Q_p = \sum_{n=1}^{m=9} Q_{pn} \tag{2}$$

$$Q_s = C_{d2} \cdot A_v \sqrt{\frac{2P_s}{\rho}} \tag{3}$$

$$Q_{pn} = C_{d1} A_{dn} \sqrt{\frac{2\left|P_{pn} - P_s\right|}{\rho}} \operatorname{sgn}(P_{pn} - P_s) \tag{4}$$

$$\dot{P}_{pn} = \frac{B}{V_0 - A_p S_{pn}}(Q_{kpn} - Q_{pn} - Q_{lpn}) \tag{5}$$

$$Q_{kpn} = \omega \frac{\pi \cdot d^2 R_p}{4} \tan\beta \cdot \sin(\omega \cdot t - (n-1)\alpha) \tag{6}$$

$$Q_{lpn} = \frac{\pi r h_g^{\ 3}}{6\mu L}(P_{pn} - P_c) \tag{7}$$

$$S_{pn} = R_p \tan\beta \cdot (1 - \cos(\omega t - (n-1)\alpha)) \tag{8}$$

Appendix B

Data Normalization Method

In order to facilitate analysis and comparison of the data it was decided to normalize the pressure ripple data. The method using in this study is called Min-Max normalization. Assuming the data is taken at the same point in the cycle, and at the same operating conditions, the definition of the "pressure normalized data" is:

Pressure normalized data P_{nd} = (Normal pump pressure P_{np} -faulty pump pressure P_{fp})/Normal pump pressure P_{np} , or

$$P_{nd} = \frac{P_{np} - P_{fp}}{P_{np}}$$

where: P_{nd} = Pressure normalized data

P_{np} = Normal pressure ripple [Pa]

P_{fp} = Faulty pressure ripple [Pa]

Proceedings of IMECE2005
2005 ASME International Mechanical Engineering Congress and Exposition
November 5-11, 2005, Orlando, Florida USA

IMECE2005-80126

DESIGN OF AN ENGINE LOAD SIMULATOR

T. Radpukdee
Department of Mechanical Engineering,
Faculty of Engineering, Khon Kaen University,
Khon Kaen 40002 Thailand
tthanar@hotmail.com

P. Jirawattana
Department of Mechanical Engineering,
Faculty of Engineering, Khon Kaen University,
Khon Kaen 40002 Thailand
piyoro@kku.ac.th

ABSTRACT

This paper discusses the design details of an engine load simulator. A proportional valve is used to regulate a resistant to the flow from a hydraulic pump. Due to the fact that pressure drop across an orifice is proportional to the square of the flow which is similar to an aerodynamic drag that is proportional to the velocity square. While this is true, more sophisticated control methods can be incorporated. However, the dynamic characteristics of the valve can not be included in designing a controller with an adaptive control method. As will be discussed, the valve effect can be eliminated mechanically or electronically. With the dynamic pressure compensation from the main valve to the pilot valve, the inertia effect of the spool can be reduced. On the other hand, the pressure in the pressure line can be read and compensated for the signal to the pilot valve directly. In addition, a variable structure control for a load simulator will be presented.

INTRODUCTION

A hydrostatic dynamometer has been introduced in the UK since the 1980's [1]. Absorbing the energy hydraulically has many advantages. Its high power density makes the hardware compact. A regenerative testing which requires lower power supply is also possible [2]. For a more complex dynamometer, the stiffness of the hydraulic system offers faster transient responses. Therefore, a more precise dynamic loading can be generated to the power source.

To learn about an engine and a drivetrain, many researchers have proposed simulated engines [3, 4] and vehicles [5, 6]. The objective of their designs for simulated vehicles is to emulate a specific load to the engine with actual hardware components and simulation components. If the engine must overcome the road load power due to rolling resistant, grade resistant and aerodynamic drag, then the load simulator can be designed to meet the following requirements; (1), a constant friction force must be added for a suitable rolling resistant and grade resistant, (2), the inertia of a flywheel should be equivalent to the real vehicle, and (3) the aerodynamic drag should be generated as a function of speed. An example of a vehicle road load function is shown in Figure 1 and Figure 2 is a schematic diagram of a load simulator.

The physics of the flow through an orifice allows us to match the load torque required to deliver fluid with the load torque due to an aerodynamic drag of the car. With the aid of control technology such as the work done by Zhang and Alleyne [7] which use a model reference load controller, they successfully emulated the vehicle load even for a system with different inertia. Their design is attractive because the hardware is simple. However, the transient response of the system is dominated by the actuator dynamics and it can not be included in the controller parameter. In order to refine the dynamometer further for transient responses, we have designed a proportional valve with dynamic pressure compensation, as will be explained in this paper.

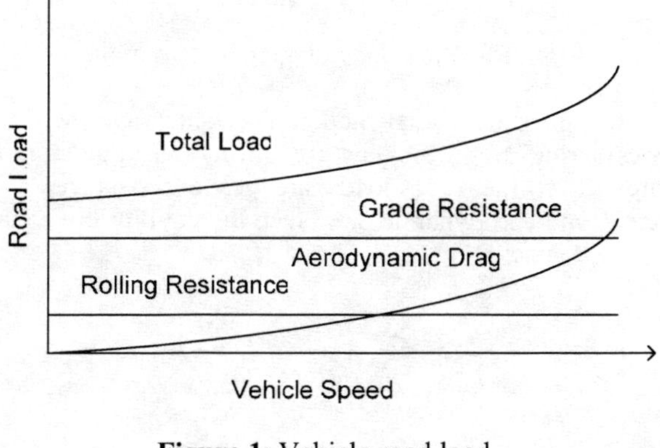

Figure 1: Vehicle road load.

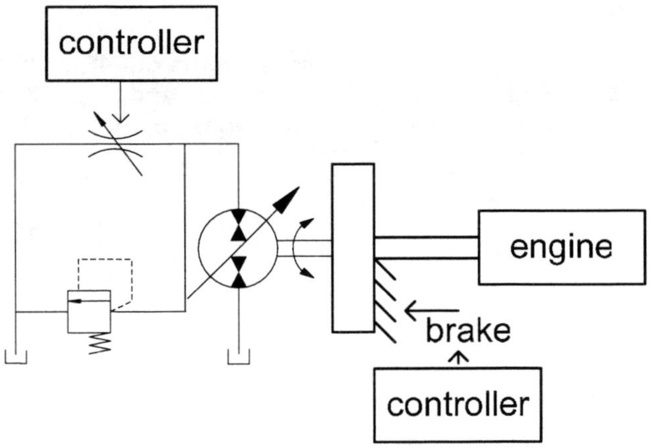

Figure 2: Schematic diagram of a load simulator.

Another control structure such as a variable structure control appears attractive for the application of a non-linear resistive control. This is because the ability of the variable structure control in stabilizing complex and highly non-linear flow-pressure relationship in the valve. The complete vehicle reference model can be simulated online without having to be linearized. In addition, despite the complexity of the controller, the valve dynamic can be taken into account with this type of control.

To discuss the addressing design problems, the paper is divided into three sections: First an example of a mathematical model of a vehicle and a simulated vehicle followed by the controller and valve designs, and finally, the conclusion will be given.

MATHEMATICAL MODELING

Vehicle Dynamics

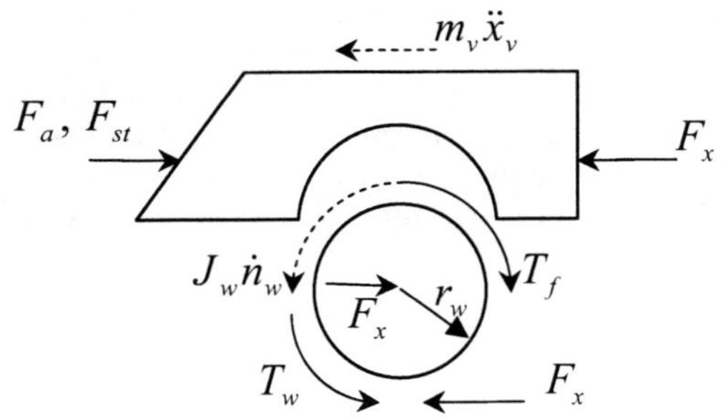

Figure 3: Free-body diagram of a vehicle.

The mathematical model of a vehicle can be derived by considering Newton's law for linear and angular motion (Figure 3). If the road loads are grade resistance, rolling resistance and aerodynamic drag, then the equilibrium equation can be written as

$$m_v\ddot{x}_v + \frac{1}{2}\rho_{air}C_D A_v \dot{x}_v^2 = F_x - F_{st} \tag{1}$$

Where m_v is the vehicle mass, x_v is the vehicle displacement, $F_a = \frac{1}{2}\rho_{air}C_D A_v \dot{x}_v^2$ is the aerodynamic drag, and C_D, F_x, and F_{st} are the drag coefficient of vehicle, the tire longitudinal force, and the climbing resistant force, respectively.

Considering the torque equilibrium equation at the wheel, we have

$$J_w\dot{n}_w + b_w n_w = T_w - F_f r_w - F_x r_w \tag{2}$$

Where J_w is the wheel moment of inertia, b_w is the wheel damping, r_w is the wheel radius, n_w is the wheel speed, T_w is the driving torque, T_f is the constant rolling resistance torque, and F_f is the constant rolling resistance force .

For simplicity we assume that no slipping occurs at the wheel. Therefore, combining equation (1) and (2) using $\dot{x}_v = r_w n_w$ and $T_f = F_f r_w$, we obtain

$$\dot{n}_w = \left(\frac{1}{J_w + m_v r_w^2}\right)\left(-\frac{1}{2}\rho_{air}C_D A_v r_w^3 n_w^2 + T_w - b_w n_w - \left(F_f + F_{st}\right)r_w\right) \tag{3}$$

Where $J_{car} = m_v r_w^2$ is the equivalent moment of inertia of the car.

An example of a dynamic response of the simulated vehicle is illustrated in Figure 5. This result corresponds to the driving torque in Figure 4.

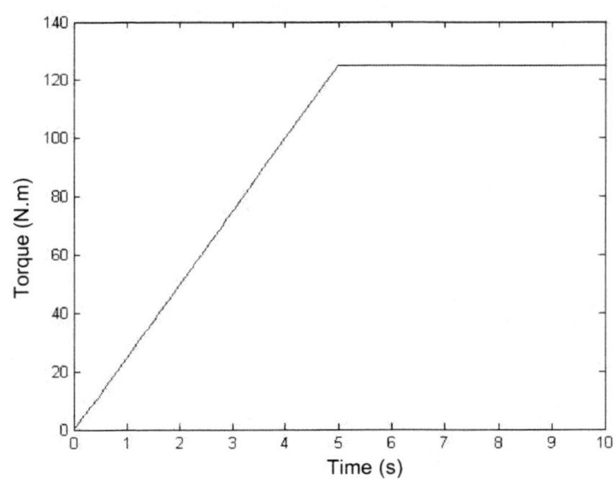

Figure 4: Driving torque.

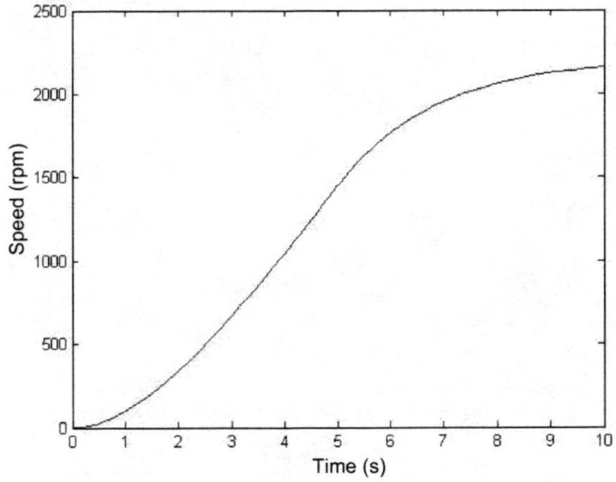

Figure 5: Speed response for a simulated vehicle.

Hydraulic Pump and Pressure Control Valve

Instead of driving a wheel, the engine is connected to a hydraulic pump (Figure 7) and the torque balance of the system is

$$\dot{n}_s = \left(\frac{1}{J_t}\right)\left(T_s - \left(b_p + b_e\right)n_s - T_l\right) \qquad (4)$$

Where n_s is the system speed, J_t is the equivalent moment of inertia of the system, $T_l = T_{p,l} + T_b$ is the load torque, $T_{p,l}$ is the load torque of pump, T_b is the load torque due to the grade and rolling resistances, T_s is the applied torque, and b_p, b_e are the pump and engine damping respectively.

For a positive displacement unit, the torque and pressure relation can we defined as

$$T_{p,l} = D_P P_L \qquad (5)$$

Where D_P and P_L are the pump displacement and load pressure, respectively.

The flow-pressure relationship of the hydraulic fluid flow across a pressure control valve can be described as

$$Q = C_d A_0 \sqrt{\frac{2}{\rho_{oil}} P_L} \qquad (6)$$

Where Q is the flow rate, ρ_{oil} is the oil density, finally, C_d, and A_o are the discharge coefficient and the effective cross sectional area of the valve, respectively.

Using equation (4), (5), (6) and the relation $Q = n_s D_P$, the torque balance equation for the load simulator can be written as

$$\dot{n}_s = \left(\frac{1}{J_t}\right)\left(-(b_e + b_p)n_s + T_s - \frac{1}{2}\frac{\rho_{oil} D_P^3 n_s^2}{C_d^2 A_o^2} - T_b\right) \qquad (7)$$

Equations (3) and (7) have the same mathematical form but the latter acts as a fast under-damp system. The speed response due to the torque trajectory in Figure 4 is shown in Figure 6. In order to make the load simulator system act as a vehicle model which has higher inertia, some means of control technique should be utilized.

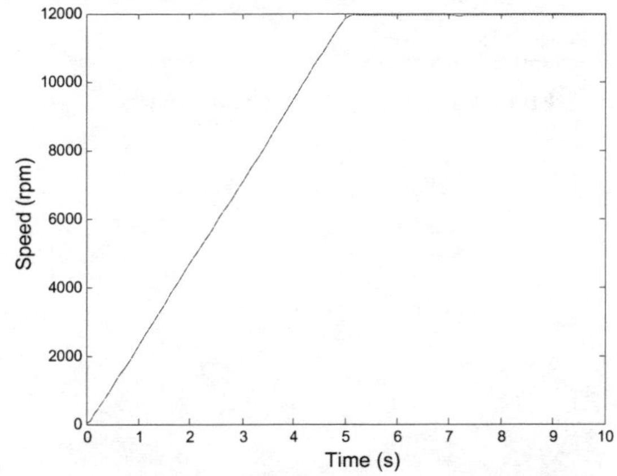

Figure 6: Speed response for a load simulator.

Figure 7: Engine load simulator test rig.

CONTROLLER AND VALVE DESIGNS

Valve Design for a Model Reference Controller

A model reference controller design for a resistive load application has been introduced in [7]. A thorough performance analysis is also addressed in [6]. For an adaptive control in Figure 8 (outlined by Astrom and Wittenmark [8]), even the actuator bandwith is high compare to that of the engine-pump, the effect of the transient behavior of the actuator can clearly be illustrated in Figure 10. Figure 11 is the simulation case where the actuator dynamics has been neglected for the same driving torque (Figure 9).

Note that this controller consists of a feedforward $T(s)/R(s)$ and a feedback $S(s)/R(s)$. Where T, R, and S are the polynomial obtained by solving the Diophantine equation.

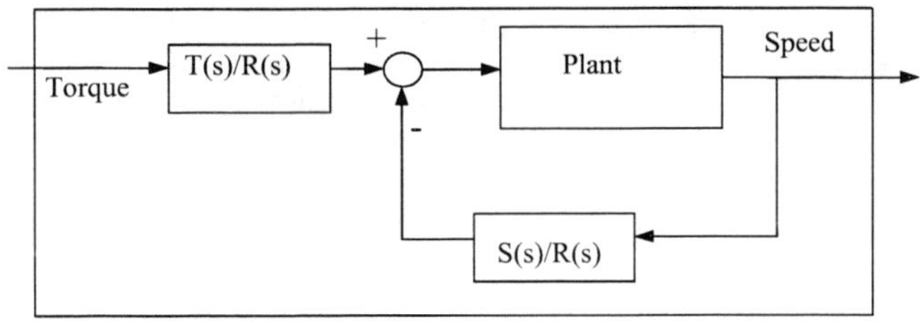

Figure 8: Model reference controller.

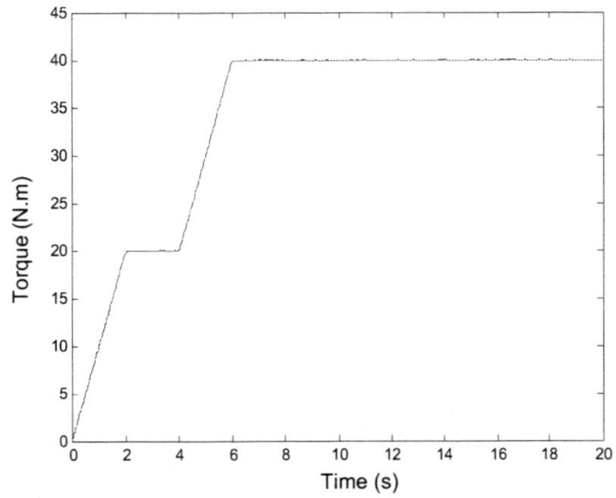

Figure 9: Driving torque.

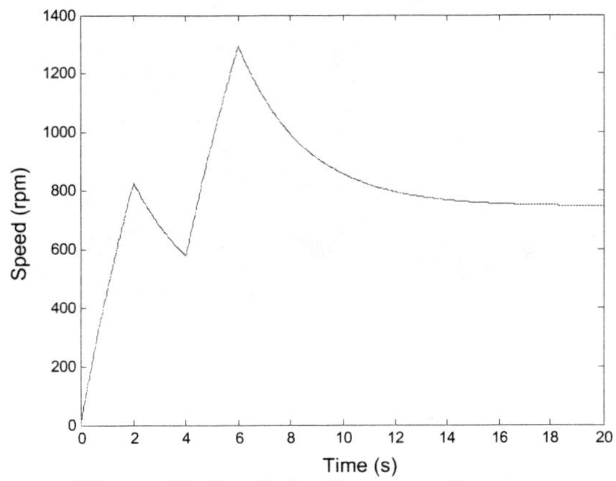

Figure 10: Load simulator speed response.

As concluded by Zhang and Alleyne [6], this configuration (referred to as "one-DOF feedback") preserves the load model and is easy to use with a non-linear load model. We wish to take an advantage of this configuration. However, the valve bandwidth should be extended. Therefore, we have designed a valve with dynamic pressure compensation to be used in our laboratory. This valve has been fabricated with 2 special chambers intended to sense the valve transient responses. The schematic diagram of the valve is shown in Figure 12 and the main valve picture is in Figure 13.

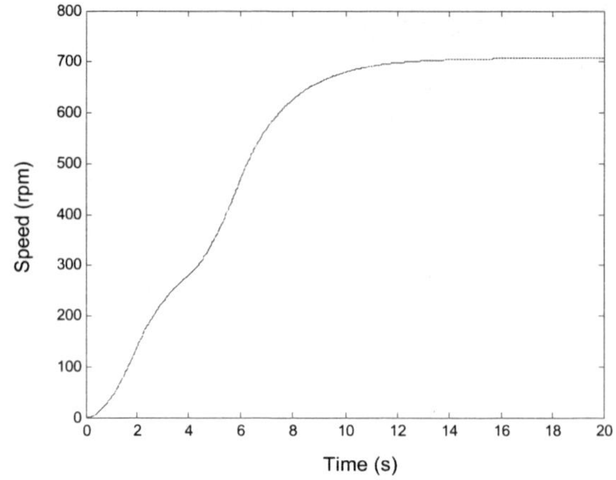

Figure 11: Speed reference.

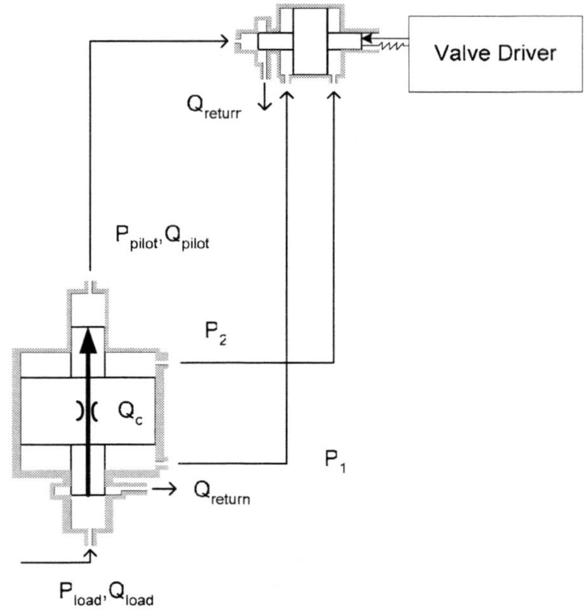

Figure 12: The 2-stage pressure control valve with dynamic pressure compensation.

Figure 13: Main valve with dynamic pressure compensation.

The overall operation of the valve will be explained in this section. Begin with an electrical signal moves the spool of the pilot valve, this results in a pilot flow from the main to the pilot unit. The pilot pressure chamber and the load pressure line are connected via an orifice restrictor in the spool. Therefore, the pilot flow causes the flow via the restrictor. This makes the spool to move, increases the flow path from the load line to the reservoir and reduces the load pressure. The spool will stop when the force due to pilot pressure from the top of the spool balanced with the pressure force from the load line.

Obviously, the load simulator speed fluctuation that occurred due to the valve transient responses can simply be reduced by increasing the valve damping. This can be done by increasing the effective cross sectional area of the restrictor control. However, this modification will result in a reduction of the valve bandwidth especially for the low flow operation. Therefore, in order to avoid imposing any other limitation to the actuator, the valve has been built with the chambers for dynamic pressure output.

At steady state, the spool has no motion, and pressure in the upper (P_2) and lower (P_1) chambers are equal. Therefore, they have no influence on the system. On the other hand, when the pilot pressure is suddenly dropped, the spool is then forced to move up. The compliance of the fluid in the pilot and load chamber acts as a spring. While the flow across the restrictor results in a damping of the system. The vibration of the spool then causes pressure fluctuation in the load line. Now, with pressure compensation chambers, a sudden motion of the spool squeezes the oil in the chamber. This squeeze pressure force created pushes the pilot spool against the electric signal for the required pressure, while the opposite is true for the downward motion of the main spool. As expected, the spool will reach steady state faster with no overshoot. If the compliance of the fluid in the lines connecting dynamic pressure chambers is neglected, the block diagram representation of the valve with pressure compensation can be simplified and shown in Figure 14. Where K is the pressure gain and the simulation result is shown in Figure 15.

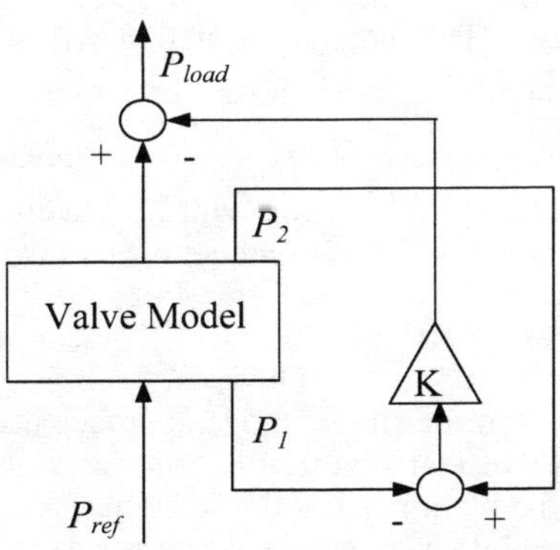

Figure 14: Simplified block diagram representation of a valve with dynamic pressure compensation.

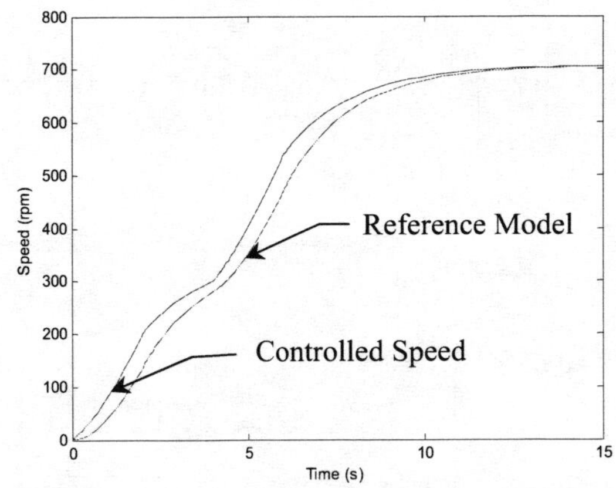

Figure 15: Load controller with dynamic pressure compensation performance.

Variable Structure Control for a Resistive Load Controller

As discussed in the previous section, the effects of the transient response of the valve can be reduced by means of dynamic pressure compensation. However, it depends on the objective of the design whether this fluctuation is important. If the load simulator is designed to test the engine for transient operation, it is necessary to eliminate the actuator effect to an acceptable level.

Variable structure control is another controller that could be used for a resistive load control application. It also has the ability to take the actuator effects into account. Variable structure control or sliding mode control comes from the derivation process of the control law. The controller is derived by the condition that the controlled output slides on the surface that makes the error vanished. To achieve this, there must be a discontinuity in the controller.

The conventional sliding mode controller is quite simple to begin. The control law for the engine load simulator is determined from the sliding mode method (outlined in [3]). The speed difference between the pump and the reference model is used to calculate an error. Then, the desired resistive pressure can be obtained. With this approach, the trajectory error is forced to zero by satisfying Lyapunov function. The actual controlled input (resistive pressure) will reach the desired pressure when the pressure control valve reaches its steady state. The control structure for the sliding mode control is shown in Figure 16.

According to the structure shown below, the first sliding surface (s_1) which defined from the speed difference, is given as

$$s_1 = n_p - n_{pd} \quad (8)$$

Where n_p and n_{pd} are the actual and desired speed of the pump, respectively.

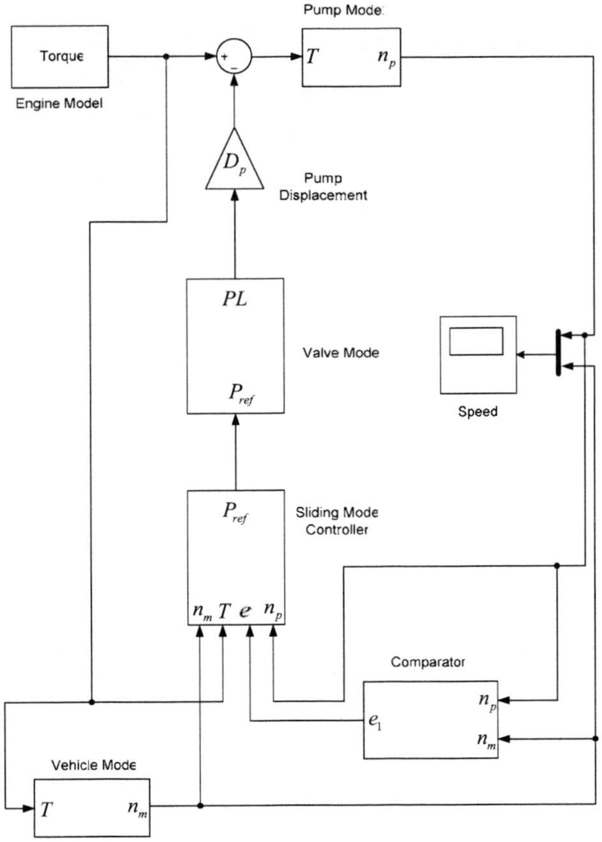

Figure 16: Structure of the sliding mode control for resistive load controller.

Differentiate equation (8) with respect to time, then utilizing equation (7) and the sliding mode formulation, the following relationship can be obtained

$$\frac{ds_1}{dt} = -\eta_1 sat\left(\frac{s_1}{\phi_1}\right) = \left(\frac{-b_t}{J_t}\right)n_p + \frac{T_s}{J_t} + \left(\frac{-D_p}{J_t}\right)\frac{\rho_{oil}D_p^2 n_p^3}{2C_d^2 A_o^2} - \dot{n}_{pd} \tag{9}$$

Where the saturation function is given by

$$sat\left(\frac{s}{\phi}\right) = \begin{cases} \frac{s}{|s|}, & for\ |s| \geq \phi \\ \frac{s}{\phi}, & for\ |s| < \phi \end{cases}$$

with the sliding gain

$$\eta_1 > 0.0$$

and the boundary layer thickness

$$\phi > 0.0$$

From the relationship between the pump flow rate and oil flow across the sharp edged orifice, the load pressure equation can be written as

$$P_L = \frac{\rho_{oil}D_p^2 n_p^2}{2C_d^2 A_o^2}$$

Equation (9) can be rearranged as

$$-\eta_1 sat\left(\frac{s_1}{\phi}\right) = \left(\frac{-b_t}{J_t}\right)n_p + \frac{T_s}{J_t} + \left(\frac{-D_p}{J_t}\right)P_L - \dot{n}_{pd} \tag{10}$$

The control law can be obtained by solving equation (10), therefore

$$P_L = \left(\frac{-b_t}{D_p}\right)n_p + \frac{T_s}{D_p} - \frac{J_t}{D_p}\dot{n}_{pd} + \eta_1 \frac{J_t}{D_p} sat\left(\frac{s_1}{\phi_1}\right) \tag{11}$$

Where b_t is the equivalent damping of the system.

The control law (equation 11) will be used as the command input to the valve which will generate the resistive pressure to the engine.

In general, there are the mechanical losses in a hydraulic pump due to viscous and coulomb frictions. While the flow losses are due to leakages and compressibility. These losses result in the difference between an ideal and actual analysis of the torque, pressure and speed relationships. However, for mathematical simplicity, the system equation is analyzed with an ideal analysis. The reason for this assumption is that, the variable structure control allows us to estimate the model with an ideal equation which bounded by its variation due to efficiencies. This bounded equation will be taken into account by the sliding gain η_1 in the Lyapunov guarantee of reaching the sliding surface $S(t)$ [11]. Nevertheless, including the pump efficiencies to the model will improve the controller performance. Since more precise model offers lower chattering in system response.

The next part of this section will present the performance of this controller. Obviously, simulation result in Figure 17 shows that the sliding mode approach can reduce the valve effects on transient significantly but the controlled speed chatters before the plant reaches the reference value. Since the error is near the sliding surface, the control calculation could drive the error across the surface [3]. In addition, there is the steady state error for this configuration.

The experimental result of controlled speed for such a controller and its driving torque are shown in Figures 18 and Figure 19, respectively. It is clear how the driving torque affects the controlled speed.

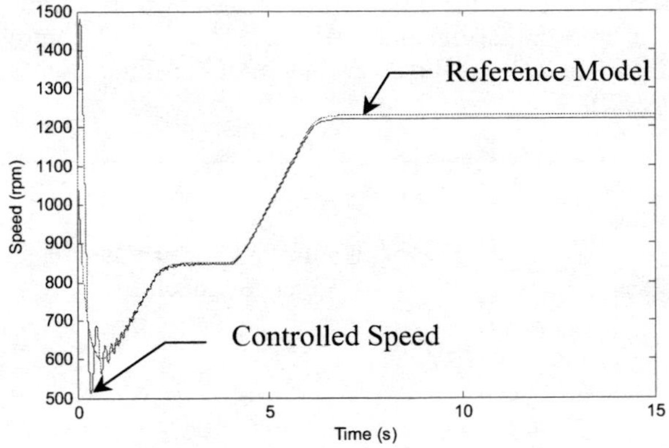

Figure 17: Load controller with sliding mode control performance

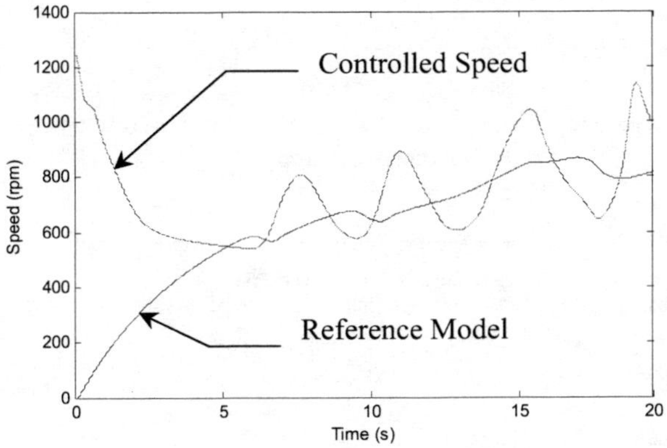

Figure 18: Experimental result of load controller with sliding mode control

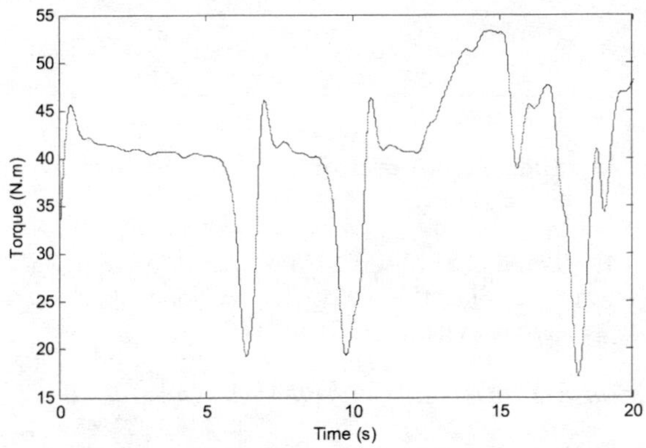

Figure 19: Driving torque

It should be noted here that the maximum bandwidth of the load model for this control configuration is 0.05 rad/s. While those of the actuator and engine-pump are 18 and 0.5 rad/s, respectively. The major reason that imposed the limitation to the load model bandwidth in our system is the actuator saturation. Instead of incorporating an anti-wind up scheme which is beyond the scope of this study, the load model is chosen to be able to control within the operation range of the actuator. The experimental result of the engine load simulator when the load model bandwidth is beyond that of the system is shown in Figure 20 (torque is varied from 0-25% of throttle valve opening).

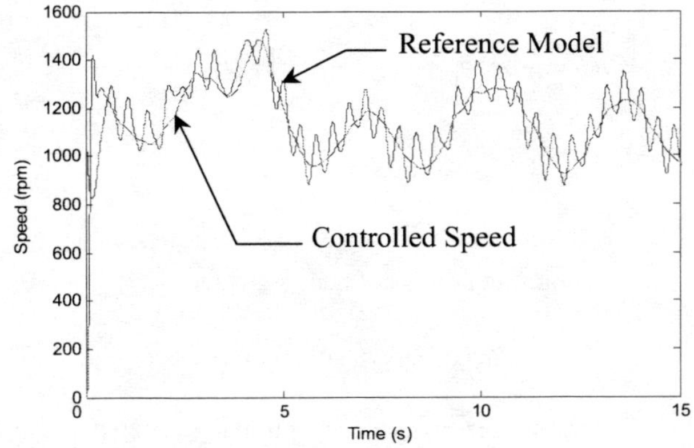

Figure 20: Sliding mode control for a load resistive controller performance when the bandwidth of the load model is too high.

Variable Structure Control for a Resistive Load Controller with Integral Control and Full-State Valve Feedback

As reported by Chern and Wu [9], incorporating an integral control can eliminate steady state error for a direct control of the robot manipulator. Similarly, the derivation of the variable structure controller which includes integration begins with the speed difference (equation 13 and equation 14). However, the control structure for indirect control is shown in Figure 21.

As can be seen from Figure 21, the dynamic characteristics of the valve are taken into account in the derivation of the controller as will be outlined in this section. If the state equation of the load simulator can be written as

$$\dot{n}_{p1} = n_{p2} \tag{12}$$

$$\dot{n}_{p1} = -a_{p1}n_{p1} - \frac{G_v D_P P_{ref}}{J_{sp}} + \frac{D_p P_d}{J_{sp}} \tag{13}$$

Where P_{ref} is the reference pressure, P_d is the pressure in the system, J_{sp} is the equivalent moment of inertia of the plant system, finally, G_v and a_{p1} are the transfer function of the valve and a parameter of the plant, respectively.

The state equation of the reference model is

$$\dot{n}_{m1} = -a_{m1}n_{m1} + \frac{D_p P_d}{J_{sm}} \tag{14}$$

Where n_m is the speed of the model, J_{sm} is the equivalent moment of inertia of the model system, and a_{m1} is the parameter of the model.

Subtracting equation (14) from equation (13), and applying integration to sliding condition. The system can be described as

$$\dot{e}_1 = e_2 = -a_{p1}e_1 + (a_{m1} - a_{p1})n_{m1} - \frac{G_v D_P P_{ref}}{J_{sp}} + \frac{D_p P_d}{J_{sp}} - \frac{D_p P_d}{J_{sm}} \quad (15)$$

While the sliding function is defined as

$$s = c_1(e_1 - k_I(-e_1)) + e_2 \quad (16)$$

In the similar manner as the conventional sliding mode, the derivatives and trajectory of the error has to be vanish. Therefore, equation (15) must satisfy,

$$s\dot{s} < 0 \quad (17)$$

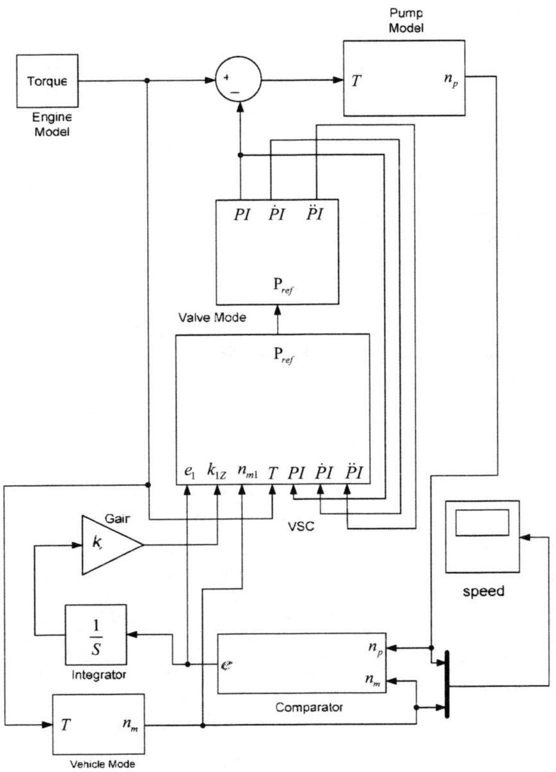

Figure 21: Variable structure control for resistive load controller with integral control and full-state valve feedback.

The control law (pressure reference) can be obtained (outlined in [9]) and the simulation result of the variable structure control for the load simulator can be illustrated in Figure 22. As seen from equation (15), the dynamic characteristics of the actuator are included in derivation of the control law. Therefore, the transient responses of the valve can be significantly reduced. In addition, due to the integral combination, the steady state response is also improved. The experimental result for this controller is shown in Figure 23. Note that the initial condition of the experimental controlled speed (Figure 23) is different from that of the simulation speed (Figure 22).

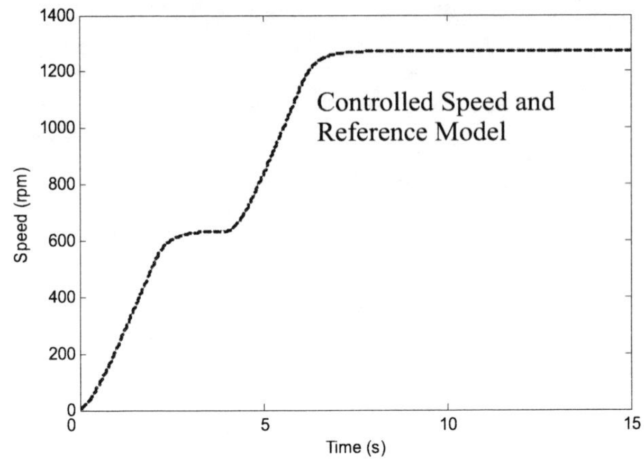

Figure 22: Variable structure control for a load resistive controller performance.

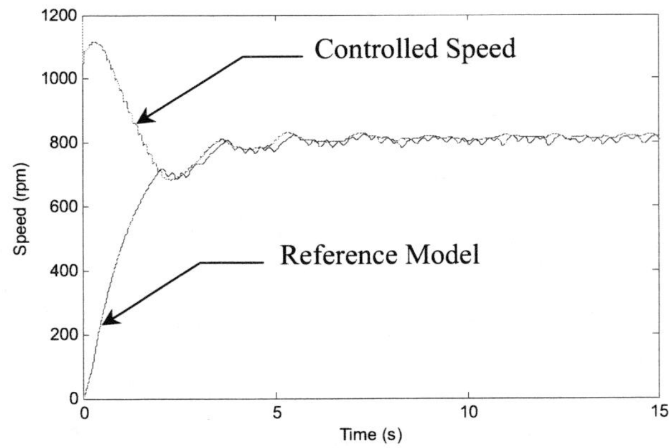

Figure 23: Sliding mode control for a load resistive controller performance.

It should be observed that an extension of this controller for non-linear models can be done by using non-linear expression in equation (13) and (14).

Non-Linear Resistive Load Controller with Integral Control

There is always a trade-off between performance and complexity. Therefore, every design is suitable for a specific circumstance. The following discussion will present a model reference control with integral control for a resistive load controller. For this, the transient effects of the valve can not be eliminated but the controller is simple. In addition, there is no need to linearize the load and the vehicle reference models. The block diagram representation is shown in Figure 24 and the simulation results are in Figure 25.

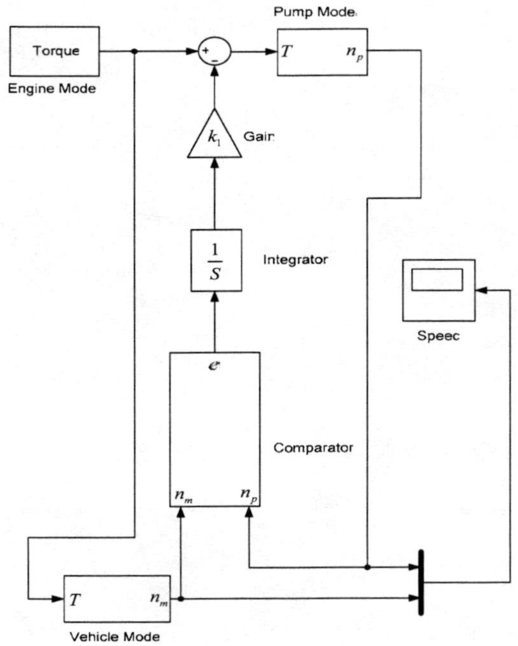

Figure 24: Non-linear resistive load controller with integral control.

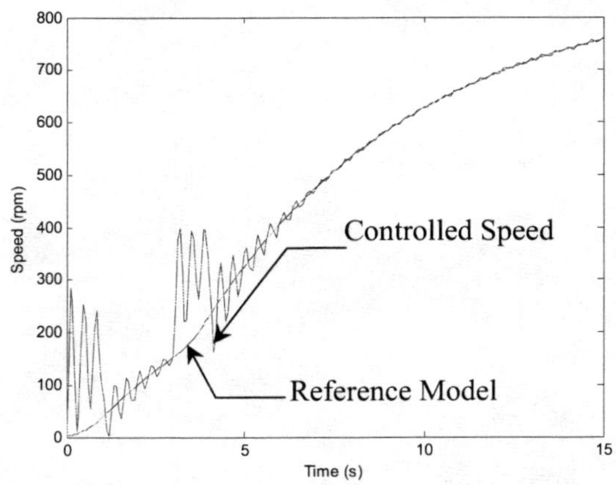

Figure 25: Non-linear resistive load controller with integral control performance.

4. Conclusions

The design details of an engine load simulator have been presented. With a resistive load controller, we can simply connect the load simulator to the engine to be tested. The simulator will continuously generate the dynamic load to the engine for any position of the drive pedal and engine speed corresponding to the reference vehicle model. To refine the simulator for the model reference control approach, the pressure control valve with dynamic pressure compensation has been presented. With pressure compensation, speed fluctuation due to transient response of the valve can be reduced. In case that machining facility is limited, a more complex controller such as a variable structure control can be derived and can take the actuator dynamics into account. By combining the advantages of the sliding mode and the integral control, the transient and steady state error can be reduced. Even though the variable structure control is complicated, it becomes attractive if the transient response is an important consideration. However, for a steady state testing, a model reference control with an integration would perform satisfactory.

ACKNOWLEDGMENTS

The authors would like to thank National Metal and Material Technology Center (MTEC) of Thailand for their research funding.

REFERENCES

[1] Dorey, R.E., Wang, D., Sept. 1989, "*A Hydraulic Dynamometer for Engine Testing*", 2nd Bath International fluid Power Workshop.

[2] Babbitt, G.R., Lumkes, J., Luceir, P., Fronczak, F.J., and Beachley, N.H., 1994 , "*Regenerative Testing of Hydraulic Pump/Motor System*", SAE Paper 941750.

[3] Moskwa, J.J., 1987, "*Automotive Engine Modeling for Real Time Control*", MIT Ph.D. Thesis.

[4] Lewendowski, L., Benson, D., Fronczak, F.J., and Beachley, N.H., 1987, "*Test Rig Design for Hydrostatic Accumulator Energy-Storage Automotive Research*", ASME Publication DE-Vol.11, Trends in Vehicle Design Research.

[5] Babbitt, G.R. and Moskwa, J.J., August 22-27, 1999, "*Implementation Details and Test Results for a Hardware in the Loop Vehicle Model*", Proceeding of the 1999 IEEE, International Symposium on Computer Aided Control System Design, Kohala Coast-Island of Hawaii, Hawaii, USA.

[6] Zhang, R. And Alleyne, A.G., November 17-22, 2002, "*Dynamic Emulation Using a Resistive Control Input*", Proceeding of IMECE2002, ASME International Engineering Congress & Exposition, New Orleans.

[7] Zhang, R. And Alleyne, A.G., 2001, "*A Model Reference Load Controller with Adaptation Using a Two Stage Pressure Relief Valve*", American Control Conference, Arlington, VA.

[8] Astrom, K. And Wittenmark, B., 1995, "*Adaptive Control*", Massachusetts, Addison-Wesley Reading.

[9] Chern, T.L. and Wu, Y.C., 1992, "*Integral Variable Structure Approach for Robot Manipulators*", IEE Proc., Pt D, 139, pp 161-166.

[10] Chern, T.L. and Wu, Y.C., 1992, "*An Optimal Variable Structure Control with Integral Compensation for Electrohydraulic Position Control System*", IEEE Trans. Industrail Electronics, 39, no.5, pp 460-463.

[11] Slotine, Jean-Jacques E. and Weiping, L., 1991, "*Applied Nonlinear Control*", Prentice-Hall International, Inc.

Proceedings of IMECE2005
2005 ASME International Mechanical Engineering Congress and Exposition
November 5-11, 2005, Orlando, Florida USA

IMECE2005-80138

DEVELOPMENT OF HYDRAULIC-ELECTRIC POWER UNITS FOR MOBILE ROBOTS

Justin W. Raade, Kurt R. Amundson, and H. Kazerooni
Department of Mechanical Engineering
University of California, Berkeley, California, 94720, USA
exo@berkeley.edu, http://bleex.me.berkeley.edu

ABSTRACT

Energetic autonomy of a hydraulic-based mobile robot requires a power source capable of both hydraulic and electrical power generation. The hydraulic power is used for locomotion and the electric power is used for the control computer, sensors and other peripherals. In addition, the power source must be lightweight and quiet. This study presents several designs of internal combustion engine-based power units. Each power unit is evaluated with a Ragone plot which shows its performance over a wide range of operation times. The best-performing design of the hydraulic-electric power units (HEPU), based upon the Ragone plot analysis, is described in detail. This HEPU produces constant pressure hydraulic power and constant voltage electric power. The pressure and voltage are controlled on board the power unit by a computer. A novel characteristic of this power unit is its cooling system in which hydraulic fluid is used to cool the engine cylinders. The prototype power unit weighs 27 kg and produces 2.3 kW (3.0 hp) hydraulic power at 6.9 MPa (1000 psi) and 220 W of electric power at 15 VDC.

INTRODUCTION

Most human scale and field robotic systems are currently powered by tethers or heavy battery systems. In order for a robotic device to obtain energetic autonomy free from tethers and heavy batteries, a compact, portable power unit providing both mechanical power for actuation and electrical power for computation and control is essential. The NiMH battery pack in ASIMO, Honda's humanoid walking robot, is one example of a battery-based power unit [1]. However, batteries have a low specific energy (energy per unit mass): 150 W-hr/kg for a high performance lithium ion battery [2]. Due to this low specific energy, batteries become large and heavy unless the operation time is short or the robotic system requires little power.

A fuel with a higher specific energy than batteries is desirable in a mobile robotic system. Internal combustion (IC) engines utilize the high specific energy of gasoline: 12000 W-hr/kg. The power units described here utilize IC engines to produce compact, lightweight power sources. This is primarily motivated by the fact that IC engines have been the primary source of power for automobiles, earthmoving machinery, motorcycles, and other wheeled vehicles. We envision mobile field robots as another class of these field vehicles that operate outdoors for periods of hours. In fact several field and service robotic systems have already experimented with IC engines as their prime mover [3]-[4].

This paper describes the basic design challenges of a generic hydraulic-electric power unit (HEPU) for robotic applications. First, an analysis of the HEPU systems based on Ragone plots is presented. Then the specifications of several versions of the HEPU are described. Finally, the final architecture of the HEPU, including its hydraulic and electric power generation, cooling system, and control, is described in detail.

RAGONE PLOT ANALYSIS

Ragone plots are useful for evaluating the performance of a power unit for a wide range of operation times [5]. They plot the power unit's specific power (power divided by total mass) versus the specific energy (energy divided by total mass). The specific energy $\hat{E}$ can be expressed by

$$\hat{E} = \frac{E_{tot}}{m_{tot}} = \frac{P_{tot}\tau}{m_{fuel} + m_{tank} + m_{eng}} \quad (1)$$

where E_{tot} is the total energy required for the operation time τ of the system, m_{tot} is the total power unit mass including the mass of the fuel m_{fuel}, fuel tank m_{tank}, and engine m_{eng}, and P_{tot} is the total output power (hydraulic power plus electrical power). The mass of the fuel is

$$m_{fuel} = \frac{E_{tot}}{\eta \hat{h}} \tag{2}$$

where $\hat{h}$ is the specific energy of the fuel and η is the overall efficiency. The overall efficiency is defined as the total output power divided by the fuel power (measured fuel flow rate $\dot{m}_{fuel}$ times specific energy $\hat{h}$ of the fuel) flowing into the engine.

$$\eta = \frac{P_{tot}}{\dot{m}_{fuel}\hat{h}} \tag{3}$$

The mass of the fuel tank is proportional to the mass of the fuel, and can be expressed by

$$m_{tank} = \frac{m_{fuel}}{\beta} \tag{4}$$

where β is a constant defined as the ratio of fuel mass to the mass of the tank required to hold the fuel. Using Equations (2) and (4) in Equation (1) and simplifying produces

$$\hat{E} = \left[\frac{1}{\eta \hat{h}}\left(1+\frac{1}{\beta}\right)+\frac{m_{eng}}{P_{tot}\tau}\right]^{-1}. \tag{5}$$

The specific power $\hat{P}$ can be similarly expressed as

$$\hat{P} = \frac{P_{tot}}{m_{tot}} = \frac{\hat{E}}{\tau}. \tag{6}$$

Several valuable trends can be learned by examining a Ragone plot. As the operation time τ becomes very long, the specific power $\hat{P}$ approaches zero while the specific energy $\hat{E}$ tends toward a finite value. This can be shown by examining Equations (5) and (6) as the operation time approaches infinity.

$$\lim_{\tau\to\infty} \hat{E} = \eta \hat{h}\left(\frac{\beta}{\beta+1}\right) \qquad \lim_{\tau\to\infty} \hat{P} = 0 \tag{7}$$

Alternatively, as the operation time becomes very short, the specific energy approaches zero while the specific power approaches a finite value. As the operation time approaches zero, Equations (5) and (6) become:

$$\lim_{\tau\to 0} \hat{E} = 0 \qquad \lim_{\tau\to 0} \hat{P} = \frac{P_{tot}}{m_{eng}} \tag{8}$$

If one is concerned with very long operation times for a mobile robotic power unit, then Equation 7 shows that the most critical parameters of the power unit are the *overall efficiency*, the *specific energy of the fuel*, and the *fuel storage ratio*. For a power unit designed for very short operation times, Equation 8 shows that the most critical parameters are the *total power* and the *engine mass* (the fuel and fuel tank mass are negligible).

Mobile robotics require lightweight power units in order to operate effectively in the field. Therefore, it is imperative to find the lightest power unit for a given operation time. Using a Ragone plot to evaluate the power units described in this work enabled us to determine the lightest choice for a given application. Figure 1 shows a Ragone plot of each HEPU system developed at the University of California, Berkeley. All the HEPU systems use gasoline with a fuel storage ratio $\beta = 4$ (based on a sturdy plastic tank holding the gasoline). The diagonal lines correspond to constant operation times on a logarithmic scale. Based on the information illustrated in Figure 1, for operation times less than approximately ten hours the lightest power unit is the ZDZ80-based HEPU due to its high specific power. For very long operation times greater than ten hours, the lightest system is the GX31-based HEPU due to its high overall efficiency. In general, two-stroke engines have more power per unit mass (specific power) than four-stroke engines, making the two-stroke HEPU systems lighter for short operation times. Four-stroke engines, on the other hand, have a higher efficiency than two-strokes, making the four-stroke HEPU systems lighter for long operation times.

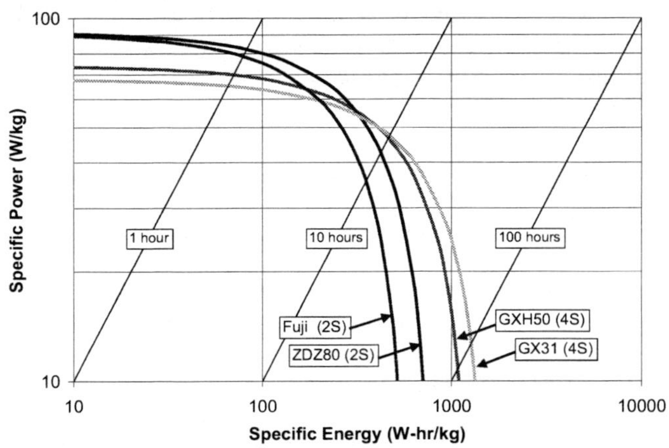

Figure 1: Ragone plot of HEPU systems, labeled by engine name. The two-stroke power units are labeled "2S" and the four-strokes are labeled "4S."

SEQUENCE OF HEPU DESIGNS

GENERAL ARCHITECTURE

All of the HEPU designs presented in this work have the same general architecture as shown in Figure 2. The engine provides the shaft power which turns a hydraulic pump to produce hydraulic flow for the actuators of the robotic system. The engine also spins an alternator which provides electrical power for the sensors, control computers, and ancillary equipment such as cooling fans or pumps. The power units described herein can be utilized by any hydraulic robotic system.

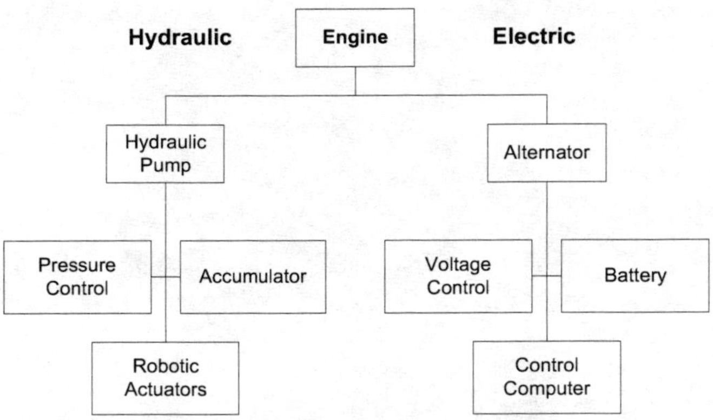

Figure 2: HEPU system architecture.

HONDA GX31-BASED HEPU

The Honda four-stroke engine GX31 serves as the prime mover in the first generation HEPU. A hydraulic schematic for the GX31-based HEPU is shown in Figure 3.

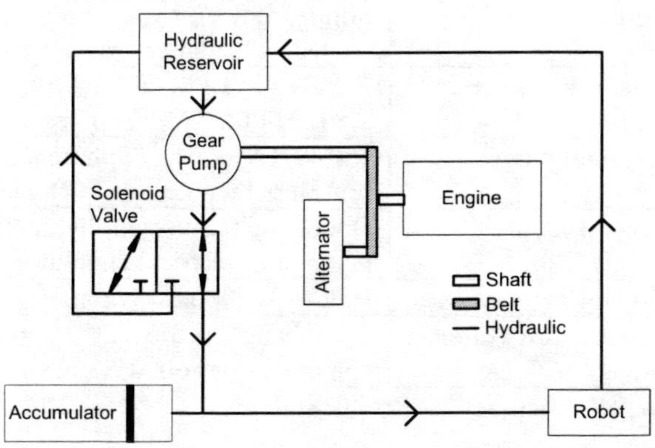

Figure 3: Hydraulic schematic of GX31 HEPU.

The engine spins a gear pump and alternator via a belt drive. A solenoid valve maintains the hydraulic pressure by rerouting the flow from the pump back to the reservoir when the pressure in the accumulator reaches the desired value. An independent speed governor on the engine maintains the electrical voltage by maintaining a constant engine speed. A photograph of the system is shown in Figure 4. The system specifications are listed in Table 1. The GX31 HEPU proved to be a lightweight power unit suitable for a robotic system with low hydraulic flow requirements.

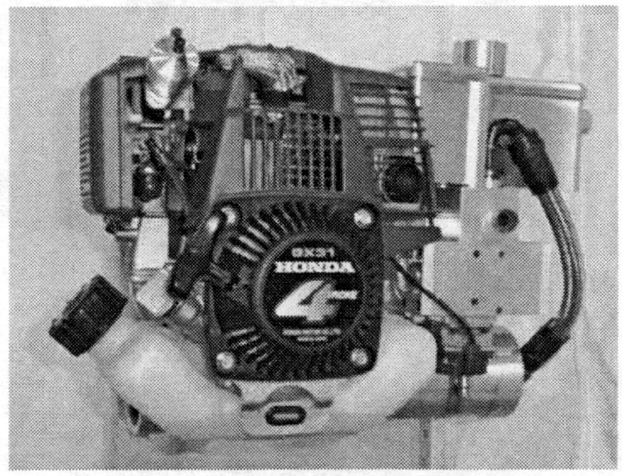

Figure 4: Honda GX31 HEPU.

Description	First generation HEPU
Engine	Honda GX31 4-stroke, air cooled, centrifugal clutch
Hydraulic power	750 W (5.3 LPM @ 6.2 MPa) 1.0 HP (1.4 GPM @ 900 psi)
Electrical power	120 W (Aveox three phase brushless motor/alternator)
Transmission	Double belt drive with hydraulic pump and electrical alternator
Hydraulic pump	Haldex W600 gear pump
Pressure and voltage control	Solenoid hydraulic bypass valve, bang-bang throttle control
Overall efficiency (η)	16%
Mass without fuel or tank (m_{eng})	13 kg (28 lbs)

Table 1: GX31 HEPU specifications.

HONDA GXH50-BASED HEPU

The second generation HEPU has the same hydraulic and electrical architecture as the first generation HEPU (Figure 3), but utilizes the Honda four-stroke engine GXH50 for a significant power increase over the GX31. This power unit was used extensively for field testing of the Berkeley lower extremity exoskeleton (BLEEX) [6]-[8]. Figure 5 shows a photograph of the system. The system specifications are listed in Table 2.

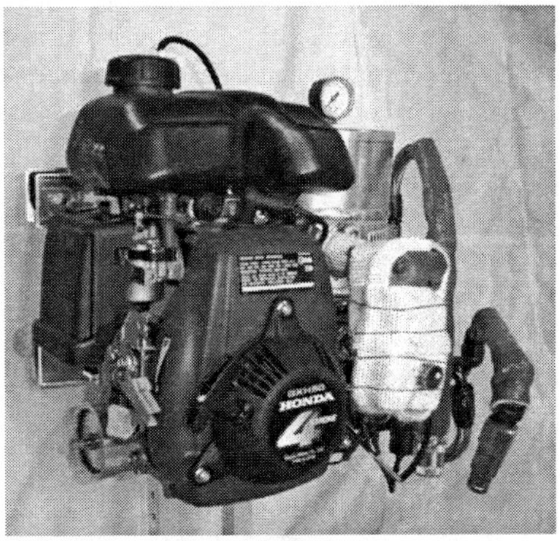

Figure 5: Honda GXH50 HEPU.

Description	Second generation HEPU
Engine	Honda GXH50 4-stroke, air cooled, centrifugal clutch
Hydraulic power	1.0 kW (9.1 LPM @ 6.9 MPa) 1.4 HP (2.4 GPM @ 1000 psi)
Electrical power	100 W (Faulhaber three phase brushless motor/alternator)
Transmission	Double belt drive with hydraulic pump and electrical alternator
Hydraulic pump	Haldex W600 gear pump
Pressure and voltage control	Solenoid hydraulic bypass valve, stock governor throttle control
Overall efficiency (η)	13%
Mass without fuel or tank (m_{eng})	15 kg (34 lbs)

Table 2: GXH50 HEPU specifications.

FUJI BT50SA-BASED HEPU

A two-stroke engine (the Fuji BT50SA) was chosen for a new HEPU design due to its light weight and high power output [9]. Figure 6 shows a hydraulic schematic of the Fuji-based HEPU. In contrast to the previous HEPU systems, an electrically-actuated clutch maintains hydraulic pressure by disconnecting the engine from the pump when the desired pressure is reached in the accumulator. A CAD image of the system is shown in Figure 7. The system specifications are listed in Table 3. The Fuji-based HEPU was evaluated on a test stand but not fully constructed due to its high noise level.

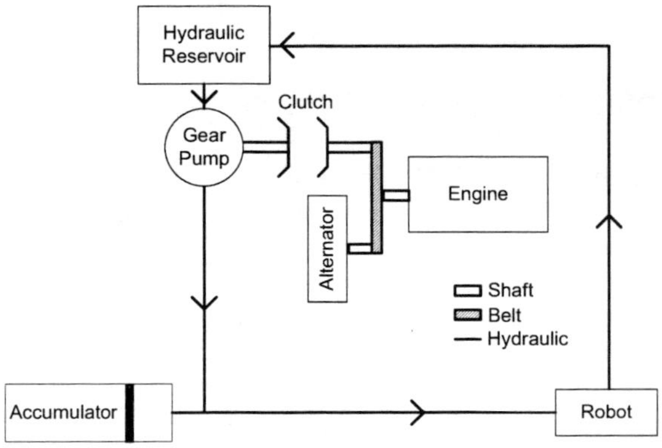

Figure 6: Hydraulic schematic of Fuji BT50SA HEPU.

Figure 7: CAD model of Fuji BT50SA HEPU.

Description	Third generation HEPU
Engine	Fuji BT50SA 2-stroke, liquid cooled, solenoid activated clutch
Hydraulic power	1.1 kW (15 LPM @ 4.1 MPa) 1.5 HP (4.0 GPM @ 600 psi)
Electrical power	120 W (Aveox three phase brushless motor/alternator)
Transmission	Double belt drive with hydraulic pump and electrical alternator
Hydraulic pump	Haldex W600 gear pump
Pressure and voltage control	Bang-bang electric clutch bypass, computer-controlled engine speed
Overall efficiency (η)	6.0%
Mass without fuel or tank (m_{eng})	14 kg (30 lbs)

Table 3: Fuji BT50SA HEPU specifications.

ZDZ80-BASED HEPU

A HEPU system with integrated noise deadening measures was realized with a ZDZ80 two-stroke engine. Figure 8 shows a CAD image of the system. The system specifications are listed in Table 4. See the section on the ZDZ80 HEPU for a detailed description of the design of this power unit.

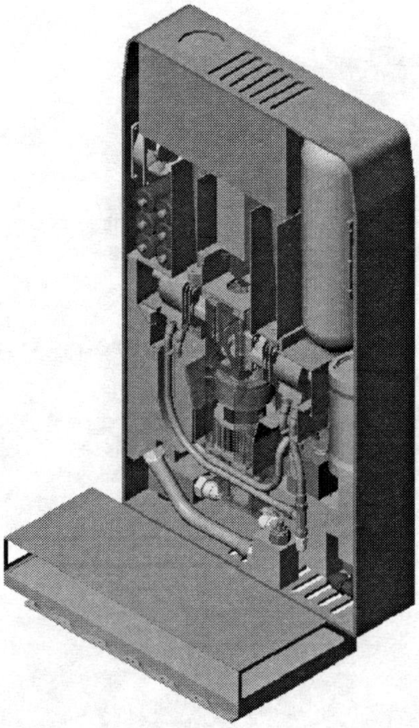

Figure 8: ZDZ80 HEPU.

Description	Fourth generation HEPU
Engine	ZDZ-80 2-stroke, liquid cooled, direct drive
Hydraulic power	2.3 kW (20 LPM @ 6.9 MPa) 3.0 HP (5.2 GPM @ 1000 psi)
Electrical power	220 W (Kollmorgen RBE series, three phase brushless motor/alternator)
Transmission	In-line direct drive of pump and alternator
Hydraulic pump	Haldex W300 gear pump
Pressure and voltage control	Solenoid hydraulic bypass valve, computer-controlled engine speed
Overall efficiency (η)	8.1%
Mass without fuel or tank (m_{eng})	27 kg (59 lbs)

Table 4: ZDZ80 HEPU specifications.

DETAILED DESIGN OF ZDZ80-BASED HEPU

The following sections summarize a more complete description of the ZDZ80-based HEPU design found in [10].

HEPU SPECIFICATIONS

The design requirements for a mobile fieldable robotic system are functions of the robot size, its maneuvering speed, and its payload capability. The main feature of many field robotic systems that affects the design of their power units is the load carrying capability in the field.

While high pressure hydraulics often lead to less power loss, we chose 6.9 MPa (1000 psi) as the system pressure. This leads to more reasonable hydraulic components for mobile systems that need to work in the field and perhaps in proximity of humans. The hydraulic flow requirements are usually calculated using the speed characteristics of the robot. High speed movements lead to large hydraulic flow requirements. In the case of the BLEEX project, the walking speed from CGA (clinical gait analysis) data resulted in 20 LPM (5.2 GPM) of hydraulic flow [6]. Our experience in building various exoskeleton systems suggest that one requires approximately 220 W of electric power for on-board robot computers and sensors as well as the power unit sensors and controller.

OVERALL HEPU ARCHITECTURE

The HEPU is designed to provide electric and hydraulic power. It uses a compact two-stroke opposed twin cylinder IC engine capable of all-angle operation. Figure 9 and Figure 10 show how the engine (1) drives a single shaft (2) to power an alternator (3) for electric power generation, a cooling fan (4) for air circulation, and a gear pump (5) for hydraulic power generation. This single shaft design elegantly avoids noisy and heavy belt drive mechanisms common in systems comprising many rotating shafts. A hydraulic solenoid valve (7) regulates the hydraulic fluid pressure by directing the hydraulic flow from the gear pump to either an accumulator (10) or to the hydraulic reservoir (13). The accumulator consists of an aluminum cylinder in which a free piston separates the hydraulic fluid from the pressurized nitrogen gas. A carbon fiber tank (11) is attached to the gas side of the accumulator as reservoir for the nitrogen gas. In general, the larger the volume of this gas reservoir is, the smaller the pressure fluctuation will be in the presence of hydraulic flow fluctuations. A pressure transducer (9) measures the pressure of the hydraulic fluid for the controller. A manifold (6) is designed to house both the solenoid valve (7) and filter (8). A novel liquid cooling scheme utilizes the returning hydraulic fluid itself to cool the engine. The hydraulic fluid from the robot actuators is divided into two paths. Approximately 38% of the hydraulic fluid is diverted to cool the engine cylinders. A heat exchanger (12) removes the heat from this hydraulic fluid before it reaches the hydraulic reservoir (13) and is mixed with the remaining 62% of the fluid.

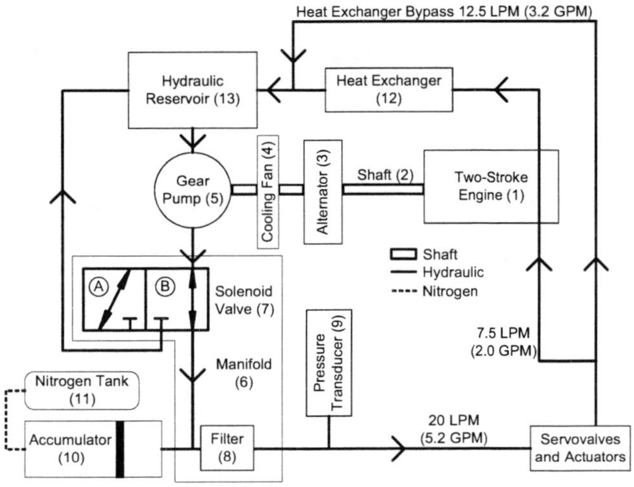

Figure 9: HEPU schematic layout. Components labeled with numbers in parentheses also correspond to Figure 10.

HYDRAULIC POWER PRODUCTION

The two-stroke opposed twin cylinder IC engine (model 80 B2 RV, manufactured by ZDZ Model Motor) capable of producing 6 kW (8.1 hp) of shaft power at 8200 rpm is used as the prime mover of this power unit. This engine has an 80 cm^3 displacement and weighs only 2 kg (4.4 lbs). Since the gear pump was limited to turn at maximum speed of 6300 rpm and since we did not intend to utilize any transmission speed reducer in this power unit, we were forced to drive the engine at speeds lower than the maximum-power speed of the engine. The engine can produce approximately 3.06 kW (4.0 hp) of shaft power at 6300 rpm which is adequate to produce the total output hydraulic and electrical power (2.5 kW or 3.4 hp). The engine is controlled with a servo motor mounted to its throttle. The engine directly drives an alternator, a cooling fan and a gear pump. The pump (model WP03-B1B-032L-20MA12, manufactured by Haldex) has a 3.2 cm^3 displacement volume per revolution.

CONTROL

The HEPU controller measures and regulates two important variables: engine speed and hydraulic pressure. A unique control scheme was needed to maintain constant operating pressure with a fixed displacement pump running at a constant speed. An accumulator at the outlet of the pump supplies the fluid to the actuators and functions like a capacitor to compensate for transient peak flows. The hydraulic pressure is read by the pressure sensor. The computer controls the solenoid valve to maintain the pressure. When the pressure reaches the desired value (6.9 MPa in this case), the computer diverts the hydraulic flow to the reservoir by moving the valve to position A in Figure 9. To prevent pressure drop in the accumulator when the hydraulic fluid in the accumulator is consumed by the servovalves and the actuators, the computer diverts the flow back to the accumulator by moving the valve to position B. The modulation of this valve based on the measured pressure allows the system to produce hydraulic power at near-constant pressure. The operating pressure in the accumulator is maintained in a band of 6.9 +/- 0.2 MPa (1000 +/- 30 psi).

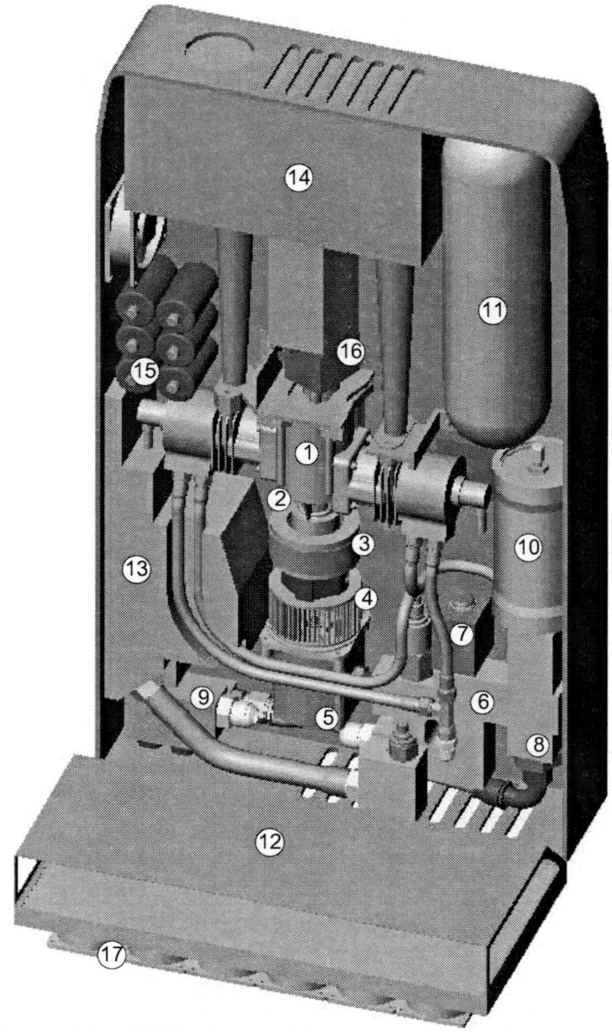

Figure 10: HEPU physical layout. Engine (1); shaft (2, not visible); alternator (3); cooling fan (4); gear pump (5); manifold (6); solenoid valve (7); filter (8, not visible); pressure transducer (9, not visible); accumulator (10); nitrogen tank (11); heat exchanger (12); hydraulic reservoir (13); muffler (14); batteries (15); carburetor and throttle (16); heat exchanger fans (17). Internal baffling around engine is not shown for clarity.

When the solenoid valve diverts the hydraulic fluid to the reservoir, the engine speed increases rapidly. The opposite is also true: when the valve diverts the hydraulic fluid to the accumulator, the engine speed decreases rapidly and the engine might even stall. The engine speed variation leads to a large voltage variation. Additionally the high engine speeds might damage the pump. For the above reasons, it is desirable to control the engine speed to a constant value. It was decided to maintain the speed at 6300 rpm (maximum allowable pump speed). The engine speed is measured by counting the pulses from a Hall effect sensor on the alternator. An on-board computer modulates the engine throttle with a servo.

COOLING

Since the engine was designed for high performance model aircrafts, it requires a large amount of air for cooling its cylinders (air is generously available when the engine is installed on aircraft models.) For the application of field robotics, it is necessary to package the engine tightly in a sound-deadening shield; therefore liquid cooling was required. A novel liquid cooling scheme was devised that uses the hydraulic fluid itself to cool the engine. The engine cylinder heads were modified to allow hydraulic fluid to pass through them and absorb heat. This makes the addition of a water-based cooling system unnecessary and results in a simplified system with fewer components. Using the hydraulic fluid as the cooling medium increases the load on the heat exchanger since the heat from the engine must be removed to prevent the hydraulic fluid from exceeding the operating temperature of any hydraulic components. The maximum temperature allowable was determined by the pump which had the lowest temperature tolerance of any component in the system (the gear pump required hydraulic fluid temperature cooler than 65° C or 149° F).

The fluid returning from the actuators is split into two separate paths, as shown in Figure 11. Approximately 62% of the hydraulic fluid returns directly to the reservoir. The remaining 38% passes first through the cylinder heads where excess heat is extracted from the engine, then through a heat exchanger where the heat in the fluid is dissipated, and finally returns to the reservoir. As shown in Figure 11, the heat exchanger must remove the heat generated from the dissipative effect of the servovalves on the actuators in addition to the heat generated in the engine cylinder heads. Increasing fluid volume in the reservoir increases convective heat transfer (cooling) to ambient air and allows longer operation times. This is a typical solution in industrial hydraulics, but is not feasible in this application where a large reservoir is undesirable. Therefore, careful sizing of the heat exchanger was critical to ensure adequate cooling at a minimum weight.

A thermal model was created (using measured data from a test stand whenever possible) to estimate the behavior of the hydraulic system and evaluate the hydraulic fluid temperature at the most sensitive component, the pump [10]. At steady state the heat exchanger removes 4.00 kW of heat and the calculated pump inlet temperature is 61°C (141°F), under the maximum allowable pump temperature, 65°C.

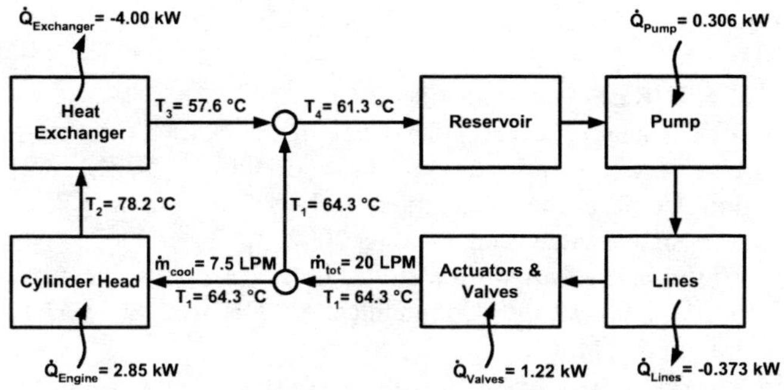

Figure 11: Cooling system schematic of the HEPU with calculated heat flow and temperature values.

ELECTRICAL POWER GENERATION

The HEPU generates electrical power for the sensors, cooling fans, and the control computer. The total electrical system power budget is 220 W, with 100 W for cooling fans and 65 W for the control computer and sensors. The remaining 55 W are expected to be consumed in losses and other peripheral components. A three phase, 12-pole frameless, brushless DC motor (model RBE-1812, manufactured by Kollmorgen) is used as an electric power alternator (3 in Figure 10). The three phases were converted to single-phase, 240 VDC by a bridge rectifier (the back EMF constant of the motor is 26.9 V/krpm so that at the operational speed of 6300 rpm the rectified voltage is 240 VDC). Two DC-DC converters are used to create two 15 VDC bus voltages to be used for two sets of components. One 15 VDC line is used to power the electrically noisy components such as solenoid valves, cooling fans, and the ignition for the engine. The second 15 VDC line is used to charge a set of batteries, power the control computer, HEPU controller, and the throttle servo. External power is used to power the system when the engine is off. A battery powers the control computer, HEPU controller, throttle servo, and sensors for a short time in case the engine shuts down. This gives the operator ample time to connect external power to the system.

CONCLUSION

Gasoline-fueled hydraulic power units are capable of producing mobile robotic power. When the system mass is the most critical parameter, Ragone plots are very useful for determining the best design of a power unit that minimizes the system mass for a given operation time. The ZDZ80-based HEPU as designed produces constant-pressure hydraulic power (2.3 kW) and constant-voltage electrical power (220 W) for a mobile robotic system.

ACKNOWLEDGMENTS

This work is partially funded by DARPA grant DAAD19-01-1-0509.

REFERENCES

[1] Y. Sakagami, et al, "The Intelligent ASIMO: System Overview and Integration," Proc. of the 2002 IEEE/RSJ Intl. Conference on Intelligent Robots and Systems EPFL, Lausanne Switzerland, October 2002, pp. 2478-2483.

[2] *High Energy Lithium-Ion Cell, VL 45 E cell*, document no. 54032-2-0603, the Communication Department, SAFT, Bagnolet, France.

[3] P. Weiss, "Dances with Robots", *Science News*, vol. 159, no. 25, June 2001, pp. 407-408.

[4] E. Colin, D. Polome, V. Piedfort, and Y. Baudoin, "AMRU 3: a teleoperated six-legged electro-hydraulic robot," In *Mobile Robots IX*, William J.Wolfe, Wendell H. Chun, Editors, Proc. SPIE 2352, 1995, pp. 192-203.

[5] J. W. Raade and H. Kazerooni, "Analysis and design of a novel hydraulic power source for mobile robots," *IEEE Trans. on Automation Sci. and Eng.*, vol. 2, no. 3, pp. 226-232, July 2005.

[6] A. Chu, H. Kazerooni, and A. Zoss "On the Biomimetic Design of the Berkeley Lower Extremity Exoskeleton (BLEEX)," Proc. of IEEE International Conference on Robotics and Automation, Barcelona, Spain, 2005.

[7] H. Kazerooni, L. Huang, J. L. Racine, and R. Steger "On the Control of the Berkeley Lower Extremity Exoskeleton (BLEEX)," Proc. of IEEE International Conference on Robotics and Automation, Barcelona, Spain, 2005.

[8] A. Zoss and H. Kazerooni "On the Mechanical Design of the Berkeley Lower Extremity Exoskeleton (BLEEX)," Proc. of IEEE International Conference on Intelligent Robots and Systems, Edmonton, Alberta, Canada, 2005.

[9] C. Grimmer, "Mechanical Design and Testing of an Exoskeleton Hydraulic Power Unit," Master of Science thesis, University of California, Berkeley, 2005.

[10] K. R. Amundson, J. W. Raade, N. Harding, H. Kazerooni, "Hybrid hydraulic-electric power unit for field and service robots," Proc. IEEE International Conference on Intelligent Robots and Systems, Edmonton, Alberta, Canada, 2005.

Proceedings of IMECE2005
2005 ASME International Mechanical Engineering Congress and Exposition
November 5-11, 2005, Orlando, Florida USA

IMECE2005-80305

SLIDING MODE CONTROLLER AND FILTER APPLIED TO A MODEL OF AN ELECTROHYDRAULIC ACTUATOR SYSTEM

SHU WANG **RICHARD BURTON** **SAEID HABIBI**

College of Engineering University of Saskatchewan
Saskatoon, Saskatchewan

ABSTRACT

A new robust state and parameter estimation strategy called the Variable Structure Filter (VSF) has recently been proposed and used for state and parameter estimation. A very common problem of linear stochastic systems is to design a combined robust control and estimation strategy, given system and noise uncertainties. Variable Structure Control (VSC) and its special form of Sliding Mode Control (SMC) show superb robustness. This paper proposes a new strategy involving the Sliding Mode Control and the Variable Structure Filter. Both the estimator and controller are based on the concepts of Variable Structure Systems (VSS). In the presence of bounded parametric uncertainties and noise, a robust stability is guaranteed. Further more, the combined strategy can be used to achieve high regulation rates or short settling times.

The object of this paper is to introduce this combined VSF and SMC strategy and to demonstrate its application to a third order model of a high precision hydrostatic system, referred to as the Electrohydraulic Actuator System (EHA).

1 INTRODUCTION

It is often an objective to force a plant to follow a prescribe path (sometimes referred to as trajectory tracking control). The presence of noise and nonlinearities in the actual plant makes this very difficult to do precisely. So an appropriate controller is introduced, which will modulate the inputs to the plant according to some algorithm which will make the error between the desired path and the actual path very small. Many controllers must have some information about the plant (i.e. a model), from which the controller can determine the appropriate signal to input to the plant. However, the measurement and plant noise combined with an inability to precisely model the plant (uncertainties) is very problematic for the control algorithm. The main problem then becomes one of developing an appropriate control strategy which reflects this uncertainty. One way to minimize the problem is to account for model uncertainties on the design of algorithm and to estimate as accurately as possible, the model states. A control strategy based on the concepts of the Variable Structure Filter (VSF) and the Sliding Mode Control (SMC) which does include uncertainties in its formulation, defined as the Sliding Mode Controller and Filter (SMCF) is introduced in this paper.

The design of Sliding Mode control (SMC) requires the specification of several "operational modes". The term operational mode refers to different techniques which relate to the choice of the type of the sliding surface or mode "*S*". There are different ways in which the surface function, *S*, can be defined such that it is easier and more effective to solve various types of problems. The most common operational modes of SMC are regulation (usually for regulator systems) [1], trajectory following (usually mechanical motion, such as robot manipulator trajectory) [2], model following (usually encountered in the application of optimal control system), [3], and observation (for systems requiring state estimation) [4].

Associated with these modes, there is typically a plant model that models the transformation of the plant state which can be represented as a linear stochastic system Sensor noise is common in stochastic systems and is often attributed to Electromagnetic interfaces. A second category of noise referred to as system noise is mainly a consequence of the fact that the plant model is not exactly known, or that only an approximation of some transformation models is available.

Many studies on modeling plant uncertainties have focused on establishing the "matching conditions" where the uncertainties lie in the "image" of the input matrix [5] [6]. What this really means is that the uncertainties can only affect the plant dynamics through the same channels as the plant's input. To illustrate this, the following definition of a nonlinear model with uncertainties may be described as:

$$\dot{X}(t) = [f(x,t) + \Delta f(x,t)] + [B(x,t) + \Delta B(x,t)]u(x,t) + d(t) \quad (1)$$

where Δf and ΔB are system uncertainties, and *d(t)* is an external disturbance. If $\Delta f, \Delta B$ and $d(t)$ lie in the image of *B(x, t)* for all x and t, they can all be lumped into a single vector function $\Delta D(x,t,d,u)$, so that:

$$\dot{X}(t) = f(x,t) + B(x,t)u(x,t) + B(x,t)\cdot \Delta D(x,t,d,u) \tag{2}$$

Thus, uncertainties in this system will be considered to satisfy the matching condition if the system equation can be written in the form of Eq. (2). Many system uncertainties, however, cannot satisfy this requirement and pose a challenge for control.

Misawa [7] proposed a robust sliding mode controller with uncertainties which did not satisfy the matching conditions. Thus the more general problem of uncertain systems may be dealt with in the stochastic environment.

A general problem in linear systems theory is to estimate the internal states from the system's outputs. One such common estimator is the Kalman filter, and has been the subject of much research in the past. There are, however, other strategies which have been applied for state and parameter estimation and are called "sliding" observers such as proposed in [8], [4], [9]. A new strategy referred to as the Variable Structure Filter was proposed in [10]. The VSF is a new model-based strategy that can be used for estimating the states of an observable system. The VSF uses concepts closely related to Variable Structure Control. However its implementation procedure is similar to Kalman Filter.

In this paper, a new combined controller and filter strategy is proposed and is referred to as sliding mode controller and filter (SMCF). This control strategy "fuses" a discrete-time filter and controller based on Variable Structure Systems theory. The SMCF demonstrates robust performance to system and measurement uncertainties. The filter and controller are smoothly fused together because the structures in both are deliberately allowed to change. The SMCF approaches the estimation and control problem by separating the filtering and control law into different function blocks unlike other sliding mode observers. This separation facilitates understanding the attributes associated with the system's performance. In addition, the SMCF can handle more general uncertainties, i.e. unmatched uncertainties.

The paper is organized as follows. The nomenclature used in this paper is listed in Section 2. The system definition and problem statement is presented in Section 3. Section 4 proposes the Sliding Mode Controller and Filter. Its application to a third order model of a Hydrostatic system referred to as Electro Hydraulic Actuator (EHA) system is provided in Section 5. Further remarks on SMCF are given in Section 6. The conclusions are provided in Section 7.

2 NOMENCLATURE

Table 1 gives all parameters and coefficients and their comments and dimensions used in this paper.

Table 1 Nomenclature

Symbol	Comments	Dimensions
$\hat{\ }$	Denotes uncertain values	
$\| \ \|_{ABS}$	Absolute value	
A, B, ΔA, ΔB	Matrix	
C	Sliding coefficient	$1\times n$
e	State estimation error	$n\times 1$
$e_{z_{k\|k}}, e_{z_{k\|k-1}}$	Output estimation error calculated by using the a posteriori and a priori output estimates	$m\times 1$
f, F	Vector and its upper bound	$n\times 1$
G,	Input matrix	$n\times p$
G_h	Transfer function	
H	Output matrix	$m\times n$
i, j	Subscripts used to identify elements of matrices and vectors	1×1
k	Calculation step index	1×1
K	Sliding gain	1×1
K_k	VSF gain	$n\times 1$
k_{pos}	Proportional gain	1×1
m	Number of measurements	1×1
n	Number of states	1×1
p	Number of inputs	1×1
S	Switching function	$m\times 1$
sat	Saturation function	
sgn	Sign function	
T_s	Sampling period	
u, u_{eq}	Control input and equivalent	1×1
v, $v_{\max}$	Measurement noise and its upper bound	$m\times 1$
w, $w_{\max}$	System noise and its upper bound	$n\times 1$
x, x_o	System states and their initial condition	$n\times 1$
X, X_d, X_e	States vector and desired state vector and their error vector	$n\times 1$
$x_{k\|k}, x_{k\|k-1}$	A posteriori and a priori state estimates	$n\times 1$
z	Measured output	$m\times 1$
$\hat{z}_{k\|k}, \hat{z}_{k\|k-1}$	A posteriori and a priori output estimates	$m\times 1$
δ, σ	Arbitrary positive number	1×1
Φ	System matrix	$n\times n$
γ	Constant diagonal gain matrix with elements $\gamma_{ij} \geq 1$	$n\times n$
Ψ, Ψ_c, Ψ_f	Boundary layer	$n\times 1$

3 SYSTEM DEFINITION AND PROBLEM STATEMENT

In the following sections, an introduction to SMCF is presented. It is recognized that many of the concepts are well established, but since this is the first time that the VSF is integrated to a VSC and because of the mixed background of the audience this paper has been directed towards, it was deemed prudent to included more detail than would normally be necessary.

In practice, there are many application-specific approaches in which control algorithms require the estimation of some internal states from a limited set of measurements. However, many of such proposed methods do not inherently take into consideration the uncertain nature of the system and measurements. The fundamental source of information is the same: estimates are derived from noisy measurements using mechanical, inertial, optical, acoustic, or magnetic sensors. These uncertainties can be statistically characterized, leading to stochastic methods for addressing the problems. To illustrate this, consider the following continuous-time plant with uncertainties ΔA, and ΔB:

$$\dot{X}(t) = (A + \Delta A)X(t) + (B + \Delta B)u(t) + Dv(t) \tag{3}$$

where, $X(t) \in \Re^n$ $u(t) \in \Re^p$ and $v(t) \in \Re^l$ are, respectively, vectors of the state, control and external disturbance; the pair (A, B) are controllable.

A discrete-time mathematical model of the plant of (3) is given as,

$$X(k+1) = \Phi X(k) + Gu(k) + w(k). \tag{4}$$

where:

$$\Phi = e^{AT_s};\ G = \int_0^{T_s} e^{A\tau} B d\tau;\ w(k) = w_v + w_A + w_B;$$

$$w_v = \int_0^{T_s} e^{A\tau} Dv((k+1)T_s - \tau)d\tau;$$

$$w_A = \int_0^{T_s} e^{A\tau} \Delta A((k+1)T_s - \tau)X((k+1)T_s - \tau)d\tau;$$

$$w_B = \int_0^{T_s} e^{A\tau} \Delta B((k+1)T_s)d\tau. \tag{5}$$

and T_s is the sampling time. The pair (Φ, G) is controllable because (A, B) is controllable.

In many studies, the parameter variations $\Delta A, \Delta B$ and external disturbances in Eq. (3) have to be assumed to be matched, i.e. the $rank[B \mid \Delta A \mid \Delta B \mid D] = rank[B]$ is satisfied [11]. The control strategy of SMCF in this paper is developed for more general systems, i.e. it assumes that the discrete controller and filter are robust without satisfying the above matching condition.

In trajectory following problems, the sliding surface is defined in terms of the error rather than the states as is the case in regulation control. Assume that the control problem is to constrain the states X to follow a prescribed trajectory X_d. Let:

$$X_d(k) = [x_d(k), \dot{x}_d(k), \cdots, x_d^{(n-1)}(k)] \tag{6}$$

Consider Fig. 1 in which a second order sliding surface is shown. In setting up the describing conditions for the Sliding Mode Controller or Filter, it is useful to refer to this figure for visual understanding.

Definition 2.1: Define the sliding hyperplane to be:

$$S(k) = CX_e(k) = 0 \tag{7}$$

where $X_e(k) = X_d(k) - X(k)$. $S(k)$ is referred to as the switching function and C is a real-valued constant vector such that the product $CG \neq 0$, where G is the input matrix as defined in Eq. (4). The sliding surface is, physically, a surface in which the states are forced to move along in some prescribed fashion. If $S(k) = 0$, then all states follow exactly the desired path. See Fig. 1.

A "reaching condition" (a condition which guarantees that the states will reach $S = 0$) can be stated as [12], [13]:

$$S_i \dot{S}_i < 0 \quad i = 1, \cdots, m. \tag{8}$$

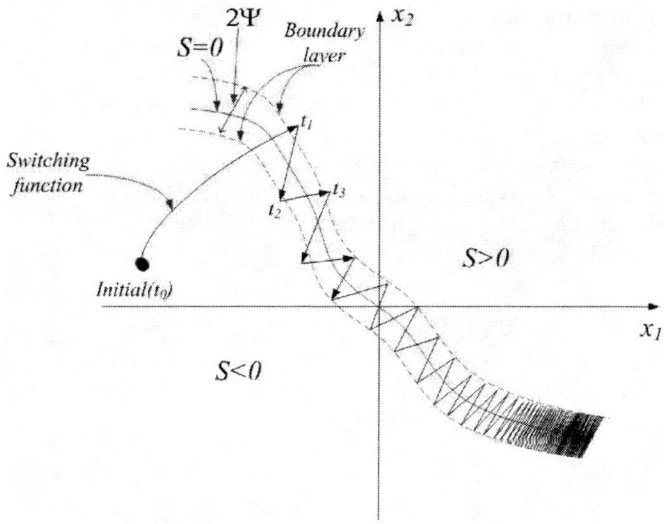

Fig. 1 Sliding mode of discrete-time second order systems

Corollary 2.1: For the discrete-time system, the pseudo-sliding mode (the discrete version of the continuous sliding surface) is said to exist if the following condition is satisfied.

$$\nabla S(k) S(k) < 0$$

$$where\ \nabla S(k) = S(k+1) - S(k) \tag{9}$$

Another "sufficient" but not "necessary" condition can also be defined as [14]:

$$|S(k+1)| < |S(k)| \tag{10}$$

Definition 2.2: Introduce an "attractive" boundary layer around the sliding surface S (note, in this layer, the reaching condition must be satisfied, in that the states must approach the surface $S = 0$ at least asymptotically [5]). This boundary layer is defined as:

$$S_\Psi(k) = \{X_e(k) \| S(X_e(k)) | \le \Psi\}, \tag{11}$$

in which the switching function $S(X_e(k))$ is not greater than some value of Ψ.

4 SLIDING MODE CONTROLLER AND FILTER

Using the concepts and the definition in section 3, it is now possible to introduce the SMCF structure. Consider Fig. 2. From the actual plant, only some of the outputs may be physically measured, and those that can, contain measurement uncertainties. These outputs are fed into the Variable Structure Filter to extract and estimate the internal states of the system. If the estimated states follow the actual plant states within acceptable boundaries, then these states may be integrated with a trajectory following Sliding Mode Controller. The Sliding mode controller uses the estimated states and the reference input r to produce a discontinuous control signal to attain trajectory tracking.

Sliding mode control has an inherent switching action as shown in Fig. 1. This switching and discontinuous control action can lead to a phenomenon known as chattering. This problem is addressed in the SMCF through the introduction of a boundary layer [15]. In addition, since the VSF is a special form of Observer and as such, chattering can be further alleviated by a bypass high frequency loop, which is formed by SMC and VSF shown in Fig. 2. [16]

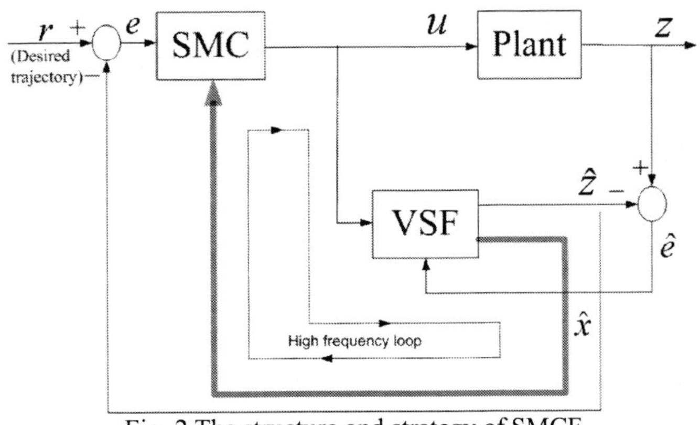

Fig. 2 The structure and strategy of SMCF

Consider the discrete-time linear time-invariant system with the unmatched uncertainties

$$X(k+1) = \Phi X(k) + Gu(k) + w(k) \tag{12}$$

$$Z(k) = HX(k) + v(k) \tag{13}$$

where $X = [x_1, x_2, \cdots, x_n]^T$ is the state vector, $Z = [z_1, z_2, \cdots, z_m]^T$ is the measurement vector, ($m \le n$), $u(k) \in \Re^{p\times 1}$ is the control input, and $w(k) \in \Re^{n\times 1}$ and $v(k) \in \Re^{m\times 1}$ are random variables representing the system and measurement noise respectively, which are both upper bounded. Φ, G and H are the system input and output matrices respectively. The pair of (Φ, G) is assumed to be completely controllable, and the pair of (Φ, H) is assumed be completely observable. The index k indicates the k-th sampling step and sampling period is indicated as T_s. It is for this general system that a controller and observer are to be designed.

Sliding Mode Controller

To facilitate the introduction of the definitions for sliding mode control, consider a simple case which is a second order system and where only one control input is involved. The sliding surface $S = 0$ about which the control signal changes, was illustrated graphically in Fig. 1. "Sliding motion" will occur when the system state repeatedly crosses and immediately re-crosses the sliding or switching surface $S = 0$, because all motion in the neighborhood of the surface is directed inwards towards the surface. From some initial point on the hyperplane, the states are confined to a trajectory on the hyperplane and are said to be in the "sliding mode".

Because of imperfections in controllers such as delays in the switching action, real sliding mode may occur. In a real sliding mode, the states would chatter about the switching hyperplane. This high-frequency chattering effect is synonymous to artificial noise, highly undesirable, and can excite unmodelled high-frequency plant dynamics. To avoid control discontinuities and switching action in the control loop, the smoothing boundary layer solution [17] is applied to the control law such that: (i) outside the smoothing boundary layer, a discontinuous control action is applied, but (ii) within the layer, the a continuous control action is used as a function of the distance of the states from the hyperplane. This means within the boundary layer, the switching action is smooth as opposed to an abrupt bang-bang action. The saturation function $sat(S, \Psi)$ is used to replace the sign function *sgn(S)* in the control law such that a continuous sliding mode can be achieved when the switching function is inside the boundary layer, i.e. $S(k) \le \Psi$, such that:

$$sat(S, \Psi) = \begin{cases} \text{sgn}(S/\Psi), & \text{for} |S/\Psi| > 1 \\ S/\Psi, & \text{for} |S/\Psi| \le 1 \end{cases} \tag{14}$$

The discrete-time system in the sliding mode involves two control subsystems which is the same as its continuous counterpart: a slow and fast subsystem. The slow subsystem is the system when it is in the sliding mode (i.e. on the S=0 surface) yielding "equivalent" control. The fast subsystem is the high switching control system in the vicinity of the sliding surface maintaining the system but without leaving the surface. The equivalent control can be obtained when the system is in the sliding mode, i.e.

$$S(k+1) = S(k) = 0 \tag{15}$$

Consider the ideal system model without uncertainties

$$X(k+1) = \Phi X(k) + Gu(k) \tag{16}$$

Substituting (7), (16) into (15), the equivalent control can be determined as:

$$u_{eq}(k) = (CG)^{-1} C(X_d(k+1) - \Phi X(k)) \tag{17}$$

Thus, equivalent control can be considered as the perfect control signal which forces the error between all output states and the desired states to zero, if no uncertainties existed. However, in the presence of uncertainties, the actual states will oscillate or switch about this ideal condition and hence the equivalent control signal will be modified by a fast acting switching control signal and as discussed previously, gives rise to "chattering" as the control signal switches about u_{eq}. A boundary layer neighboring the sliding surface S=0 is defined as

$$\{X_e(k) \| S(k) |=| CX_e(k) | \le \psi_c, \psi_c > 0\} \tag{18}$$

whose thickness is $2\Psi_c$.

Theorem 4.1. A sufficient condition for the existence of the discrete sliding mode is that there exists a k_0 such that

$$(a). \quad | S(k+1) | < | S(k) |, \ k \ge k_0 \tag{19}$$

$$(b). \quad S(k)\Delta S(k) < -\frac{1}{2}\Delta S^2(k), \ k \ge k_0 \tag{20}$$

The above two conditions are equivalent.

Proof: Choose a Lyapunov function as

$$V(k) = S^2(k) = (S(X_e))^2. \tag{21}$$

It has been shown in [18] that the condition $V(k+1) - V(k) < 0$ guarantees the existence of sliding motion, that is:

$$S^2(k+1) < S^2(k) \tag{22}$$

Thus, if condition (a) is satisfied, the existence of a sliding mode is guaranteed.

Let $\Delta S(k) = S(k+1) - S(k)$, then Eq. (22) may be rewritten as

$$\begin{aligned} &[S(k+1) + S(k)][S(k+1) - S(k)] \\ &= \Delta S(k)[\Delta S(k) + 2S(k)] < 0 \end{aligned} \tag{23}$$

which is equivalent to Eq. (20). Thus, condition (b) is also a sufficient condition for the existence of a sliding mode. Conditions (a) and (b) are equivalent.

For the case of $S(k)$>0, from Eq. (19)

$$S(k+1) > S(k) \tag{24}$$

where

$$\begin{aligned} S(k+1) &= CX_e(k+1) \\ &= C(X_d(k+1) - \Phi X(k) - Gu(k) - w(k)) \end{aligned} \tag{25}$$

Substituting ((17) into (25)

$$S(k+1) = CGu_{eq}(k) - CGu(k) - Cw(k) \tag{26}$$

Thus Eq. (26) may be rewritten as:

$$CGu_{eq}(k) - CGu(k) - Cw(k) < S(k) \tag{27}$$

Solving for $u(k)$, the control law is given as [19]:

$$u(k) = u_{eq}(k) - (CG)^{-1} S(k) + (CG)^{-1} K \cdot sat(S(k)/\psi_c) \tag{28}$$

where the adaptation gain $K = (Cw)_{\max} + \delta$,

$\psi_c > (Cw)_{\max} + \sigma$.

$$sat(S(k)/\psi_c) = \begin{cases} S(k)/\psi_c & | S(k) | \le \psi_c \\ sign(S(k)/\psi_c) & \text{otherwise} \end{cases}$$

in which $(Cw)_{\max}$ is the upper bound of $Cw(k)$, δ and σ are arbitrary positive number. It should be noted that K is a constant.

For the case of $S(k)$<0, the control law can be obtained in the same way.

Variable Structure Filter

The function of the variable structure filter is to estimate the states of a system when subjected to discontinuous inputs. The VSF uses concepts closely related to VSC. As discussed in reference [16], sliding observers use a constant feedback gain and a discontinuous vector to correct the predicted estimates of the system dynamics. On the other hand, the VSF uses a discontinuous component called a VSF gain, to correct state estimates. The form of the VSF gain, however, is different from the sliding observer gain K. This will be illustrated as follows.

Consider a linear system with unmatched uncertainties as described in Eq. (4). Define an observer switching function of the system

$$\hat{S} = e_z = z_d - \hat{z} \tag{29}$$

where $e_z = z - \hat{z}$ is the estimation error for the output and $e = x - \hat{x}$ is the estimation error for the states, and $\hat{x}$ and $\hat{z}$ are estimates of the states and output from the filter. Thus for the estimation error to be zero, the switching hyperplane associated with the filter should be

$$\hat{S} = e_z = 0. \tag{30}$$

This switching hyperplane is identical in concept to that in the VSC. VSF uses a similar calculation process for state estimation as the Kalman filter. A priori state ("crude") estimate ($\hat{x}_{k|k-1}$) is refined into an a posteriori state ("refined") estimate ($\hat{x}_{k|k}$) by using a weighted difference of e_z that is the output estimation error. Thus $\hat{x}_{k|k}$ is a linear combination of $\hat{x}_{k|k-1}$ and a function including a weighted difference of e_z, defined as the VSF gain. In the VSF, the structure of the gain changes with the error such that:

$$K_k = f(e_z, \mathrm{sgn}(e_z), v_{\max}, w_{\max}) = \begin{cases} K_k^+ = f(e_z, +|\mathrm{sgn}(e_z)|, v_{\max}, w_{\max}) & S = e_z > 0 \\ K_k^- = f(e_z, -|\mathrm{sgn}(e_z)|, v_{\max}, w_{\max}) & S = e_z < 0 \end{cases} \tag{31}$$

where **sgn** is the sign function of e_z, v_{max} and w_{max} are the upper bounds of system noise and measurement noise respectively.

A block diagram of the model for the VSF estimation process is shown in the Fig. 3, where K_k is the VSF gain and is applied to obtain the a posteriori state estimate such that:

$$\hat{x}_{k|k} = \hat{x}_{k|k-1} + K_k \tag{32}$$

The a priori output estimate for the next step (which in fact provides a new crude estimate for the next step) is predicted by using a model of the plant as:

$$\hat{z}_{k+1|k} = H(\Phi \hat{x}_{k|k} + G u_k) \tag{33}$$

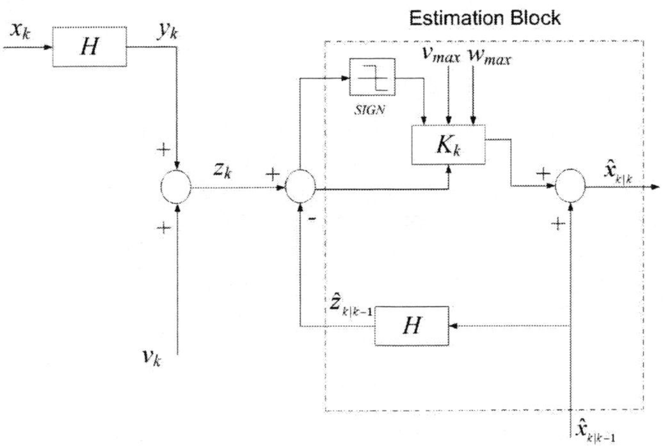

Fig. 3 The model for the estimation phase in VSF

The essential feature of the variable structure filter is the use of the VSF gain K_k. In deriving K_k., a switching hyperplane is used to restrict the estimated states to within a subspace of the actual system states. The VSF gain K_k. can be derived by the asymptotic stability condition, which is based the existence condition described by Corollary 3.1:

$$|e_{z_{ik|k}}| < |e_{z_{ik|k-1}}|, \tag{34}$$

where i is the element number of states vector, k and k-1 is the calculation step index.

Condition (34) is sufficient for stability. A further condition that is NOT sufficient for stability, but is instrumental to the derivation of the VSF gain is:

$$e_{z_{ik+1|k}} \mathrm{sgn}(e_{z_{ik|k-1}}) < 0 . \tag{35}$$

Substituting Eqs. (29), (32) and (33) into (35), and making the system noise w_k and measurement noise v_k bounded by v_{max} and w_{max} respectively, then the VSF gain can be obtained as (detailed derivation is provided in [10]):

$$K_k = \Phi^{-1} H^{+} \{ |[|H\Phi|_{ABS} (\gamma |H^{+}|_{ABS} |e_{z_{k|k-1}}|_{ABS} + (\mathrm{I} + |\Phi^{-1}|_{ABS}) |H^{+}|_{ABS} v_{\max} + |\Phi^{-1}|_{ABS} w_{\max})] |_{ABS} \mathrm{sgn}(e_{z_{k|k-1}}) \}, \tag{36}$$

where γ is a diagonal matrix with elements $\gamma_{ii} \geq 1$, and the superscript "+" on H denotes the pseudoinverse [20].

The selection of γ is a trade off in that the larger the value of γ is, the faster the estimation of the states can occur (an important factor for real time estimation in controls applications). However, a large γ can lead to the filter instabilities. The value of γ is very similar to the adaptation rule of K in Eq. (28), which is the constant of corrective term of the Sliding Mode Controller. In the absence of noise, the VSF may be regarded as predictor-corrector form of sliding mode observer.

To avoid chattering, from Definition 2.2, the boundary layer is also used for VSF as:

$$\{ e_{z_{ik}} \, \| S(k) | = | e_{z_k} | \leq \psi_f, \psi_f > 0 \} \tag{37}$$

where $e_{z_{ik}}$ are the elements of outputs, and the elements of ψ_f is $\psi_{f_i} (i = 1, \cdots m)$

The term of $\mathrm{sgn}(e_{z_{k|k-1}})$ in (36) may be replaced by saturation function as follows:

$$sat(e_{z_{k|k-1}} / \psi_f) = \begin{cases} e_{z_{k|k-1}} / \psi_f & |e_{z_{k|k-1}}| \leq \psi_f \\ sign(e_{z_{k|k-1}} / \psi_f) & \text{otherwise} \end{cases} \tag{38}$$

A small boundary layer width in (28) and (38) improves the control and estimation accuracy of the filter, but it may not be sufficient in width to overcome the chattering.

5 APPLICATION ON EHA SYSTEM

It was of interest to initially apply the SMCF to a model of a system which represented a physical plant. Simulation studies are extremely important to establish strengths and limitations of concepts before actually implementing them. Since the authors have been doing research on an ElectroHydraulic Actuator (EHA) in parallel to the SMCF [21], it was decided to use this system as a focus of the study. Further, since the authors have completed extensive studies on the EHA system and have established accurate models and a good understanding of the uncertainties which can exist in the parameters and model order, it was felt that this would be an appropriate system to consider in this first simulation study.

The ElectroHydraulic Actuator (EHA) is a high performance actuation system based on the principle of closed circuit hydrostatic transmission. It is capable of moving large loads with sub-micron precision. A mathematical model of the dominant dynamics of the EHA system can be represented by a third order transfer function as follows [21]:

$$G_h(s) = \frac{28}{s(s^2 + 54.5s + 24794)} \tag{39}$$

Assuming the presence of system noise and a sampling period $T_s = 0.001s$, the discrete state-space equation of the system may be approximated to:

$$X(k+1) = \begin{bmatrix} 1 & 0.001 & 0 \\ 0 & 1 & 0.001 \\ 0 & -24.65 & 0.9435 \end{bmatrix} X(k) + \begin{bmatrix} 0 \\ 0 \\ 0.0279 \end{bmatrix} u(k) + w(k) \tag{40}$$

$$Z(k) = \begin{bmatrix} 1 & 0 & 0 \\ 0 & 1 & 0 \end{bmatrix} X(k) + v(k) \tag{41}$$

where the system uncertainties *w(k)* are bounded by the maximum amplitude

$$w_{\max} = \begin{bmatrix} 0.001 \\ 0.001 \\ 0.001 \end{bmatrix},$$

and all the measurement uncertainties *v(k)* are bounded by:

$$v_{\max} = \begin{bmatrix} 0.1 \\ 10 \end{bmatrix}$$

For the controller, the sliding surface is chosen using the Linear Quadratic approach of [22] as:

$$S(k) = CX_e(k) = [358.0318 \quad 196.2450 \quad 35.8423] X_e(k) \tag{42}$$

The adaptation gain K is chosen as 10, which satisfies the condition of $K > (Cw)_{\max}$, and boundary layer of Sliding Mode Controller $\psi_c = 5$ which makes $\psi_c > (Cw)_{\max} + \sigma$.

For the filter, the boundary layer width is evaluated as the VSF gain reaches the maximum value such that all states can be contained within the boundary,

$$\begin{aligned} \psi_f &= |\Phi^{-1}H^+|_{ABS} \{|[|H\Phi|_{ABS}((I + |\Phi^{-1}|_{ABS})|H^+|_{ABS} v_{\max} \\ &\quad + |\Phi^{-1}|_{ABS}| w_{\max})]|_{ABS}\} \\ &= [0.2305 \quad 19.4918 \quad 509.2441]^T \end{aligned} \tag{43}$$

and the diagonal matrix γ is chosen as

$$\gamma = \begin{bmatrix} 1 & 0 & 0 \\ 0 & 1 & 0 \\ 0 & 0 & 1 \end{bmatrix}.$$

In this example, the objective of the EHA design was to achieve an acceleration of 5G. The reference input (i.e. the desired trajectory of all states) was chosen to be as illustrated in Fig. 4.

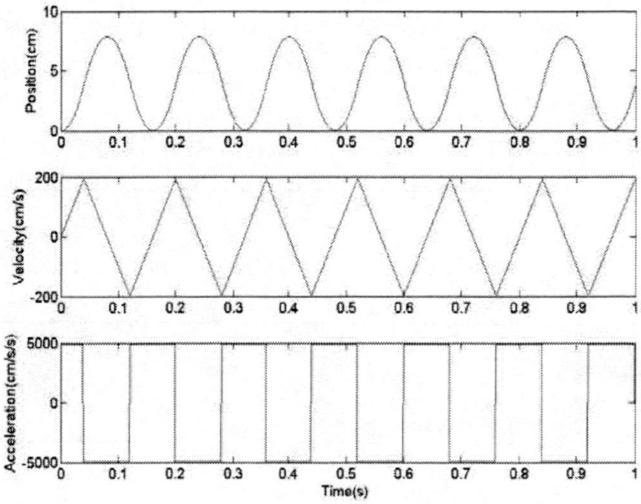

Fig. 4 The reference input of SMCF

The initial states of the EHA were defined as: $X_0 = [0 \quad 0 \quad 0]$. In this case, it is assumed that the position and velocity of the EHA are measured..

The controller receives the measured and estimated system states from the VSF and subsequently forces the system to follow the desired trajectory as shown in Fig. 4. The estimated state (acceleration), the desired state trajectories, and the actual states produced by the SMCF are shown in Fig. 5. The errors between the desired and actual system states are shown in Fig. 6. The output from the control law generated by Eq. (28) is presented in Fig. 7. It is evident that the magnitude of the errors between the desired and the actual states are very small and thus it can be observed that in the presence of uncertainties, the SMCF can obtain a high degree of robustness and accurate control without an actual measurement of acceleration. The chattering that is evident in the error plots can be further suppressed by adjusting the boundary layer designed for the SMCF.

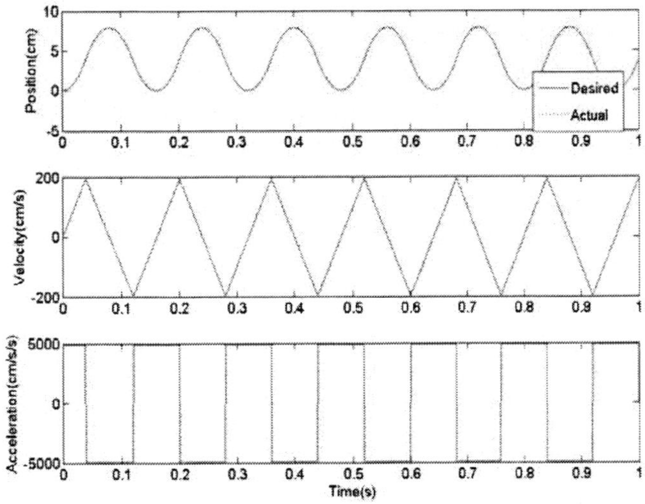

Fig. 5 The desired states trajectory and actual states of SMCF

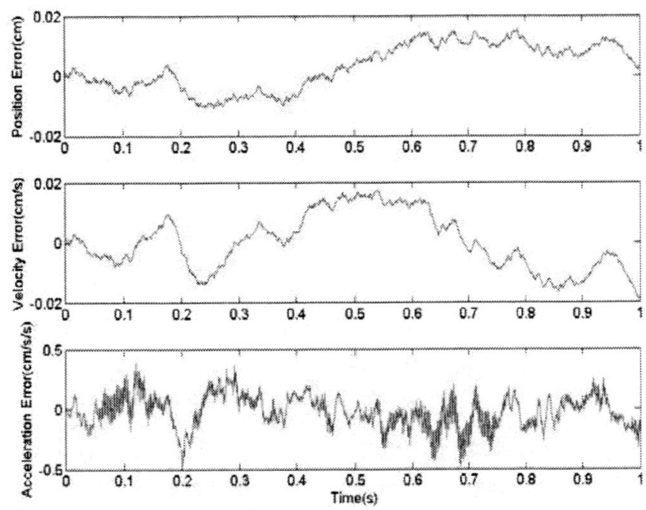

Fig. 6 The error between desired and actual states

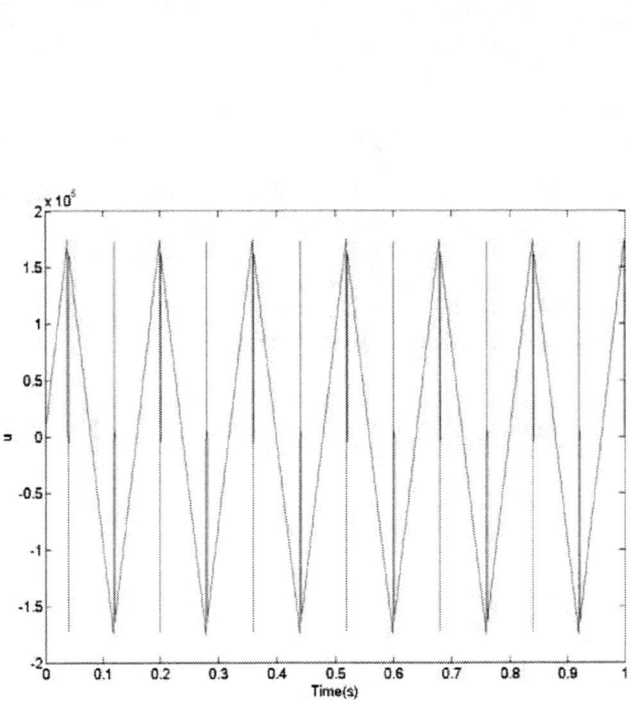

Fig. 7 The control law, *u*, of SMCF

6 ANALYSIS Study

In the previous sections, the concept of the SMCF and its application to the EHA system have been presented. From the theoretical equations and numerical simulation results, several remarks are now forwarded:

Remark 6.1. Both the SMC and VSF are based on the same basic concepts: discontinuous control theory. The Sliding Mode Controller and Variable Structure Filter described in Section 4 are both derived based on the same definitions and existence conditions, such as Definition 3.1, Definition 3.2 and Corollary 3.1. These allow the SMC and the VSF to be smoothly fused together in order to achieve a satisfactory control objective.

Remark 6.2. It is of interest to apply a well known Proportional gain type controller to the EHA model subject to the desired position trajectory only as shown in Fig. 8 [21].

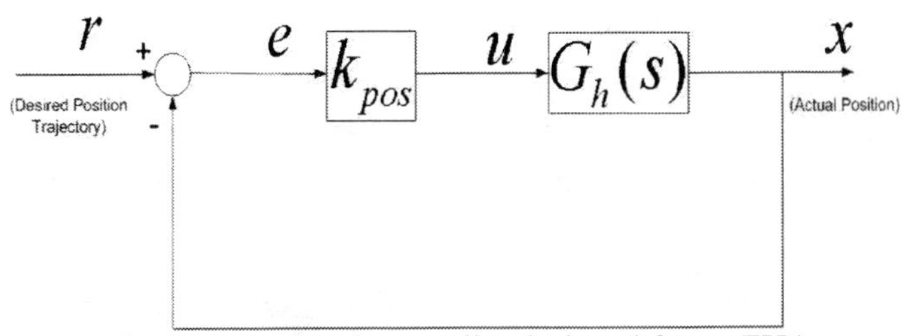

Fig. 8 Proportional controller designed for an EHA

The actual output position of the EHA system is shown in Fig. 9 using the Proportional controller and the same reference position trajectory given in Fig. 4. From this simulated result, it is obvious there is a delay between the actual state and desired state trajectory. Comparing this result to those of the SMCF (Fig. 5), the SMCF can achieve high regulation rates or short settling times for the system for similar inputs. In other words, the SMCF can obtain a shorter settling time than a Proportional controller [11].

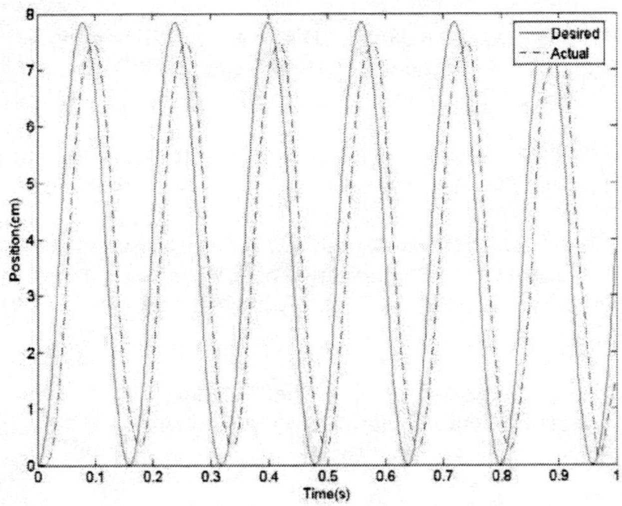

Fig. 9 Desired and actual position of EHA produced by a Proportional controller

Remark 6.3. In the SMCF, there are two groups of states which should be analyzed. The first state was those estimated by the VSF; the second were the actual plant states controlled by the SMCF. In this case, it was assumed that the acceleration was not known but position and velocity were. Thus, the VSF was designed to estimate the acceleration. In Fig. 10, two errors, are shown such as one between the reference states and the estimated states (that from the VSF), and another one between the reference states and actual states (that from the plant). It is evident that the SMCF displays a small discrepancy from the desired trajectory. In addition, chattering does not excite the unmodelled dynamics in controller and filter, because the VSF, as a special-form of asymptotic observer, serves as a high frequency filter, as described in [11].

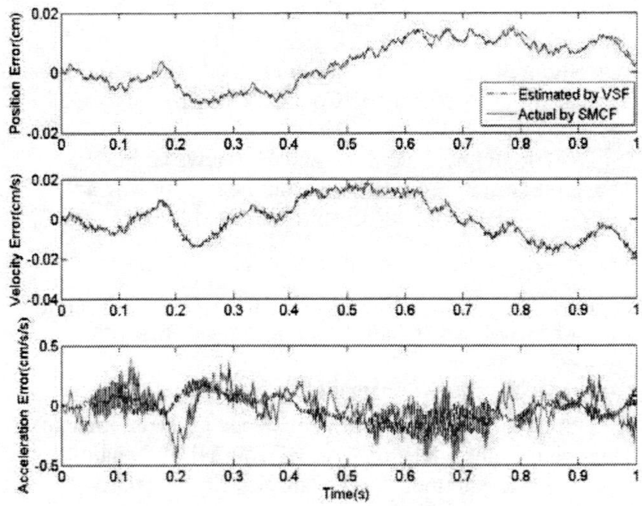

Fig. 10 Estimated and actual acceleration errors of EHA

7 Conclusions

In this paper, a new control strategy, referred to as the SMCF is proposed based on the concepts of Variable Structure Control and Filter. Some advantages, such as robustness, low control cost (see Remark 5.2.) and chattering suppression, are inherent in the design of the SMCF. The SMCF is applied to a model of the EHA and the results are very encouraging. It is known from earlier studies on the EHA that the system is nonlinear and that the third order model (lateralized) is valid at specified operating points. It is now necessary to extend the SMCF concept to nonlinear systems and then to apply the controllers to a physical EHA model. This is now being done.

ACKNOWLEDGEMENTS

The author would like to acknowledge the financial support of the University of Saskatchewan and National Science and Engineering Research Council of Canada.

REFERENCES

[1] Utkin, V. I, 1977, "Variable structure system with sliding modes." IEEE Transactions on Automatic Control, AC-22, pp. 212-222.

[2] Corradini, M. Letizia and Leo, T. and Orlando, G., 2002, "Experimental Testing of a Discrete-time Sliding Mode Controller for Trajectory Tracking of A Wheeled Mobile Robot in the Presence of Skidding Effects", J. of Robotic systems 19(4), pp. 177-188.

[3] Zinober, A.S.I, El-Ghezawi, O.M.E. and Billings, S.A., 1982, "Multivariable-Structure Adaptative Model-following Control Systems", Proceedings of IEEE, 129, pp. 6-12.

[4] Slotine, J. –J. E., Hedrick, J.K.and Misawa, E. A., 1987, "On Sliding Observers for Nonlinear Systems", J. of Dynamic Systems, Measurement and Control, 109, pp. 245-252

[5] Decarlo, R. A., Zak, S. H. and Matthews G. P., 1988, "Variable Structure Control of Nonlinear Multivariable Systems: A Tutorial", Proceedings of the IEEE, 76, No. 3, 212-232.

[6] Furuta,K., 1990, "Sliding Mode Control of A Discrete System," System & Control Letters, 14, pp. 145-152.

[7] Misawa, E. A., 1997, "Discrete-Time Sliding Mode Control for Nonlinear Systems With Unmatched Uncertainties and Uncertain Control Vector", ASME Journal of Dynamic Systems, Measurement, and Control, 119, pp. 503-512.

[8] Utkin, V. I., 1992, "Sliding Modes in Control Optimization,", Springer-Verlag Berlin.

[9] Walcott, B. L. and Zak, S. H., 1987, " State Observation of Nonlinear Uncertain Dynamical Systems," IEEE Transaction on Automatic Control, 32, pp. 166-170.

[10] S. R. Habibi and R. Burton, 2003, "The Variable Structure Filter" ASME Journal of Dynamic Systems, Measurement and Control, 125, pp. 287-293.

[11] Sabanovic, A. and Fridman, L. and Spurgeon, S., 2004, "Variable Structure Systems: From Principles to Implementation," The Institution of Electrical Engineers, London.

[12]S.V. Emelyanov, 1967, "Variable Structure Control Systems", Moscow: Nauka (in Russian).

[13]Utkin, V. I., 1978, "Sliding Mode and Their Application in Variable Structure Systems", English translation, Mir Publication, Moscow.

[14] S. Z. Sarpturk and Y. Istefanopulos and O. Kaynak, 1987, "The Stability of Discrete-time Sliding Mode Control Systems IEEE Transactions on Automatic Control, Vol. 32, pp.930-932.

[15] Slotine, J. –J. E and Sastry, S. S, 1983, "Tracking Control of Nonlinear Systems Using Sliding Surfaces, with Application to Robot Manipulators", International Journal of Control, 38, pp. 465-492.

[16] V. Utkin and J. Guldner and J. Shi, 1999, "Sliding Mode Control in Electromechanical Systems", Taylor & Francis London.

[17] Slotine, J. –J. E., 1984, "Sliding Controller design for nonlinear Systems", International Journal of Control, 40, pp. 421-434.

[18] Itkis, Y., 1976, "Control Systems of Variable Structure", New York: Wiley.

[19] Misawa, E. A., 1997, "Discrete-Time Sliding Mode Control: The Linear Case", ASME Journal of Dynamic Systems, Measurement, and Control, 119, pp. 819-821

[20] Campbell, S.L. and Meyer, C.D. Jr, 1991, "Generalized Inverses of Linear Transformations", New York: Dover.

[21] Habibi, S. R. and Singh, G., 2000, "Derivation of Design Requirements for Optimization of A High Performance Hydrostatic Actuation System", International Journal of Fluid Power, 2, pp. 11-27.

[22] Zinober, A. S. I.(Ed.), 1994, "Variable Structure and Lyapunov Control", Springer-Verlag London Limited.

Proceedings of IMECE2005
2005 ASME International Mechanical Engineering Congress and Exposition
November 5-11, 2005, Orlando, Florida USA

IMECE2005-80562

AN APPLICATION OF THE SELF-EXCITED OSCILLATION METHOD TO A HYDRAULIC MOTOR ANGULAR VELOCITY SYSTEM

Takayoshi Ichiyanagi
National Defense Academy
1-10-20 Hashirimizu
Yokosuka Kanagawa JAPAN
ichiyana@nda.ac.jp

Takao Nishiumi
National Defense Academy
1-10-20 Hashirimizu
Yokosuka Kanagawa JAPAN
nishiumi@nda.ac.jp

ABSTRACT

It has been known that a hydraulic servo system can be generally dealt with a second order delay element for designing an adequate controller. In the previous work, our research group developed a simple method utilizing the self-excited oscillation of a hydraulic servo system to directly estimate the dynamic parameters such as the damping ratio and undamped natural frequency. The advantage of this method is the "on-line" ability that is able to identify these parameters instantaneously with changing the operating conditions. Although this method was found to be very useful, it is available only when the spool valve is close to the neutral position. This paper proposes an online parameter estimation method for the hydraulic motor angular velocity system. In this method, the self-excited oscillation around the displaced position from the neutral spool position is utilized. In order to demonstrate the effectiveness, the experimental work is conducted by giving the various input voltages, system pressures and angular velocities. The resulting frequency characteristics of these identified transfer functions are then compared with those of the measured data by the frequency characteristics method in order to confirm coincidence.

INTRODUCTION

A hydraulic motor that is controlled by a servovalve has been used as a rotary actuator in various industry fields because it has several advantages such as a high power density, its compactness, and so on, over other actuator systems. For the development of this hydraulic servo motor system, it is required to know a precise knowledge of the system dynamic behavior and the corresponding mathematical descriptions. The hydraulic servo motor system has non-linear characteristics in their dynamic behavior. However, it is often treated as simplified descriptions by linearized approximation for the design of a controller or analyzing the actuator dynamics. Their dynamic characteristics are generally dealt with as a second order delay element [1].

When it comes to design a controller adequately, an estimation of the dynamic parameters, which are the damping ratio ζ and the undamped natural frequency ω_n in the second order delay element, is necessary. One of the practical ways to obtain these parameters is the use of an experimental identification method. In the previous works, our research group developed a simple method utilizing the self-excited oscillation of the hydraulic servo system to estimate these parameters directly. The usefulness and the validity of the method were fully described in the previous reports. Firstly, this method was applied to a servovalve in the case where the output detector delay was negligible [2]. Then, the effects of the output detector delay on the identified results were assessed by using an electric circuit, which simulated the detector delay as a first order delay element [3]. Furthermore, this method was applied to an on-line identification system, which is realized by using a digital signal processor [4] and an application example to hydraulic servo motor system is also reported [5-6].

Even though the previous reports proved the usefulness of the proposed method, it identifies the dynamic parameters only at the condition that the servovelve spool is close to the neutral position. In other words, the method was only available for the hydraulic motor angular position system. This research addresses an application of the self-excited oscillation method to a hydraulic motor angular velocity system. The angular velocity self-excited oscillation around a constant rotational speed is utilized to identify the dynamic parameters. When the motor rotates at a constant speed, the servovalve spool is displaced from the neutral position. The paper describes the fundamental principle of the proposed angular velocity self-excited oscillation method and its experimental set up. In order to demonstrate the usefulness, the on-line experimental identification for the hydraulic servo motor system is carried out by providing the various system pressure and angular velocity. The identified results derived from the conventional frequency response method and the proposed self-excited oscillation method is also compared to confirm the validity of the proposed method.

NOMENCLATURE

- e deviation signal
- e_a setting voltage of non-linear element
- G_L transfer function of hydraulic servo motor system
- G_N transfer function of non-linear element
- K_L gain of hydraulic servo motor system
- K_N variable gain of non-linear element
- P_s supply pressure
- P_t tank pressure
- V servo amplifier input voltage
- V_a amplitude of self-excited oscillation wave
- v_i input voltage
- x spool displacement of servovalve
- Γ amplitude correction factor
- θ angular position
- ζ dumping ratio
- ω angular velocity
- ω_n undumped natural frequency
- ω_s angular frequency of self-excited oscillation wave
- ξ frequency correction factor

SELF-EXCITED OSCILLATION METHOD

A schematic diagram of an identified hydraulic servo motor system is illustrated in Fig.1. It mainly consists of an internal gear type hydraulic motor, load inertia, a servo amplifier and servovalve. The linear transfer function from the servo amplifier input voltage V to the angular velocity of the hydraulic motor $\dot{\theta}$ can be described by Eq.1 as a second order delay element, provided that the connecting pipes are treated as a lumped element and the dynamic characteristics of the servovalve is negligible, i.e. the servovalve dynamics is considerably faster than the hydraulic motor response.

$$G_L(s) = \frac{\dot{\theta}(s)}{V(s)} = \frac{K_L \omega_n^2}{s^2 + 2\zeta\omega_n s + \omega_n^2} \tag{1}$$

where K_L, ω_n and ζ are the gain of the hydraulic servo motor system, the undamped natural frequency and the damping ratio, respectively.

The self-excited oscillation method is able to directly identify the system dynamic parameters ζ and ω_n of the second order delay element by using the self-excited oscillation of the identified system. Figure 2 (a) shows the block diagram of the angular position self-excited oscillation system that was utilized to identify the linear transfer function of the hydraulic servo motor system where the sevovalve spool is at the neutral

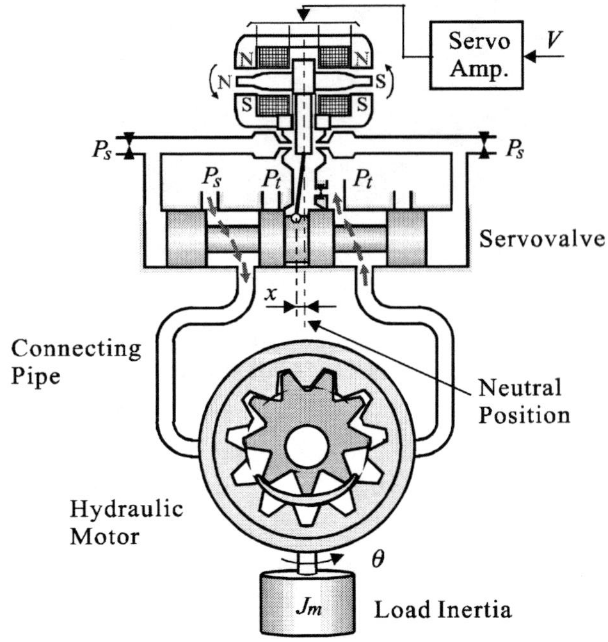

Figure 1. Schematic diagram of hydraulic servo motor system

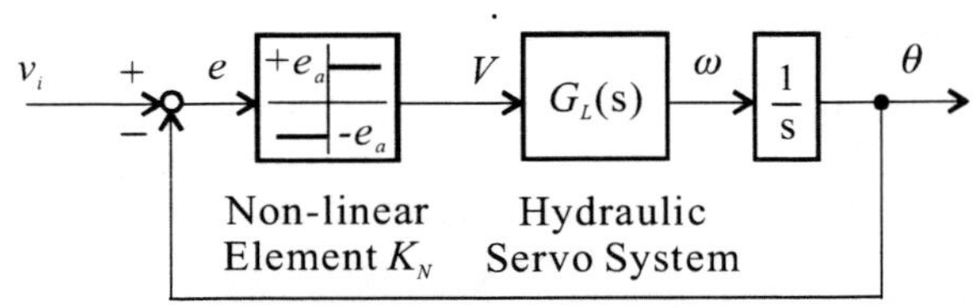

(a) Angular position self-excited oscillation system

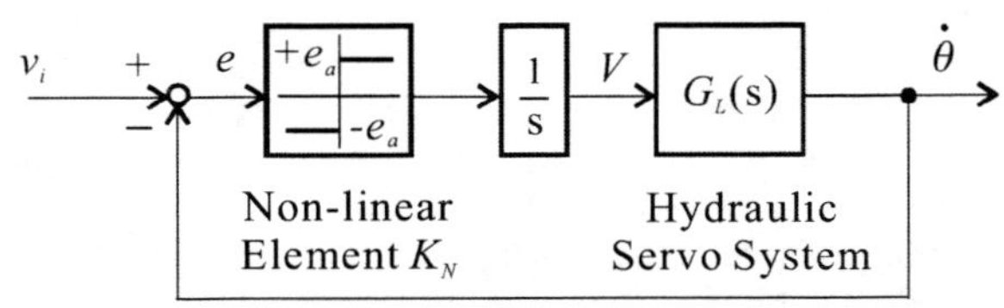

(b) Angular velocity self-excited oscillation system

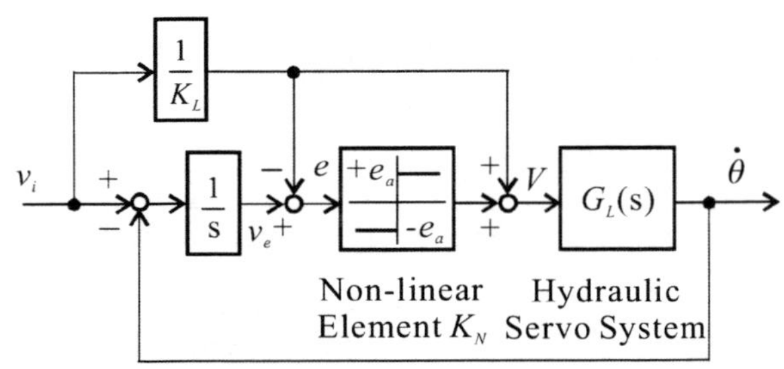

(c) Revised angular velocity self-excited oscillation system

Figure 2. Self-excited oscillation system

position. This system is composed only by arranging a non-linear element G_N in forward side of the hydraulic servo motor system G_L. Since the non-linear element, which is expressed by Eq.2, has variable gain characteristics, the system causes the limit cycle at the point of stability limit.

$$K_x = \frac{e_a}{|e|} \tag{2}$$

In this identification method, two dynamic parameters can be easily obtained as Eq.3 and Eq.4 by measuring the amplitude V_s and the angular frequency ω_s of the measured self-excited oscillation wave.

$$\omega_n = \frac{\omega_s}{\xi} \tag{3}$$

$$\zeta = \frac{\Gamma K_L e_a}{2V_s \omega_n} \tag{4}$$

where ξ and Γ are the frequency correction factor and the amplitude correction factor. These correction factors are expressed by the following equations that were obtained and interpolated by simulation analysis.

$$\left.\begin{aligned} \xi &= 1.0 - 0.0315\zeta + 0.00415\zeta^2 - 0.000185\zeta^3 \\ \Gamma &= 1.27 + 0.0647\zeta - 0.00762\zeta^2 + 0.000307\zeta^3 \end{aligned}\right\} \tag{5}$$

Consider a condition where the hydraulic motor is driving at a constant rotational speed. In this case, the servovalve spool is displaced from the neutral position (see the displacement x in Fig.1) and the dynamic characteristics of the identified system are varied. In order to apply the self-excited oscillation method to this hydraulic motor angular velocity system, the angular velocity ω is used for the feedback signal to make up the angular velocity self-excited oscillation system shown in Fig.2 (b). It is clear from Fig.2 (a) and (b) that the open loop transfer functions of both the angular velocity self-excited oscillation system and the angular position self-excited oscillation system are the same. Therefore, all equations described above can be applied to obtain the dynamic parameters. However there is a problem in the measurement of angular velocity. The noise of high order harmonic frequency is often imposed in the measured angular velocity. This noise component yields an input signal, which is not based on the proper angular velocity self-excited oscillation, into the servo amplifier by passing through the non-linear element. Consequently it causes the inaccurate estimation result of the dynamic parameters. Hence, in this research, the integration element is placed forward to the non-linear element to reduce the influence of this noise component. The self-excited oscillation system shown in Fig.2 (b) is transformed into the system shown in Fig.2 (c). It is the revised angular velocity self-excited oscillation system. The feed-forward signals are added into the input signal to the non-linear element v_e- (v_i/K_L) and to the servo amplifier $\pm e_a+(v_i/K_L)$. The former signal is oscillated around zero so that the non-linear element outputs $\pm e_a$ periodically. The later block diagram realizes the self-excited oscillation at a reference rotational speed.

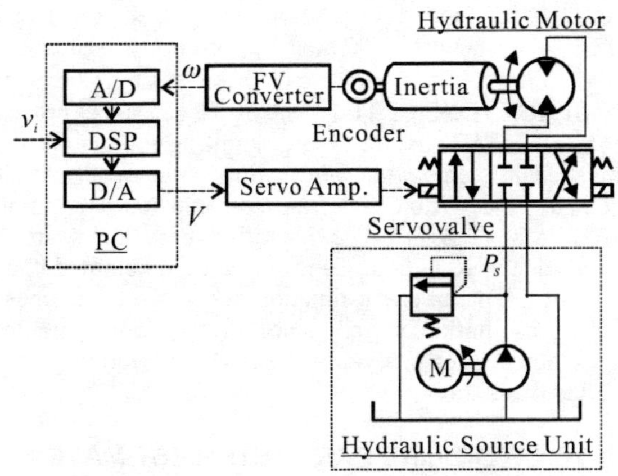

Figure 3. Experimental apparatus

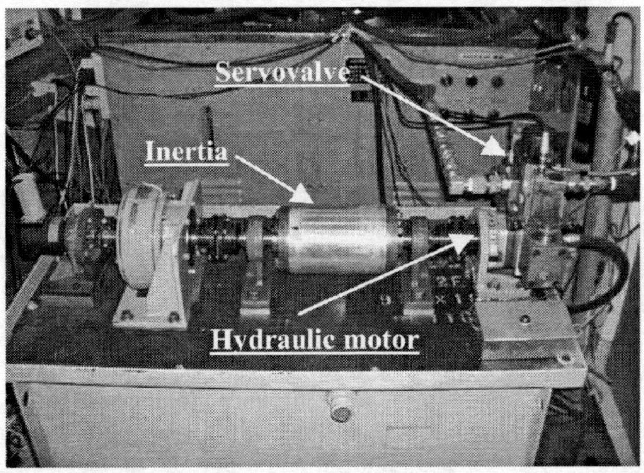

Figure 4. Hydraulic servo motor system

EXPERIMENTAL SETUP SYSTEM

Figure 3 and 4 show the experimental apparatus used in this study. The hydraulic source unit consists of a pump, an electric motor, a relief valve and a main tank. The system supply pressure is adjusted by this relief valve. The identified hydraulic servo motor system consists of the servovalve, the hydraulic motor and the load inertia, which is mentioned in the previous chapter. The main specification of this system is shown in Table 1. The gain of the hydraulic motor system K_L has been obtained in advance by measuring the static characteristics. This gain shows the dependency characteristics on the supply pressure as expressed in Eq. 6.

$$K_L = 4.25 + 2.72 \times 10^{-6} P_s - 0.134 \times 10^{-12} P_s^2 \tag{6}$$

where the supply pressure unit of this equation is [Pa]. The angular velocity ω is detected by an encoder and sent to the PC through a FV converter as a feed-back signal. For the on-line parameter estimation, a digital signal processer is used to analyze the self-excited oscillation wave and calculate the dynamic parameters. The program written in MATLAB/Simulink® is implemented in the single board type DSP (dSPACE DS1102:TI-TMS320C31 25MHz processor with 16bit AD/DA converter). The sampling time is chosen to 0.5 [ms]. Data acquisition and editing parameter sets are managed by the virtual instrument panel on PC monitor (dSPACE Control desk). The identified parameters are also stored and displayed in this window. It should be noted that the least time of the parameter estimation in this work becomes the self-excited oscillation period, since the dynamic parameters can be obtained after analyzing the angular frequency of one self-excited oscillation wave.

RESULTS OF ON-LINE PARAMETER ESTIMATION

Figure 5 shows the example plot of the self-excited oscillation wave. The input/output signals of the non-linear element shown in Fig.2 (c) are examined. It can be seen from this figure that the noise component imposed on the angular velocity oscillation wave is completely attenuated and the non-linear element outputs the setting voltage e_a corresponding to the angular velocity self-excited oscillation period without suffering from the influence of the noise.

Figure 6 shows the results of the on-line parameter estimation with changing the supply pressure P_s continuously from 5 MPa to 10 MPa at the reference rotational speed $K_L v_i$=32.5 rad/s and the setting voltage e_a=1.5 V. It is obvious that these dynamic parameters are estimated continuously by the proposed identification method. This indicates the realization of real-time identification. The results also show the identified dynamic parameters are notably affected by the supply pressure. In particular, the undamped natural angular frequency ω_n increases almost in proportion to the supply pressure P_s.

The identified dynamic parameters on the condition that the servovalve spool is displaced from the neutral position are shown in Fig.7. The reference rotational speed of the hydraulic motor $K_L v_i$ is varied from 0 rad/s to 120 rad/s continuously at the supply pressure P_s=8 MPa and the setting voltage of the non-linear element e_a=1.5 V. This rotational speed range approximately corresponds to the displacement range of the servovalve spool. The maximum rotational speed in this identified hydraulic servo motor system is 140 rad/s at the maximum spool displacement. The result at $K_L v_i$=0 rad/s is the same condition as the angular position self-excited oscillation system reported previously. It is confirmed that the dynamic parameters at the servovalve spool is displaced are also identified continuously as well as the spool is in the neutral position.

Finally, in order to investigate the validity of the on-line identification results, a comparison between the proposed self-excited oscillation method and the frequency response method is carried out. Figure 8 shows the bode plot of the identified hydraulic motor angular velocity system at the supply pressure P_s=8 MPa, the setting voltage of the non-linear element e_a=1.5 V and the reference rotational speed $K_L v_i$=0, 70, 133 rad/s. In this figure the line shows the self-excited oscillation method results obtained from Eq.1 using the identified dynamic

Table 1 Specification of hydraulic servo motor system

Motor displacement	1.75×10^{-6} m^3/rad
Load inertia	1.92×10^{-2} kgm^2
Servovalve flow rate	30 L/min
Hydraulic fluid	MIL-H5606C
Fluid temperature	40±2℃

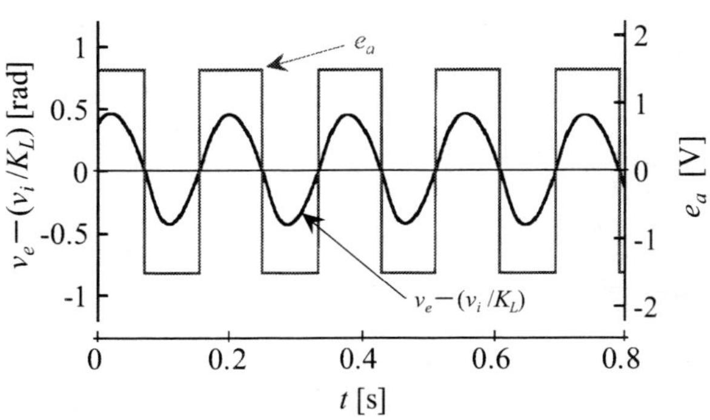

Figure 5. Example plot of self –excited oscillation wave

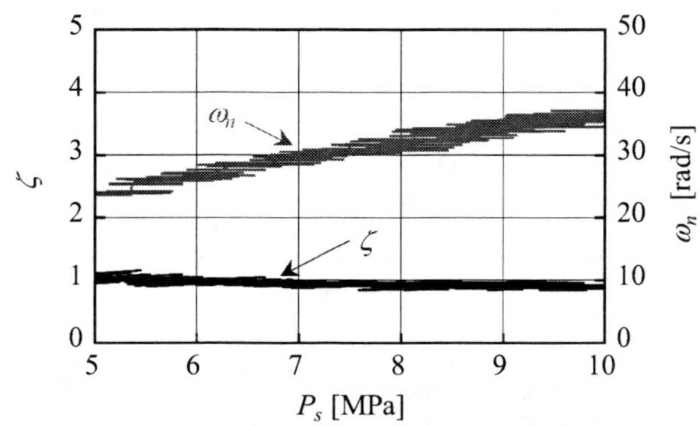

Figure 6. Identified dynamic parameters ($K_L v_i$=32.5rad/s, e_a=1.5V)

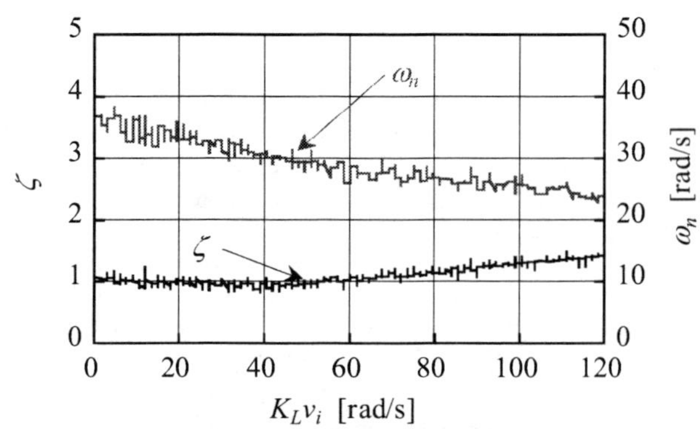

Figure 7. Identified dynamic parameters (P_s=8MPa, e_a=1.5V)

parameters ζ and ω_n, on the other hand, the plotted points show the experimental results from the frequency response. A good agreement is achieved between these two methods. This suggests the developed on-line identification method is capable of practical usage.

CONCLUSIONS

The self-excited oscillation method was revised for the hydraulic servo motor system which operates at a constant rotational speed. In this case, the sevovalve is always opened to delivery the flow and its spool is displaced from the neutral position. The angular velocity self-excited oscillation system was utilized for this system instead of the angular position self-excited oscillation system used in the previous report. Then it was clarified that the present on-line identification method was enable to identify the dynamic parameters of the hydraulic servo motor system, where the rotational speed are varied continuously. It was also confirmed that the on-line identified results coincide with the measured data obtained from the frequency response method. From the results examined in this report, it was verified that the present identification method makes possible the on-line parameter estimation of the hydraulic motor angular velocity system.

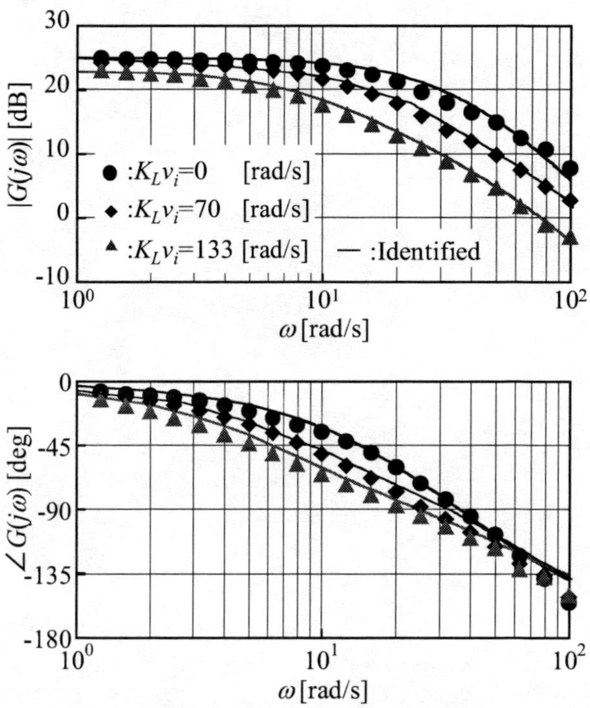

Figure 8. Identified dynamic parameters (P_s=8MPa, e_a=1.5V)

REFERENCES

1. Watton, J., Fluid power system, Prentice Hall, ISBN 0-13-323197-6, 299-301,1989
2. Konami, S., Nishiumi, T. and Hata, K., Identification of linearized electro-hydraulic servovalve dynamics by analyzing self-excited oscillations (First report, a case in which flow-rate detector delay is negligible), Journal of Hydraulics & Pneumatics, 27-4, 143-149, 1996 (in Japanese)
3. Konami, S., Nishiumi, T. and Hata, K., Identification of linearized electro-hydraulic servovalve dynamics by analyzing self-excited oscillations (Second report, a case in which flow-rate detector delay must be consider), Journal of Hydraulics & Pneumatics, 28-3, 88-94, 1997 (in Japanese)
4. Ichiyanagi, T., Nishiumi, T. and Katoh, H., Real-time identification using A self excited oscillation method to an electro-hydraulic servo system, 1st International Conference on Computational Method in Fluid Power Technology, 423-431, 2003
5. Nishiumi, T., Konami, S. and Watton, J., An application of a neural network controller to a hydraulic servo cylinder system, Journal of Hydraulics & Pneumatics, 31-4, 115-122, 2000
6. Hwang, U-K. and Cho, S-H., Application of a neural network for identification and control of a servovalve controlled hydraulic cylinder system, Proc. of 5th JFPS International Symposium on Fluid Power NARA 2002, Vol. 1, 205-210, 2002

Proceedings of IMECE2005
2005 ASME International Mechanical Engineering Congress and Exposition
November 5-11, 2005, Orlando, Florida USA

IMECE2005-80609

STABILITY AND PERFORMANCE ANALYSIS OF A METERING POPPET VALVE

Roger Fales
Mechanical & Aerospace Engineering Dpt.
University of Missouri – Columbia
Columbia, MO 65211
falser@missouri.edu

ABSTRACT

Poppet type metering vales have many benefits including low leakage and an economical design. These benefits make the poppet valve an appealing alternative to spool valves in a valve stack. The fact that the metering element is not hydrostatically balanced as in a spool valve leads to control design challenges. In this work, a model of an electro hydraulic metering poppet valve is considered. Due to design compromises, the response of production metering poppet valves tends to be too slow to maintain a desired flow rate when there are fast upstream pressure variations. Re-designing to speed up the response of the valve may lead to stability issues which can be traced to plant uncertainty. Frequency response analysis of the valve model shows that the model varies greatly depending on the operating point chosen for the linearization. The analysis presented will help define the problem of designing hardware and control systems for higher performance but still reliable metering poppet valves.

INTRODUCTION

In this work, the stability and performance are examined for metering poppet valve with variations in operating conditions and parameters. Specifically, the supply pressure is varied and the area of a metering slot is varied to increase performance. A model was created to simulate a metering poppet valve system. In this work, a model of the Vickers Valvistor is used for the metering poppet valve. The Valvistor has a unique flow amplifying hydro mechanical control system which employs a metering slot to control the position of the main metering element. The Valvistor modeling effort was based on the work of previous researchers. Nonlinear simulations show that flow control is poor and there are stability issues at higher pressures. Various operating conditions are considered in the simulations including fast variations of supply pressure. Varying supply pressure is a concern when multiple valves are connected to a single variable displacement pump where the pump pressure is controlled by sensing the load pressure at the valve largest load (i.e. a load sensing system). In a load sensing system, the supply pressure is maintained at a level which is larger than the highest load pressure in the system. The supply pressure can vary from some minimum pressure setting up to the maximum pressure of the system. Linear analysis is used to explore the valve's stability issues due to wide pressure fluctuations.

LITERATURE REVIEW

Metering poppet valve related research appears in many publications. In the literature, there are works related to modeling of poppet valves based on detailed analysis of fluid mechanics as well as complete dynamic system models. There are also works which explore the control performance of poppet valves.

In work by Johnston, Edge, and Vaughan[i] and Vaughan, Johnston, and Edge[ii], the design and models of poppet valves are discussed. These works focus on the geometry of the poppet and the consequences to flow and forces using both experiments and simulations. This work focuses only on the fluid mechanics and does not focus on the dynamics of poppet motion. In another work focused strictly on fluid mechanics, Yang[iii] determined that flow forces on poppet valves could be predicted with reasonable accuracy using CFD models. The study was restricted to steady state flow forces.

Work has been done by Zhang, Alleyne, and Prasetiawan[iv] in which the limitations on performance of the same Vickers Valvistor that is considered in this work. The valve which was the subject of the research is one of few metering poppet valves in industrial production. A linearization and simplification of the valve model was analyzed. A pilot poppet and a feedback slot on the main poppet control the main poppet position by controlling the pressure on one side of the main poppet (See Figure 2 and the Appendix). The valve was found to have a fundamental limitation in bandwidth which would preclude its use in many applications due to low response speed. The system's transfer function was found to have a zero which limited the ability of a feedback controller to improve the speed of response. It was concluded that a re-design of the valve would be necessary to improve its performance.

The model and a discussion of the control system for another type of metering poppet valve are given by Opdenbosch, Sadegh, and Book[v]. In this study a detailed model is given. The valve configuration is different from others in that the valve actuation stage contains a pressure compensation device. A controller was designed based on Nodal Link Perceptron Networks. A simulation showed that the valve with the control system was capable controlling flow. There was no discussion of how the physical characteristics of the valve might be changed to alter performance.

The above cited works are not focused on examining the changes in performance and stability as parameters such as the supply pressure and valve areas are varied.

NOMENCLATURE

Symbol	Description	Units
P_A	Inlet pressure	Pa
P_B	Outlet pressure	Pa
P_p	Pilot Pressure (pressure applied to the top of the main poppet)	Pa
Q_A	Flow into the valve	m^3/sec
Q_B	Flow out of the valve	m^3/sec
A_{tp}	Area of the top of the main poppet	m^2
A_{bt}	Area of the bottom of the main poppet	m^2
x_m	Main poppet position	m
k_m	Main poppet spring rate	N/m
b_m	Main poppet damping rate	N/m/sec
m_m	Main poppet mass	Kg
x_p	Pilot poppet position	m
k_p	Pilot poppet spring rate	N/m
b_p	Pilot poppet damping rate	N/m/sec
m_p	Pilot poppet mass	Kg
F_p	Solenoid force on pilot poppet	N
K_e	Slope of Force vs. current relationship for the pilot solenoid	N/Amp
V_p	Pilot fluid volume above the main poppet	m^3
K_p	Pilot poppet flow gain	$m^4/(\sec\sqrt{N})$
K_s	Feedback slot flow gain	$m^4/(\sec\sqrt{N})$
K_m	Main poppet flow gain	$m^4/(\sec\sqrt{N})$

MODEL

The model consists of the Vickers Valvistor and a simple load (Figure 1). The Valvistor model is based on previous work by other researchers[vi]. Consult the references given in this work and the Appendix for details on the Valvistor model. The load is modeled as a constant outlet pressure of 2 MPa for part of this work and then allowed to vary in one case. The model's input is pilot solenoid current and the output is the total flow to the load.

The valve uses a novel method for controlling the main poppet position. To open the valve to causes a flow through the main poppet metering area, a force, F_p, first pushes downward on the pilot poppet (Figure 2). This allows fluid to flow from the volume above the main poppet to the outlet of the valve and thus decreasing the pressure on the area, A_{tp}, which decreases the force which is in a downward direction on the main poppet. This force is balanced by the spring, and pressures on areas, A_{bt} and A_{tp}-A_{bt}. Therefore the supply pressure, P_A, and downstream pressure, P_B, have an affect on the force balance both causing and upward force. As the pilot poppet opens, the upward forces become greater than the downward forces and the main poppet moves upward. As it moves upward the metering slot begins to open and fluid from the supply flows into the volume above the main poppet thus increasing P_p. The pressure rise rate in the volume above the main poppet is controlled by the balance of flows or the difference, Q_2-Q_p. As the pilot pressure increases and settles to a steady value, the poppet slows and stops moving. The outlet flow, Q_B, is the sum of the pilot flow, Q_p, and the main flow Q_1.

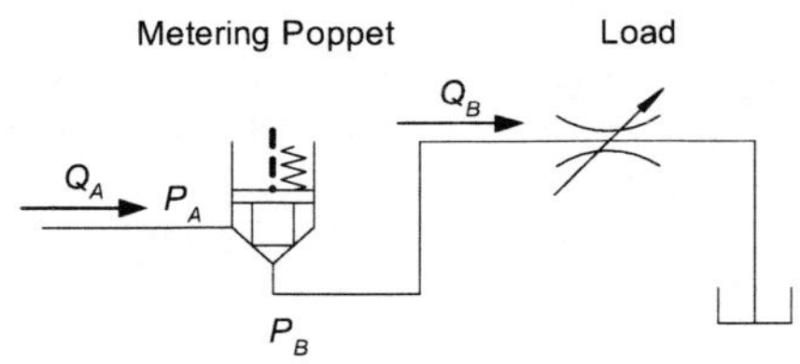

Figure 1 Metering poppet valve, and load

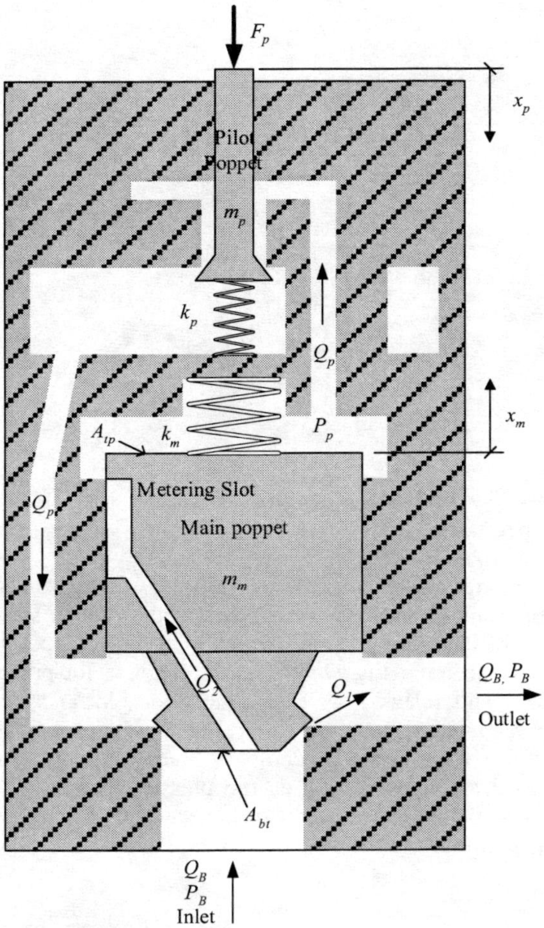

Figure 2 Valvistor diagram vi

SIMULATIONS

A set of plots (Figure 3) of nonlinear simulation responses have been made. These plots are with the original feedback slot and with supply pressures varying from 5 to 35 MPa in increments of 5 MPa. The outlet pressure (or load pressure) is held constant at 2 MPa. At 0.2 seconds the input was increased so that the output flow increased from slightly less than 5 liter/min to about 7 liter/min with the 5 MPa supply pressure. Flows increase as the pressure is increased so the lowest trace is for 5 MPa and the highest trace is for 35 MPa. The speed of response becomes faster as the supply pressure is increased. As expected the flow control accuracy is poor since the flow increases greatly as the supply pressure increases.

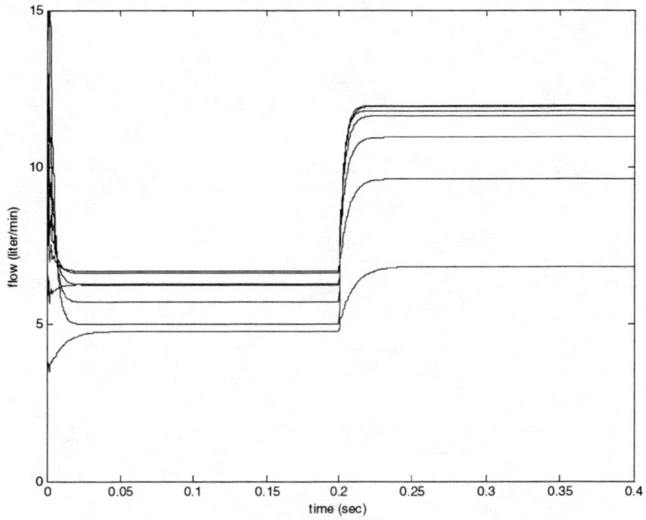

Figure 3 Flow control accuracy

The question of what would happen if the feedback slot was made larger to increase the speed of response of the system will now be considered. A set of simulations were used to explore this possibility with supply pressure increasing from 5 to 35 MPa. A plot of the result is given in Figure 4. A step input is given at 0.2 sec. to the pilot solenoid. At low pressure, the result is good and the speed of response is much inproved which was the goal of redesigning the valve with the modified slot. However, as the supply pressure is increased the output becomes oscillatory and flow control accuracy is poor.

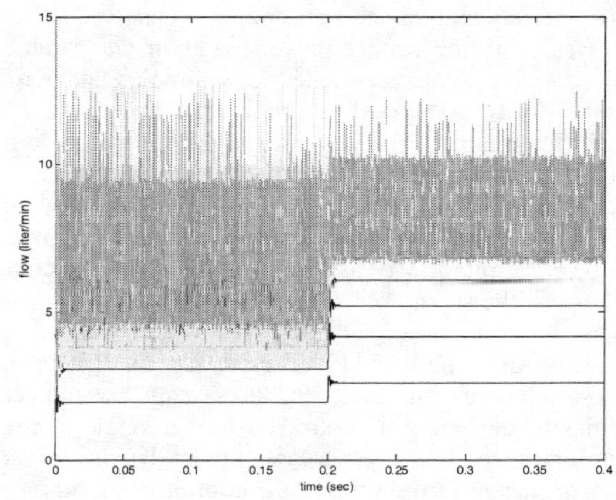

Figure 4 System flow control performance with modified feedback slot

LINEAR ANALYSIS

To further study the dynamics of the system, a linear model was developed for the Vickers Valvistor. A block diagram of this model is given in Figure 5. The model is of the system with the input of pilot poppet position, x_p, and an output of outlet flow, Q_B with the pilot pressure dynamics included. The partial derivatives used to linearize the model, *K1* through *K5*, are defined as follows:

$$K1 = \frac{\partial Q_p}{\partial x_p} = -K_p\sqrt{(P_{0p} - P_{0B})},$$

$$K2 = \frac{\partial Q_p}{\partial P_p} - x_{0p}K_p/(2\sqrt{P_{0p} - P_{0b}}),$$

$$K3 = \frac{\partial Q_2}{\partial P_p} = -x_{0m}K_S / \sqrt{P_{0A} - P_{0p}},$$

$$K4 = \frac{\partial Q_2}{\partial x_m} = K_s\sqrt{P_{0A} - P_{0p}}, \quad \text{and}$$

$$K5 = \frac{\partial Q_B}{\partial x_m} = K_m\sqrt{P_{0A} - P_{0B}}.$$

The flows indicated above are defined as follows:

$$Q_p = x_pK_p\sqrt{P_p - P_B}, Q_2 = x_mK_s\sqrt{P_A - P_p}, \quad \text{and}$$

$$Q_B = x_mK_m\sqrt{P_A - P_B} + Q_p.$$

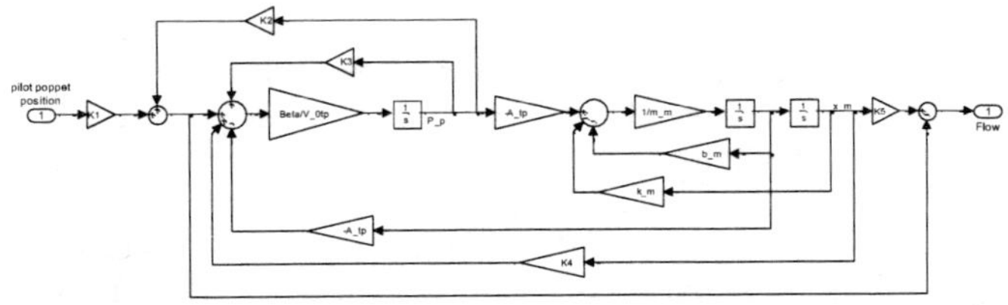

Figure 5 Linear model of the Vickers Valvistor

In this section, the linearized version of the nonlinear model is used to analyze the frequency response and the poles of the system as the supply pressure and model parameters change. An analysis of the Bode plot and poles and zeros of this system was performed. The model linearization was computed for a range of supply pressures, P_A, and Bode plots and pole/zero plots were created for each case. Again, the supply pressure ranges from 5 MPa to 35 MPa. The Bode plots are shown in Figure 6 with arrows which indicate the direction which the Bode plot moves as the supply pressure is increased. It can be seen the supply pressure affects the DC gain which is expected since the flow gain increases as the pressure drop increases. Also the bandwidth increases as the pressure increases. These variations in the system would cause difficulties in designing a controller since a single controller design is optimized for only one case of plant dynamics. As the parameters change in the system, the controller would have to change in order to maintain optimal performance using gain scheduling for example. This is not always a simple task since sensors would be required to determine the state of the system which would be used to select the appropriate controller gains.

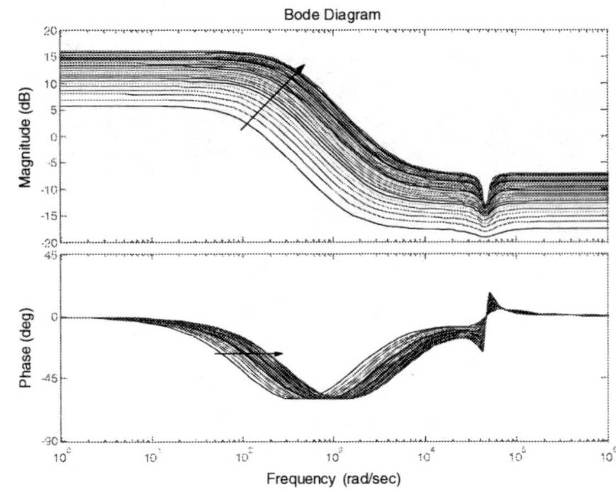

Figure 6 Set of Bode plots for the model with increasing supply pressure

In Figure 7 a root locus is given for the system as the pressure increases. Arrows are given in the plot to show the direction of increasing supply pressure. A low frequency real pole can be seen moving away from the origin as the pressure is increased. This matches well with the Bode plot and the time domain simulation results which indicated an increase in bandwidth. Also, a set of complex poles (and zeros) move closer to the imaginary axis as the pressure increases. This suggests that the system might become unstable if the supply pressure continued to increase.

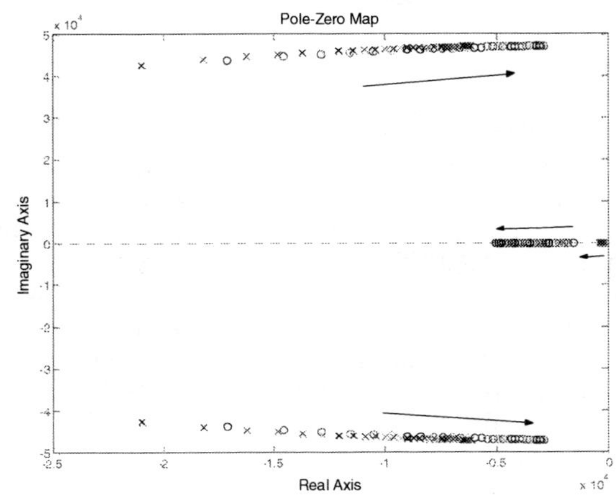

Figure 7 Pole/zero maps for the higher order model with pressure increasing

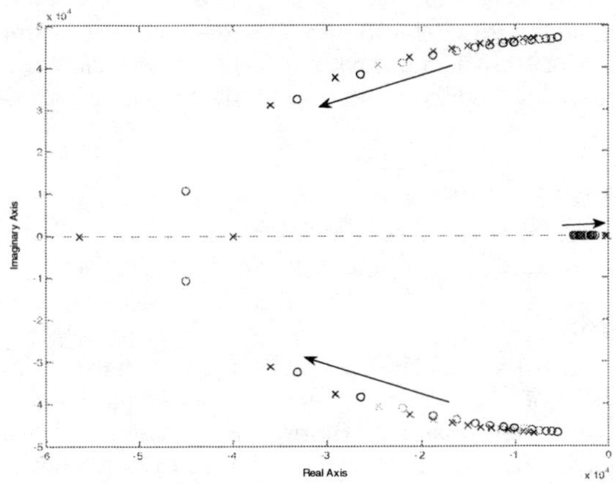

Figure 8 Locus of poles and zeros with the load pressure increasing

In the case of increasing load pressures, the dynamics of the system vary in a way opposite from the case of increasing supply pressures. Figure 8 is a root locus plot showing the movement of the poles for a case where the feedback slot has not been modified, the load pressure is increased, and the supply pressure is held constant at 20 MPa. The load pressure increases from 3 MPa to 18 MPa. The arrows indicate the direction of movement of the roots as the supply pressure is increased. The poles in the complex plain move toward the real axis as load pressure increases. The pole and zero close to the origin move closer in toward the origin along the real axis as the load pressure increases. This indicates that as load pressure increases, the system will respond more and more slowly. This is sensible because the valve actuation requires a difference between inlet and outlet pressures to operate since there is no external pilot supply pressure. Once there is no difference between the inlet and outlet pressures no valve actuation is possible. This behavior as load pressure increases holds for the entire range of supply pressure considered; therefore, the worst case scenario from the stability point of view is the case where load pressure is low and supply pressure is at a maximum.

In the next plots, the same plant (with fixed load pressure) is analyzed except with the feedback slot modified to improve performance with conditions the same as in Figure 4. The poles (Figure 9) on the real axis are moved to the left indicating improved speed of response. The pole zero plot also shows that the system becomes unstable as the supply pressure is increased from 5 to 35 MPa due to the lightly damped complex conjugate pair of poles moving across the imaginary axis at the higher pressures. It correlates well with the conditions of Figure 4 which shows that the nonlinear system under the same conditions becomes unstable. This is proof that the varying pressure drop can cause the system to become unstable.

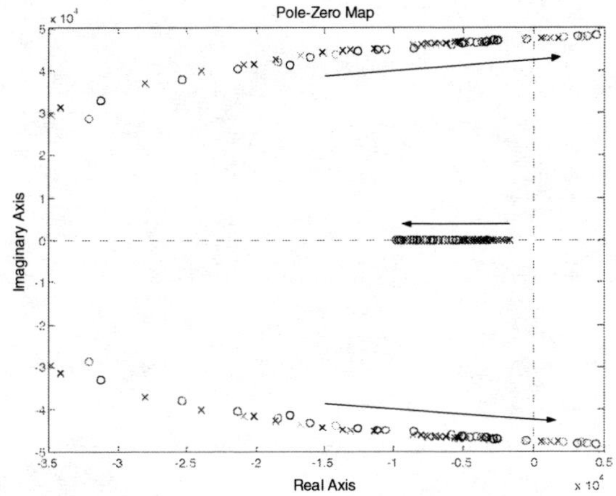

Figure 9 Root locus plot for the modified system for varying supply pressures

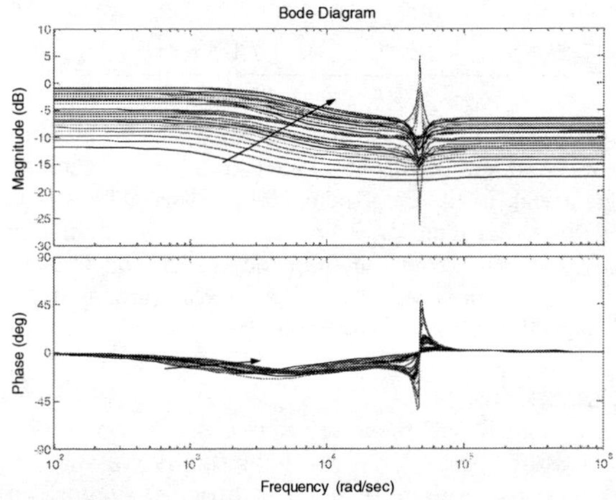

Figure 10 Bode diagram for the modified system with varying supply pressures

Turning back to the original system, the uncertainty in the system without the feedback slot modification can be used to create a model of the plant linear modeling error based on changes in supply pressure. The error in the magnitude of the frequency response of the system can be determined from the information in Figure 6. These errors are captured as a multiplicative plant uncertainty as follows[vii]. A nominal plant is chosen as the plant linearized with a supply pressure of 20 MPa. Plots are made of relative error in magnitude response between the nominal plant, G, and other linearizations (perturbed plants) at other supply pressures, G_p, an element of all possible perturbations . The error for each is given by $\frac{G_p(j\omega)-G(j\omega)}{G(j\omega)}$. The relative error in the plant model is the maximum error associated with all perturbation plants and is given as $w_i(\omega) \geq \max_{G_p \in \Pi} \left| \frac{G_p(j\omega)-G(j\omega)}{G(j\omega)} \right|$. The error, $w_i(\omega)$, is shown graphically in Figure 11 as a dotted line which is the upper boundary for the model error for all cases of supply pressure.

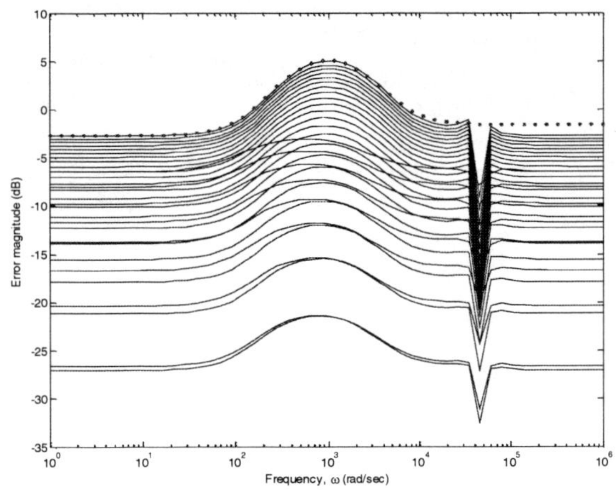

Figure 11 Modeling error in the frequency domain for supply pressures between 5 and 35 MPa

The transfer function corresponding to the error is $w_i(s) = \dfrac{0.8346\ s^2 + 5021\ s + 7.051\times10^5}{s^2 + 2785\ s + 9.632\times10^5}$. The error model shows in Figure 11 that the most uncertain frequency range for the model begins at about 100 rad/sec or about 16 Hz. There is a peak in the error near 1000 rad/sec which is greater than 0 dB (greater than 100%) modeling error. The uncertainty is significant throughout the frequency range and is especially noticeable at the low frequencies as a variation in the steady state flow as the supply pressure changes.

CONCLUSIONS

Flow control accuracy was shown to be poor when the supply pressure varies. Changes in the supply pressure alter the valve dynamics and therefore flow is difficult to predict in field conditions. Attempts to increase the flow gain of the feedback slot lead to higher performance at the expense of stability. Design of a controller to obtain accurate velocity or position control for hydraulic actuation system with this valve will be a challenge since the dynamics vary greatly with operating conditions.

An electronic feedback controller could be used to reduce the effects of the plant uncertainty especially prevalent above 16 Hz as well as at low frequencies. A mechanical solution to the problem of improving the response speed of the valve would be to somehow automatically reduce the feedback slot gain as the pressure drop increases. This would amount to a small variable orifice built into the main poppet within the feedback slot flow path so that the feedback slot gain could be increased at low supply pressures and decreased at high supply pressures.

ACKNOWLEDGMENTS

The author would like to thank Mark Vonderwell and Eko Prasetiawan for their helpful assistance during this work.

APPENDIX

Details of the Vickers Valvistor are given in the following. A diagram of the Valvistor performance data is given in Vickers technical literature. Below are the states and dynamic equations for the Vickers Valvistor adapted from reference iv.

State vector:

$$\begin{bmatrix} x_1 & x_2 & x_3 & x_4 & x_5 \end{bmatrix}^T = \begin{bmatrix} x_m & \dot{x}_m & P_p & x_p & \dot{x}_p \end{bmatrix}^T \quad (1)$$

Dynamic equations:

$$\dot{x} = \begin{bmatrix} x_2 \\ -\dfrac{k_m}{m_m}x_1 - \dfrac{b_m}{m_m}x_2 + \dfrac{A_{bt}P_A + (A_{tp} - A_{bt})P_B - A_{tp}x_3}{m_m} \\ \dfrac{\beta}{V_p}\left(K_S x_1\sqrt{P_A - x_3} - K_p x_4\sqrt{x_3 - P_B}\right) \\ x_5 \\ -\dfrac{k_p}{m_p}x_4 - \dfrac{b_p}{m_p}x_5 \end{bmatrix} x + \begin{bmatrix} 0 \\ 0 \\ 0 \\ 0 \\ K_e \end{bmatrix} u_v \quad (2)$$

Output:

$$y = K_p x_4\sqrt{x_3 - P_B} + K_m x_1\sqrt{P_A - P_B} = Q_B \quad (3)$$

The model used for the nonlinear simulations was based on these equations but includes additional detail such as saturation of the pilot and main poppet positions, leakage, and a varying volume, V_p, based on the position of the main poppet.

REFERENCES

[i] Johnston, D., K. Edge, N. Vaughan. 1991. "Experimental Investigation of Flow and Force Characteristics of Hydraulic Poppet and Disc Valves." Proceedings of the Institution of Mechanical Engineers Part A, Vol. 205, No. (A3), pp. 161-171.

[ii] Vaughan, N., D. Johnston, K. Edge. 1992. "Numerical Simulation of Fluid Flow in Poppet Valves." Journal of Mechanical Engineering Science, Vol. 206, No. (C2), pp. 119-127.

[iii] Yang, R. 2004. "Predicting Hydraulic Valve Flow Forces Using CFD." ASME International Mechanical Engineering Congress and Exposition, Anaheim, CA.

[iv] Zhang, R. , A. Alleyne, E. Prasetiawan. 2002. "Performance Limitations of a Class of Two-stage Electro-hydraulic Flow Valves." International Journal of Fluid Power, Vol. 3, No. (1).

[v] Opdenbosch, P., N. Sadegh, W. Book. 2004. "Modeling and Control of an Electro-hydraulic Poppet Valve." ASME International Mechanical Engineering Congress and Exposition, Anaheim, CA.

[vii] Skogestad, S., I. Postlethwaite. 1996. "Multivariable Feedback Control." Wiley. New York.

Proceedings of IMECE2005
2005 ASME International Mechanical Engineering Congress and Exposition
November 5-11, 2005, Orlando, Florida USA

IMECE2005-80980

CONTROL OF ELECTRO-HYDRAULIC POPPET VALVES VIA ONLINE LEARNING AND ESTIMATION

Patrick Opdenbosch
Graduate Research Assistant
G.W. Woodruff School of Mechanical Engineering
Georgia Institute of Technology
Atlanta, GA 30332
(404) 894 3256
gte608g@mail.gatech.edu

Nader Sadegh Ph.D.
Associate Professor
G.W. Woodruff School of Mechanical Engineering
Georgia Institute of Technology
Atlanta, GA 30332
(404) 894 8172
nader.sadegh@me.gatech.edu

ABSTRACT

Online learning state trajectory control applied to Electro-Hydraulic Poppet Valves (EHPV) is considered herein. The control problem is to track a desired flow conductance coefficient K_V for pressure or flow control applications. In general terms, the control methodology employed herein computes the input signal sent to the valve from the addition of three components. The first component comes from an experimentally approximated inverse input-output map of the system which gives a nominal input. The second component is computed through a neural network structure called the Nodal Link Perceptron Network that learns online the adjustment of this nominal map. The third component is an adaptive proportional feedback control input. This last component uses two system parameters known as the Jacobian and the Controllability parameter, which are estimated online via a recursive least squares algorithm with forgetting factor. The proposed controller is explored through experimental data on a pressure control application and the results are discussed.

INTRODUCTION

The advance of technology is always accompanied by challenging problems whose solutions are far from trivial. Nowadays, engineers design controllers for systems with complex nonlinearities, state constraints, and time varying characteristics. One possible approach to do so would involve the development of detailed models of such complex systems. Based on this developed model, the control scheme could then be designed using nonlinear control tools to meet the required performance specifications. In many situations, the complexity of the system and the unavailability of suitable models drive the engineer to rely on inverse input-output mapping to control the plant. This is seen in many industrial control applications, especially in areas related to motion control, where it is required that the system at hand tracks a desired trajectory without having much knowledge of the system's model [1]. In response to this challenge, new technologies such as neural networks give an alternative approach to the control design task, especially for the inverse input-output mapping learning technique. The inclusion of neural networks in the control task enables the online learning of the dynamic behavior of such complex systems.

Perceptron neural nets are a special kind of network that can be used for online identification and control of complex systems. Particularly, they can approximate and store complex nonlinear functions with the ability to self-adapt to reflect the system's changes that might arise as a result of prolonged operation [2],[3]. The neural network architecture employed herein is known as the Nodal Link Perceptron Network (NLPN) [4]. It is used to learn the corresponding control input required to follow a desired state trajectory. As a consequence, it does not need to explore the entire inverse plant dynamics map, saving memory and effort.

The tendency of including neural network technology for intelligent control purposes is rapidly increasing [2],[3],[4],[5],[6],[7] to cite a few examples. Some specific examples related to fluid power applications are included next. Pinsopon et. al. used a cerebellar model articulation controller neural network architecture to control a hydraulic actuator with an open-centre electro-hydraulic valve [5]. In particular, this type of controller is shown to improve the performance of the system compared to a conventional proportional-integral controller. In terms of online neural network adaptation, Nishiumi and Watton used an artificial neural network to improve the speed characteristic of a hydraulic motor [6]. Other examples of artificial neural networks employed for control purposes in the fluid power industry can be found at [7].

When dealing with complex systems such as electro-hydraulic valves, it is common to see that the dynamics of the valves are approximated by a second order system [8]. In other cases, the valve bandwidth is sufficiently high that it is assumed that their dynamics can be neglected [9]. By exploiting this characteristic, some researches have employed the inverse static input-output mapping [9] for control purposes. Moreover, some researches have used memory based learning techniques [10] or associative memory techniques [11] for the same.

In regards to inverse input-output mapping learning, a control scheme that uses the NLPN is proposed by Sadegh [2]. This scheme is used for state trajectory tracking by linearizing the system about a desired state trajectory. As a result, an error equation is formulated, and the control law is derived to force this error to zero. This control law employs some information about the system, namely the Jacobian and the Controllability matrix of the linearized system (linearized about the desired state trajectory). The author also points out that approximate knowledge of these matrices often suffices in practice.

The control valve considered in this paper is an Electro-Hydraulic Poppet Valve (EHPV) described in detail in [12],[13]. The flow control of hydraulic fluid through the EHPV is achieved by changing the valve flow conductance coefficient K_V via a PWM input current acting on a pilot and a poppet type orifice. Among the distinguishing features of this valve one finds an internal pressure compensation mechanism that helps maintain a consistent amount of initial input current needed to open the valve, a three-degree-of-freedom mechanical system with state constraints, zero leakage, bidirectionality, and low metering flow control.

This paper explores the idea of combining online state trajectory learning with the online identification of the Jacobian and Controllability matrices via recursive least squares for the control of electro-hydraulic valves. The control law proposed by Sadegh in [2] with the improvement of learning the needed system information online can successfully deal with the nonlinearities and uncertainties inherent in EHPV's.

This paper is organized as follows: The nomenclature is presented first. Next, the problem statement along with some definitions and assumptions is introduced. After these sections, the control design is discussed and the experimental results are given. The concluding remarks are presented at the end followed by acknowledgments.

NOMENCLATURE

i_{sol}	Input current sent to the valve's solenoid
J	Jacobian parameter
K_V	Flow conductance coefficient
q	Net flow through the valve
Q	Controllability parameter
T	Oil temperature
ΔP	Pressure differential across the valve
η	Forgetting factor for estimation
λ	Forgetting factor for NLPN training

PROBLEM STATEMENT

The system at hand is the novel electro-hydraulic poppet valve described in the introduction [12],[13]. Commonly, these valves are commanded to provide a desired flow restriction or flow conductance on pressure or flow control applications. The flow conductance of the EHPV is introduced in Eq. (1).

$$q|q| = K_V^2 \Delta P \tag{1}$$

If the valve dynamics are much faster than the dynamics of the other components in a hydraulic circuit, then knowledge of the steady state mapping of the K_V would suffice for control purposes. This steady state mapping is then inverted so that the necessary input current can be computed and be sent to the solenoid. In practice, this inverted mapping is simply stored in a lookup table and downloaded to the corresponding controller. From its definition, it can be deduced that the steady state values of the flow conductance for the EHPV are dominated by the pressure differential across the valve, the oil's temperature, and the intrinsic flow path geometry. The latter can be manipulated via the input current sent to the valve's solenoid.

$$K_V\big|_{t\to\infty} = F_{ss}(\Delta P, T, i_{sol}) \tag{2}$$

In general, temperature effects on the flow conductance coefficient are fairly small: the flow conductance slightly increases for a significant increase in oil temperature at a constant valve opening. Hence, the flow conductance is typically considered solely dependent on pressure differential and input current.

In terms of the EHPV's dynamic response characteristics, Opdenbosch and Sadegh [12] state that the flow conductance of this particular valve can be approximated by a first order nonlinear discrete-time equation as shown next, where $K_V[k]$ denotes $K_V[kT_s]$ – i.e. the flow conductance is sampled every T_s seconds (note that this can be easily extended to a second order approximation if needed).

$$K_V[k+1] = F(K_V[k], \Delta P[k], T[k], i_{sol}[k]) \tag{3}$$

The objective is then to develop a controller that learns online the input sequence $i_{sol}[k]$ that forces the flow conductance $K_V[k]$ of this valve to follow a desired trajectory. This desired trajectory would be provided by a higher level controller, omitted here for simplicity, that determines the needed flow conductance based on operator input or flow/pressure demand. Moreover, it is desired that the nonlinear mapping, both the steady state and the transient one need not be known accurately in advance. In other words, an online learning tracking controller is sought to learn the system's mappings at the same time that forces it to follow a desired K_V.

DEFINITIONS

The following definitions are used in this paper:

i) The set of admissible desired states, denoted by χ, is a bounded open subset of $\Re$ (*e.g.* $\chi \subset \{K_V \in \Re : \|K_V\| < r\}$ for some finite $r > 0$).

ii) The Jacobian of the system in Eq (3) is defined as $J[k] \equiv D_{K_V}(F)$, where $D_\varsigma(\cdot)$ is the partial derivative operator with respect to ζ.

iii) The term $Q[k] \equiv D_{i_{sol}}(F)$ is defined as the Controllability parameter which is a generalization of the well known results for linear systems [14].

ASSUMPTIONS

The following standard assumptions for controllability

commonly adopted in the literature are assumed herein for the system at hand.

i) The function F is at least a twice continuously differentiable (C^2) function

ii) The system is strongly controllable (with respect to χ)[15],[16]: $\forall x, z \in \chi$, there exists a unique input vector $u \in \Re$ such that $F(x,u) = z$.

iii) The set of admissible inputs for the system is given as $U = \{i_{sol} \in \Re : z = F(x, i_{sol}),\ x, z \in \chi\}$.

iv) The controllability matrix, Q has full rank for all $i_{sol} \in U$ and $K_V \in \chi$ [1],[14].

CONTROL DESIGN

Consider the SISO (Single-Input-Single-Output) nonlinear system given in Eq. (3). For control purposes, let the desired flow conductance be denoted by ${}^{d}K_V$ and the tracking error by

$$e[k] \equiv K_V[k] - {}^{d}K_V[k] \tag{4}$$

Linearizing the system in Eq. (3) about a given desired trajectory $\left({}^{d}K_V[k], \Delta P[k], T[k], {}^{d}i_{sol}[k]\right)$ yields the following error equation, as described in [2].

$$e[k+1] = J_d[k]e[k] + Q_d[k]\left(i_{sol}[k] - {}^{d}i_{sol}[k]\right) + o\left(e[k], i_{sol}[k] - {}^{d}i_{sol}[k]\right) \tag{5}$$

where the following definitions are applied:

$$J_d[k] \equiv \frac{\partial F\left({}^{d}K_V[k], \Delta P[k], T[k], {}^{d}i_{sol}[k]\right)}{\partial K_V[k]} \tag{6}$$

$$Q_d[k] \equiv \frac{\partial F\left({}^{d}K_V[k], \Delta P[k], T[k], {}^{d}i_{sol}[k]\right)}{\partial i_{sol}[k]} \tag{7}$$

Based on the error equation, the proposed control law introduced by Sadegh in [2] is modified as shown in Eq. (8). The control law along with the entire control scheme is portrayed in Fig. 1. It is important to point out that this control law combines adaptive proportional feedback correction, along with an online learned feedforward control action.

$$i_{sol}[k] = -\sigma \tilde{Q}^{-1}[k-1]\tilde{J}[k-1]e[k] + \bar{i}_{sol}[k] + \Delta\tilde{i}_{sol}[k] \tag{8}$$

Note that in the case where the system would need to be considered as a higher order system, the sate equation can be 'lifted' into a block-input form so that the number of inputs and outputs is the same as that of the states (see [14],[17],[18]). This procedure would then yield a square Q_d matrix. Furthermore, the careful reader should notice the fact that the control law uses a gain σ that can be chosen by the user, and also uses approximations for the Jacobian and the Controllability parameters $\tilde{J}[k-1]$ and $\tilde{Q}[k-1]$ respectively. These parameters are estimated online using the method of recursive least squares.

In order to estimate the Jacobian and the Controllability parameters, the error equation (5) needs to be considered, more specifically, approximated from available measured data. Let the approximated error be given by

$$e[k+1] \approx \Delta K_V[k+1] = \tilde{J}[k]\Delta K_V[k] + \tilde{Q}[k]\Delta i_{sol}[k] \tag{9}$$

where the definitions given by Eq. (10) and (11) apply. The 'primed' terms in these equations are the time average values of the corresponding variables.

$$\Delta K_V[k] \equiv K_V[k] - K'_V[k] \tag{10}$$

$$\Delta i_{sol}[k] \equiv i_{sol}[k] - i'_{sol}[k] \tag{11}$$

From sampled data, the following matrix equation can be easily obtained and will be used for estimation purposes:

$$\begin{bmatrix} \Delta K_V[1] \\ \vdots \\ \Delta K_V[k+1] \end{bmatrix} = \begin{bmatrix} \Delta K_V[0] & \Delta i_{sol}[0] \\ \vdots & \vdots \\ \Delta K_V[k] & \Delta i_{sol}[k] \end{bmatrix} \begin{bmatrix} \tilde{J}[k] \\ \tilde{Q}[k] \end{bmatrix} \tag{12}$$

Let the vector containing the estimates of the Jacobian and the controllability parameters be denoted by $\mathbf{V}[k]$. Furthermore, let the vector containing the measured flow conductance values be denoted by $\mathbf{B}[k+1]$. Hence, Eq. 12 can be expressed in compact form as,

$$\mathbf{B}[k+1] = \mathbf{M}[k]\mathbf{V}[k] \tag{13}$$

Let each row of the $\mathbf{M}$ matrix be denoted by $\mathbf{m}^{T}[k]$. Then, recursive equations can be formulated to estimate the Jacobian and Controllability parameters in a least squares sense. These recursive equations are presented next.

$$\mathbf{V}[k-1] = \mathbf{V}[k-2] + \frac{\mathbf{C}[k-2]\mathbf{m}[k-1]\left(\Delta K_V[k] - \mathbf{m}^{T}[k-1]\mathbf{V}[k-2]\right)}{\eta[k-2] + \mathbf{m}^{T}[k-1]\mathbf{C}[k-2]\mathbf{m}[k-1]} \tag{14}$$

$$\mathbf{C}[k+1] = \frac{1}{\eta[k]}\left(\mathbf{C}[k] - \frac{\mathbf{C}[k]\mathbf{m}[k+1]\mathbf{m}^{T}[k+1]\mathbf{C}[k]}{\eta[k] + \mathbf{m}^{T}[k+1]\mathbf{C}^{T}[k]\mathbf{m}[k+1]}\right) \tag{15}$$

$$\eta[k+1] = \alpha_e \eta[k] + \beta_e (1-\alpha) \tag{16}$$

These equations include a forgetting factor η to keep the estimation from locking the values of the Jacobian and Controllability parameters (recall that since these parameters come from linearizing the nonlinear error dynamics along a state trajectory, they are time-varying in nature). Moreover, the forgetting factor is presented as a dynamic system so that initial data can be more easily discarded if necessary in the estimation.

The proposed control law given by Eq. (8) uses a feedforward term $\bar{i}_{sol}[k]$, and a variation of this feedforward term denoted by $\Delta\tilde{i}_{sol}[k]$. The feedforward term, referenced herein also as the nominal map, can be easily implemented from inversion of the steady state input/output data of the system at hand or an approximation of the same. Moreover, the variation term $\Delta\tilde{i}_{sol}[k]$ is learned online using the NLPN [4] (see Fig. 2).

$$\bar{i}_{sol}[k] = \tilde{F}_{ss}^{-1}\left({}^{d}K_V[k], \Delta P[k], T[k]\right) \tag{17}$$

$$\Delta\tilde{i}_{sol}[k] = \tilde{\mathbf{W}}^{T}[k]\mathbf{\Phi}\left({}^{d}K_V[k], \delta K_V[k], \Delta P[k], T[k]\right) \tag{18}$$

$$\delta K_V[k] = {}^{d}K_V[k] - {}^{d}K_V[k-1] \tag{19}$$

The NLPN training is performed via recursive least squares training with forgetting factor applied as well. Note that it is possible to reset the covariance matrix $\mathbf{P}$ and the forgetting factor λ if desired. The last equation (19) is used so that the learning space for the NLPN can be divided into steady and transient.

$$\tilde{\mathbf{W}}[k+1] = \tilde{\mathbf{W}}[k] + \frac{\mathbf{P}[k]\mathbf{\Phi}[k+1]E[k]}{\lambda[k] + \mathbf{\Phi}^{\mathrm{T}}[k+1]\mathbf{P}[k]\mathbf{\Phi}[k+1]} \tag{20}$$

$$\mathbf{P}[k+1] = \frac{1}{\lambda[k]}\left(\mathbf{P}[k] - \frac{\mathbf{P}[k]\mathbf{\Phi}[k+1]\mathbf{\Phi}^{\mathrm{T}}[k+1]\mathbf{P}[k]}{\lambda[k] + \mathbf{\Phi}^{\mathrm{T}}[k+1]\mathbf{P}^{\mathrm{T}}[k]\mathbf{\Phi}[k+1]}\right) \tag{21}$$

$$E[k] = -e[k] \tag{22}$$

$$\lambda[k+1] = \alpha_l \lambda[k] + \beta_l(1-\alpha) \tag{23}$$

It is commonly known that other rules can be used to adjust the weights $\tilde{\mathbf{W}}^{\mathrm{T}}[k]$ of the NLPN. For example, one can use the delta rule training (see [2]).

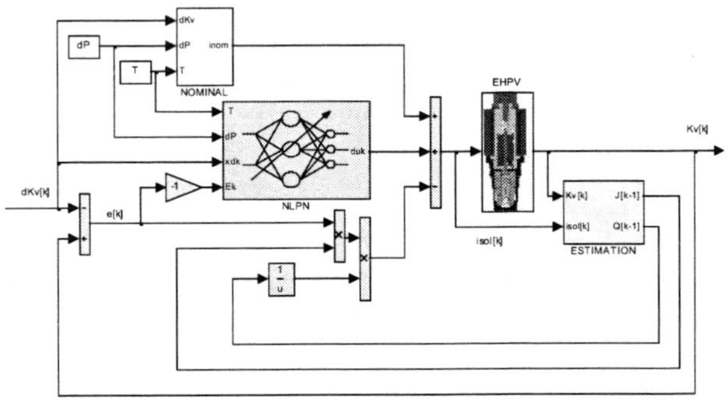

Fig. 1 Control scheme

It is worth mentioning that the nominal mapping could also be implemented by assigning an initial set of weights to the NLPN. For this, the collected steady state data is feed to the network for offline training and the learned weights are stored. Moreover, it is important to realize that this nominal mapping is added to the control input to take advantage of readily available data and to facilitate the NLPN effort. It could be omitted if such inverse mapping cannot be obtained. If this is the case, then the NLPN is left to learn the entire inverse mapping.

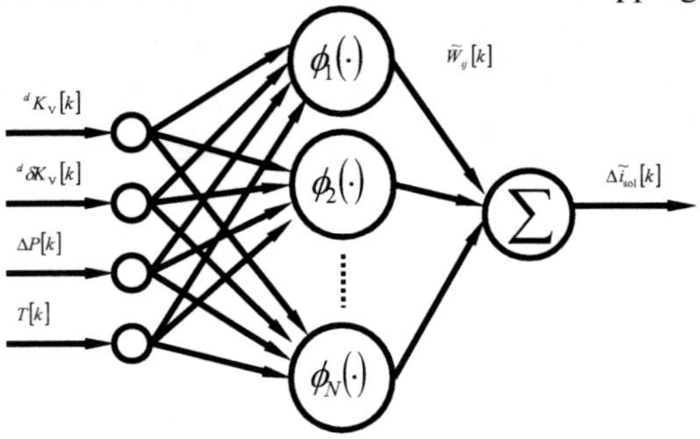

Fig. 2 Nodal link perceptron network structure

According to Sadegh [2],[3], the error signal $e[k]$ and the control input $i_{\mathrm{sol}}[k]$ are bounded sequences for a bounded functional approximation of the nodal link perceptron network ($\left|\bar{i}_{sol}[k] + \Delta\tilde{i}_{sol}[k] - {}^{d}i_{sol}[k]\right| < \delta$) provided the desired task and the network size are chosen appropriately. In addition, if the discrepancy between the estimated parameters and the true parameters is bounded, *i.e.* there are positive finite constants μ_J and μ_Q such that $\sup_k \left|J_d[k] - \tilde{J}[k-1]\right| < \mu_J$ is satisfied and also $\sup_k \left|Q_d[k] - \tilde{Q}[k-1]\right| < \mu_Q$ is satisfied, then it can be shown that the previous claim remains valid [2],[3].

EXPERIMENTAL RESULTS

The steady state characteristics of the EHPV are first investigated. Note that temperature effects are neglected herein. Since the valve is bidirectional, two steady state plots are presented. The first plot, depicted in Fig. 3(a) corresponds to the forward steady state flow conductance coefficient vs. the pressure differential across the valve. The different curves represent different input currents sent to the solenoid of the valve. These range from 984 mA up to 1504 mA. The second plot, depicted in Fig. 3(b) corresponds to the reverse steady state flow conductance coefficient vs. the pressure differential across the valve. The different curves represent different input currents sent to the solenoid of the valve ranging from 984 mA up to 1507 mA.

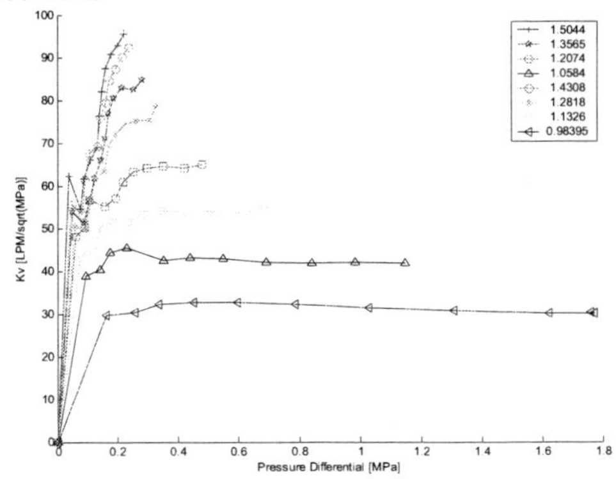

(a) Forward K_V

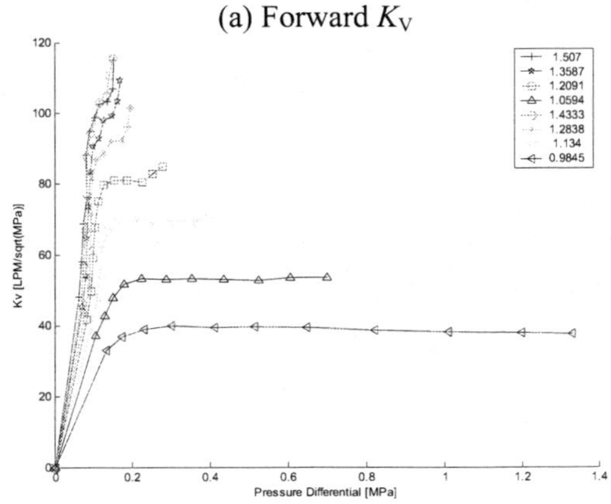

(b) Reverse K_V

Fig. 3 EHPV steady state Kv as a function of ΔP and i_{sol}

The offline learning of these mappings are presented next. For simplicity, only the inverse mappings are considered. Hence, the inverse forward mapping is given in Fig. 4(a) while that of the reverse mapping is given in Fig. 4(b).

As it can be seen, there are regions where no activation took place. This is a result of not exceeding the power envelop of the hydraulic system employed in the test. These mappings could be then implemented as lookup tables. The problem would be then how to solve the no activation regions. One solution would be to extrapolate from the nearest data points. However, this does not exempt from introducing errors. This gives rise to the need for online feedback correction.

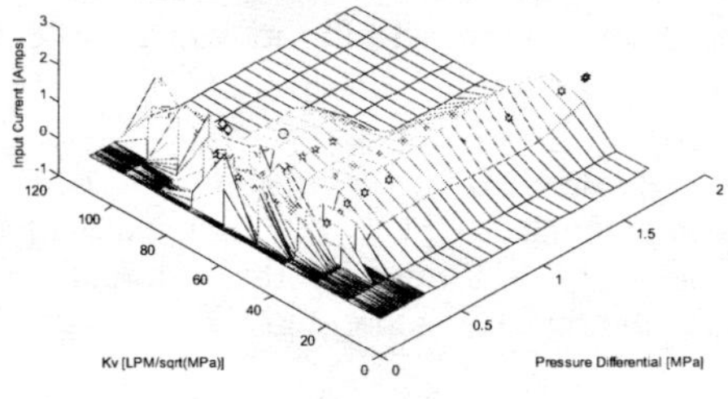

(a) Forward i_{sol}

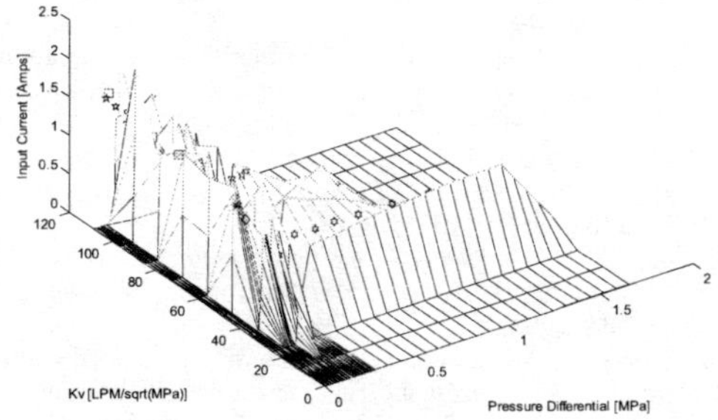

(b) Reverse i_{sol}

Fig. 4 EHPV input current as a function of ΔP and Kv

In order to demonstrate the capabilities of the NLPN, these mappings will be discarded in the online tracking task. However, they give some valuable information: the flow conductance is nearly independent of pressure differential in both flow directions for ΔP values greater than 0.4 MPa. This feature will be exploited to simplify the control task. As such, the valve is operated at a pressure differential higher than 0.4 MPa so that the ΔP term in Eq. (18) drops out. The adjustment is then given by

$$\Delta\tilde{i}_{sol}[k] = \tilde{\mathbf{W}}^T[k]\mathbf{\Phi}\left({}^dK_V[k], \delta K_V[k]\right) \tag{24}$$

The previous-point-difference input variable to the NLPN, δK_V, is used to deal with hysteresis effects. When this value is positive, the valve is ramping up; when negative, the valve is ramping down. Finally, when $\delta K_V[k] = 0$, the valve is at steady state. This way, transient mappings as well as steady state mappings can be learned.

The tracking control task was then put to the test next. For this purpose, the hydraulic system depicted in Fig. 5 was employed. In this setup, the EHPV is used as a relief valve to control the pressure on the cap side of a piston type hydraulic actuator. Note that in order to improve the aforementioned mappings, measurement of the flow conductance of the valve is needed. For this purpose, an orifice flowmeter was calibrated to measure the flow of hydraulic oil passing through the valve. The flow conductance was then computed by taking the ratio of the measured flow and the square root of the pressure differential across the valve.

The NLPN was setup to have a two dimensional input space. The grid for the desired flow conductance coefficient was partitioned as [0 22 23 46 47 72] LPM/sqrt(MPa) while the grid for the increment in flow conductance was partitioned as [-120 0 120] LPM/sqrt(MPa). The sampling period for was set to 1ms and the online weight training was performed at the same rate. The covariance matrix **P** for the NLPN was initially set to a matrix with the value of 1000 in its diagonal. The forgetting factor was fixed at a value of 0.999 and no resetting of the covariance matrix and/or resetting of the forgetting factor was used.

The estimation routine was also updated at the sampling rate. Moreover, the estimation covariance matrix was initially set to a matrix with the value of 100 in its diagonal. The forgetting factor was fixed at a value of 0.99 and no resetting of the covariance matrix and/or resetting of the forgetting factor was used. In addition, the initial estimation of the Jacobian and the Controllability parameter was arbitrarily chosen as 0.9 and 0.03 respectively. Finally, the estimation of these parameters was bounded to the interval [0.001,2] and a gain $\sigma = 0.01$ was used.

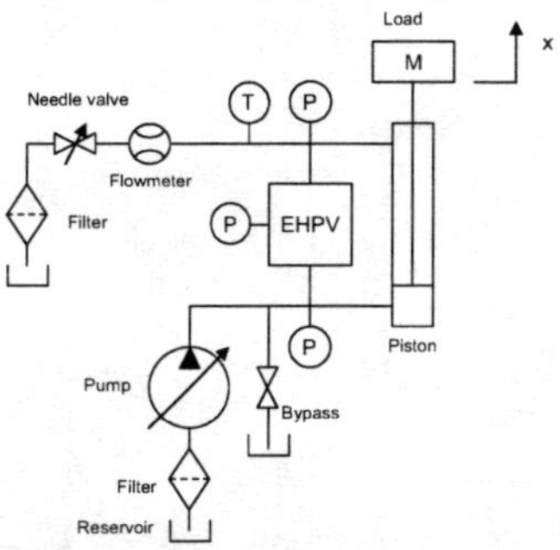

Fig. 5 Testbed hydraulic diagram

The experimental data is presented in the following figures. Initially, a nominal value of 7V is sent to the solenoid, while the feedback signal is turned off. This is seen in the first 1.2 seconds in Fig. 6(a). At this time, the feedback signal is turned on, but the NLPN learning remains off. The response in this stage is seen from 1.2 seconds to 3.5 seconds in the same figure. In terms of operating conditions, the oil temperature was near 35C, and a nominal pressure drop across the EHPV of 0.6 MPa was measured. Additionally, the approximation of the K_V employing Eq. (9) is seen in Fig. 6(a) as K_{Vappx}, while the estimated values of the Jacobian and the controllability parameters are seen in Fig 6(b).

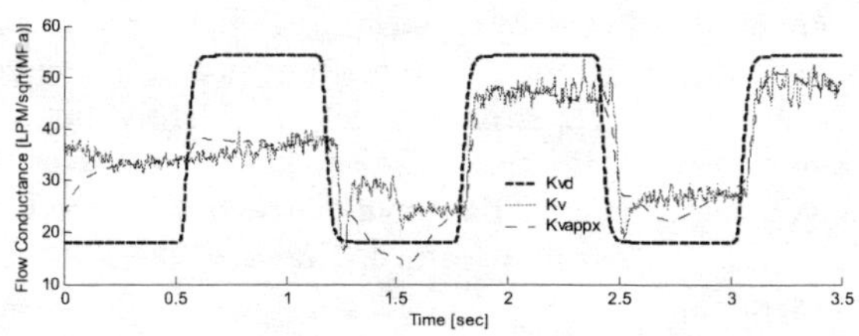

(a) Flow conductance

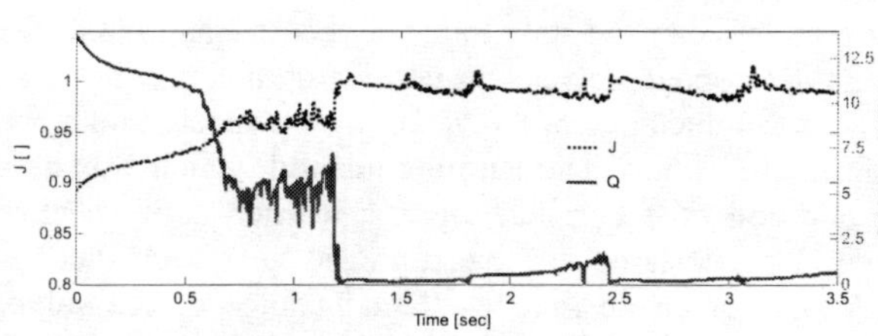

(b) Estimated parameters

Fig. 6 Initial test response: no NLPN learning

The NLPN learning is then activated at 3.5 seconds. The improved performance is presented in Fig 7(a). This figure displays the desired K_V profile along with the measured K_V and the input signal sent to the amplifier, which in turn is converted to current – it is assumed that the amplifier dynamics are much faster than those of the valve and thus neglected. This input signal is the addition of the adaptive proportional feedback signal, the nominal 7V, and the learned correction. Also, the estimated parameters are presented in Fig 7(b). Finally, the pressure differential profile along with the oil temperature is presented in Fig 7(c).

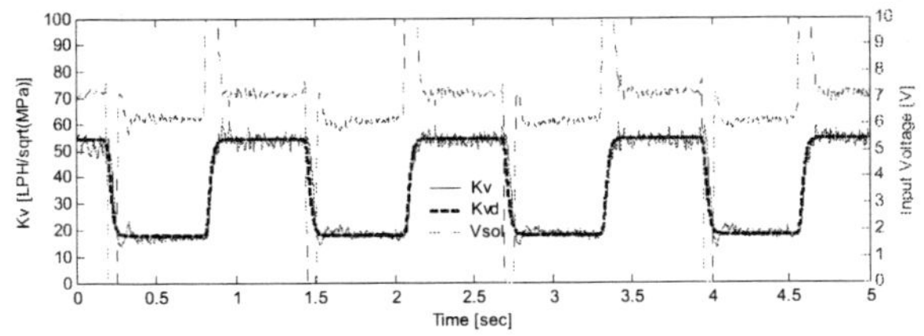

(a) Flow conductance

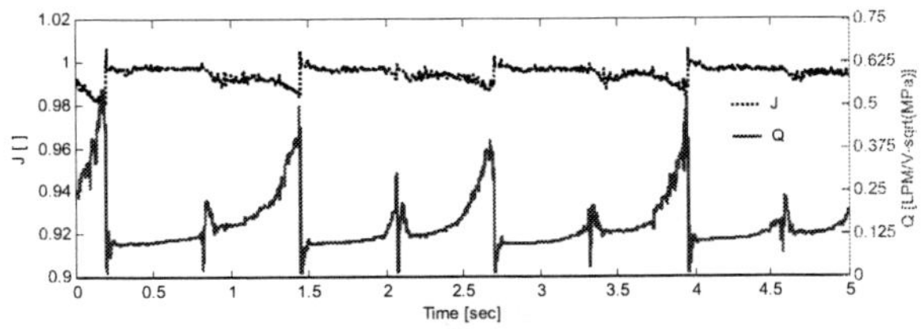

(b) Estimated parameters

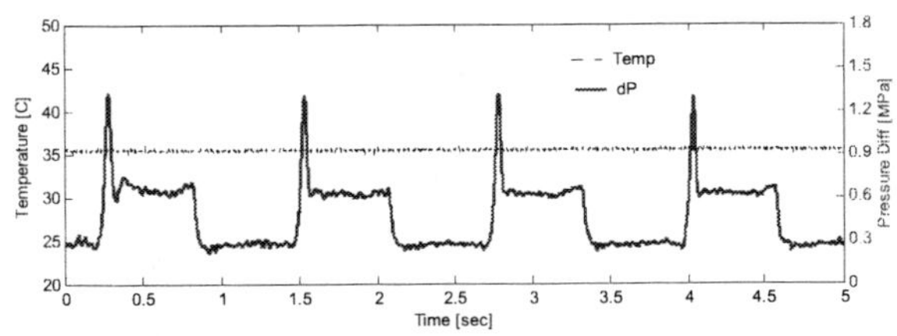

(c) Temperature and pressure profiles

Fig. 7 EHPV response with NLPN learning

It can be concluded by comparing Fig. 6 and 7, that the learning controller improved the performance of the valve. Not only is the steady state error reduced considerably, but also the transient behavior is much closer to the desired profile. Moreover, the corresponding input to reach these steady flow conductance values is now learned. The same can be said about the transient input.

CONCLUSION

An online learning state trajectory control applied to Electro-Hydraulic Poppet Valves was considered herein. The control of the flow conductance of the valve was considered on a pressure control application. The learning method, which was based on the addition of a nominal input, a learned adjustment, and a adaptive proportional feedback, proved to be very successful at tracking a given desired flow conductance of the valve. The learning was accomplished with the aid of the NLPN which used a recursive least squares training algorithm with forgetting factor included. The adaptive proportional feedback employed the estimated Jacobian and Controllability parameters. These were identified online via a recursive least squares algorithm with forgetting factor as well. This method offers an alternative approach for dealing with systems in which little is known about the dynamic behavior.

ACKNOWLEDGMENTS

The authors thank HUSCO International and the Fluid Power Motion Control Center at the Georgia Institute of Technology for the financial support given to this project.

REFERENCES

[1] N. Sadegh, "Trajectory learning and output feedback control of nonlinear discrete-time systems," in *Proc. 40th IEEE Conf. Decision and Control*, Florida, 2001, pp. 4032–37.

[2] N. Sadegh, "A perceptron network for functional identification and control of nonlinear systems," *IEEE Trans. Neural Networks*, vol. 4, No. 6, pp. 982–88, November 1993.

[3] N. Sadegh, "A nodal link perceptron network with applications to control of a nonholonomic system," *IEEE Trans. Neural Networks*, vol. 6, No. 6, pp. 1516–23, November 1995.

[4] N. Sadegh, "A multilayer nodal link perceptron network with least squares training algorithm," *International Journal of Control*, vol. 70, no. 3, pp. 385–404, 1998.

[5] U. Pinsopon et al., 'Hydraulic actuator control with open-centre electrohydraulic valve using a cerebellar model articulation controller neural network algorithm," in *Proc. Instn Mech Engrs*, vol. 213, No. 1, pp. 33-48, 1999.

[6] T. Nishiumi, and J. Watton, "Model reference adaptive control of an electrohydraulic motor drive using an artificial neural network compensator," in *Proc. Instn Mech Engrs*, vol. 211, No. 1, pp. 111-22, 1997

[7] K. Edge, "The control of fluid power systems – responding to the challenges," in *Proc. Instn Mech Engrs*, vol. 211, No. 1, pp. 91-110, 1997

[8] S. Liu, and B. Yao, "Coordinate control of energy-saving programmable valves," in *Proc. ASME Intl Mech Eng Congress*, vol. 10, pp. 123-131, 2003

[9] S. Liu, and B. Yao, "Programmable valves: A solution o bypass deadband problem of electro-hydraulic systems," in *Proc. American Control Conference*, vol. 5, pp. 4438-4443, 2004

[10] M. Krishna, and J. Bares, "Constructing hydraulic robot models using memory-based learning," Journal of Aerospace Engineering, Vol. 12, No. 2, pp. 34-42, 1999.

[11] C-W. Chung, and T-Y. Kuc, "Learning and extraction of nonlinear mappings using an associative memory, " in *Proc. IEEE International Conference on Syst, Man, & Cybernetics*, vol. 4, pp. 3430-3433, 1998

[12] P. Opdenbosch, N. Sadegh, and W. Book, "Modeling and control of an electro-hydraulic poppet valve," in *Proc. ASME Intl Mech Eng Congress*, 2004

[13] Yang, X., Stephenson, D. B., Paik, M. J., 2001, "Bidirectional Pilot Operated Control Valve," U.S. Patent 6,328,275.

[14] D. C. Foley and N. Sadegh, "Short horizon optimal control of nonlinear systems," in *Proc. 42nd IEEE Conf. Decision and Control*, Hawaii, 2003, pp. 728–33.

[15] J. Grizzle and P. Kokotovic, "Feedback linearization of sampled-data systems," *IEEE Trans. Automatic Control*, vol. 33, No. 9, pp. 857–59, 1988.

[16] A. Levin and K. Narendra, "Control of nonlinear dynamical systems using neural networks-controllability and stabilization," *IEEE Trans. Neural Networks*, vol. 4, No. 2, pp. 192–206, 1993.

[17] P. S. Albertos, "Block multirate input-output model for sampled data control systems," *IEEE Trans. Automatic Control*, vol. 35, No. 9, pp. 1085–87, September 1990.

[18] A. Khayatian, "Multirate and block methods for modeling and control of pulse modulated systems," Ph.D. dissertation, Dept. Elect. Eng., Georgia Institute of Technology, Atlanta, GA, 1993.

Proceedings of IMECE2005
2005 ASME International Mechanical Engineering Congress and Exposition
November 5-11, 2005, Orlando, Florida USA

IMECE2005-81376

SOFTWARE ENABLED VARIABLE DISPLACEMENT PUMPS *

Perry Y. Li,[†] Cassie Y. Li and Thomas R. Chase
Department of Mechanical Engineering
University of Minnesota
111 Church St. SE
Minneapolis, MN 55455
Email: {pli,yli,trchase}@me.umn.edu

ABSTRACT

Direct pump control of hydraulic systems is more energy efficient than throttle valve based methods to control hydraulic systems. This requires variable displacement pumps that are responsive and capable of electronic control. Such Electronic Displacement Controlled (EDC) pumps tend to be significantly larger, heavier and more expensive than fixed displacement counterparts. In addition, achievable control bandwidths are typically lower than throttle valve based control approaches.

We have recently begun a project to achieve the functionality of a variable displacement pump by combining a fixed displacement pump, a pulse width modulated (PWM) on/off valve under closed loop feedback control, and an accumulator. The proposed topology is the hydro-mechanical analog of the DC-DC boost converter in power electronics. Since on/off valves have little loss in either the on or the off state, this approach is potentially more efficient than throttle valve based control approaches. It has the small size/weight and low cost advantages of a fixed displacement pump. Faster response can be expected by eliminating the intervening inertias of the swash plate control system. The pump's functionalities can also be easily programmed by controlling the PWM on/off valve in different manners.

This paper presents some preliminary results from this ongoing research program. While the PWM valve based approach provides desirable features, it also introduces undesirable ripples to the system pressure and flow rate. It is shown that increasing the accumulator pre-charge pressure and the accumulator volume can decrease ripple size at the expense of response time. This apparent trade-off can be alleviated by feedback control to achieve fast response time while keeping ripple small. Feedback control using PWM control must be implemented with care since the conventional "state-space" model may not be valid when the PWM frequency is low. On the other hand, increasing PWM frequency reduces ripple size and enables the system to achieve high control bandwidths.

1 Introduction

Every hydraulic system requires at least one hydraulic pump that converts mechanical power from a prime mover (such as an I.C. engine) into hydraulic power. The lack of compact and portable hydraulic power supplies is a major impediment to applying hydraulics to many emerging robotic applications, such as personal assistance robots, robotic tools for search and rescue etc. where large forces are often needed in small or difficult to assess areas. Inability to control pump output precisely and quickly also contributes to the inefficient use of hydraulic power. Our research aims is to develop a new class of energy efficient hydraulic pumps that have 1) improved power and force densities, and 2) improved control capabilities with which a myriad of performance enhancing features can be implemented by changing software/firmware. The proposed new pumps will make hydraulic power supplies more compact. By utilizing the inherent control functionality of the pump, additional control valvings can be minimized, thus reducing the complexity of the hydraulic system and the energy loss through these valves.

Hydraulic pumps are usually positive displacement devices. By being able to vary the flow rate as required, energy loss

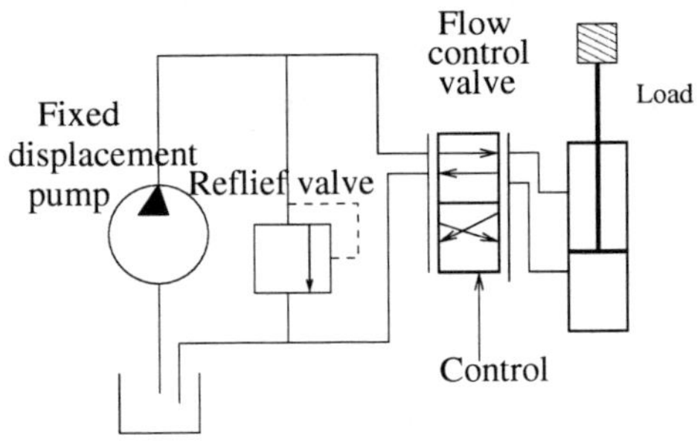

(a) A common simple open hydraulic circuit that uses a metering control valve to control the cylinder motion

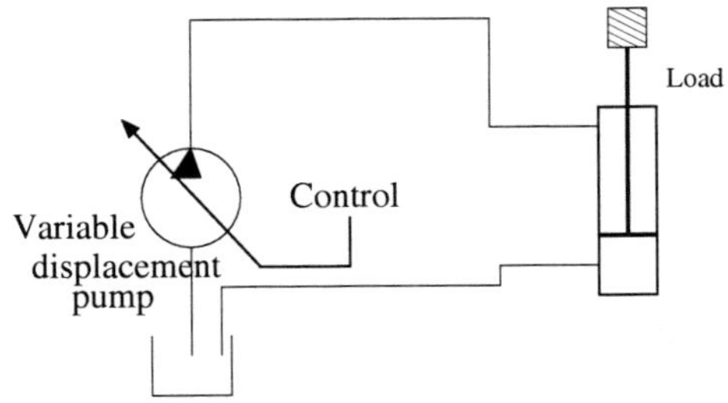

(b) An energy efficient direct pump control alternative

Figure 1. Valve controlled and direct pump controlled circuits.

through valve throttling can be minimized. For example, in the valve-controlled system in fig. 1(a), losses occur in both the relief valve and the metering valve that are partially open. If the pump flow can be controlled quickly and accurately, an alternate circuit in Fig. 1(b) can be used. In this case, since neither throttling or relief valves are used, theoretically, the efficiency will be 100%. Moreover, compared to Fig. 1(a), the hydraulic circuit in Fig. 1(b) is simpler and requires fewer components and hydraulic connections.

Since prime movers are normally operated at constant speeds, varying flow necessitates varying the pump displacement. Of the various hydraulic pumps, internal and external gear pumps are easiest to manufacture, but they have fixed displacements (Fig.2). Variable displacement pumps are typically piston pumps and are occasionally vane pumps. Both are more complicated in their designs, heavier and bulkier. Displacement in a piston pump is varied by changing the "swash plate" angle. The size, weight, complexity and cost all increase significantly as we migrate from simple fixed displacement pumps, to manually controlled displacement pumps, to the so-called electronic displacement controlled (EDC) pumps.

In this paper, we propose to combine a compact fixed displacement pump and a pulse width modulated (PWM) high speed on/off bleed-off valve to achieve a *software controllable variable displacement pump* so that the apparent pump displacement can be adjusted by varying PWM duty ratio. This approach is inspired by the advances and success of switched mode converters in power electronics to achieve compact, energy efficient, yet highly controllable devices. Just as pumps are used to convert mechanical power from the prime mover into the hydraulic power, power electronics is also concerned with the power transformation. Therefore, it is not surprising that ideas from one area can be used in the other. The advantages of the proposed approach are:

Size By eliminating the swash plate control system, significant volume and weight savings, and hence power / force density improvement, can be obtained.

Energy efficiency When the control valve is either fully on or fully off, the power loss in the bleed-off valve would be minimal (since either the pressure drop or flow is nominally zero) so that the pump operation will be efficient at all pump flows / effective displacements.

Responsiveness By eliminating the cascade of mechanical control systems, and replacing it by a direct flow control, the relative degree of the control system is reduced, and the intervening inertias is not necessary. Thus, the responsiveness of the control system is expected to improve.

Software enabled features By controlling the PWM signal, high bandwidth control of the pump output can be achieved. This in turn enables a plethora of efficiency and performance enhancing features to be implemented by programmable control software/firmware.

Two important performance issues associated with the proposed device are ripple size and responsiveness. In this paper, these are investigated through a case study. Typically, ripple size can be reduced by increasing the capacity or the pre-charge of the accumulator, however, at the expense of slower response. It is shown that this apparent trade off can be mitigated by the use of feedback control.

The rest of the paper is organized as follows. Section 2 discusses the rationale of the proposed approach based on the weight, size and performance of current pump designs. The concept of the hydraulic DC-DC converter is discussed in greater detail in section 3. Section 4 uses a design case study to illustrate the performance trade-offs of some design parameters. Fidelity of the averaged models for the PWM controlled systems will also be briefly discussed. Some concluding remarks are given in section 5.

2 Fixed and Variable Displacement Pumps

Figs. 2-3 show a gear pump and a variable displacement piston pump respectively. Gear pumps are simple, light, compact and cost effective. Unfortunately, the pump flow rate cannot be changed unless the speed of the primer mover is changed. The variable displacement piston pumps are more complicated but the pump displacement can be varied by controlling the swash plate angle.

Figure 2. Gear pumps are simple and compact, but their displacements cannot be varied.

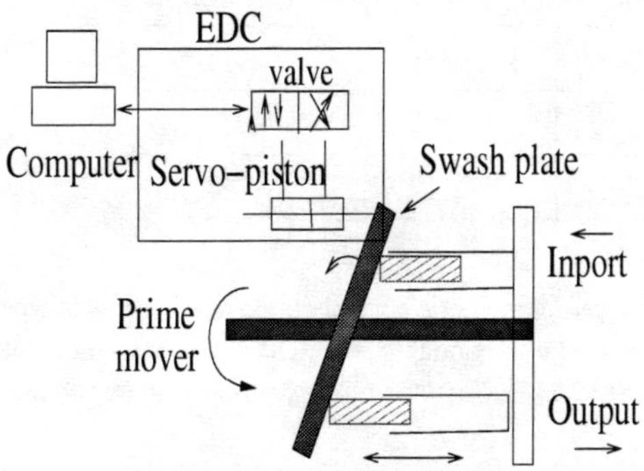

(a) Schematic of displacement control mechanism

(b) A EDC control unit

Figure 3. Piston pumps are more complex but they can vary their displacement by moving the swash plate. The EDC control unit has dimension: 215mm x 150mm x 63mm, which is fitted on the side of a Sauer-Danfoss M46 pump and constitutes 30% of the pump size.

To control the displacement of a piston pump, the swash plate angle is controlled. This can be achieved manually by using a lever (Direct Displacement Control - DDC), or via a set of additional servo pistons (hydraulic actuators) that push and pull on the swash plate. To interface with a computer, the servo pistons are in turn controlled via electrohydraulic valves which can then be interfaced to the computer. The electronic displacement control (EDC) unit can be a separate unit that is attached to a piston pump (such as the one in Fig. 3(b)) or integrated in the pump casing.

To illustrate the trade off between the weight, size and control options, a comparison of commercial pumps of various designs and features is shown table 1. It demonstrates that as control and mechatronic features are introduced, significant weight and size penalties are incurred. Notice that an EDC unit adds significantly to both the weight and size of the pump. For example, the weight and size of a fixed displacement gear pump is about 60% of those of a variable displacement piston pump. The EDC option adds an additional 30%-50% in weight and size.

In addition, the response times (neutral to full stroke) for the 2 EDC controlled pumps cited in table 1 are of the order of 1 sec, typical of other EDC pumps [1]. This is clearly inadequate for high performance direct pump control purposes. The response speed is limited by the size of the swash-plate servo hydraulic systems - larger servo-pistons will improve response speeds, but would also increase size, weight and cost of the unit.

In the current cascade control structure (Fig. 3(b)), we first control an electrohydraulic valve, which in turn controls a hydraulic actuator, the swash plate angle, and finally pump output. This increases the relative degree of the control system, and requires large forces to move the intervening inertia of the components. Both factors contribute to the low bandwidths of EDC pumps.

It would therefore be advantageous to achieve an alternate method of varying the displacement of the pump, that has significantly better control performance, but yet without the penalty on weight, size and cost that the existing option based on swash plate control suffers.

3 Main concept: Hydraulic boost (DC-DC) converter

An alternative means to vary the pump displacement is based on the hydro-mechanical analog of a DC-DC switched mode converter in power electronics [1] (Fig. 4(a)). A driving force in the advances in power electronics was the invention of power transistors that can be switched on and off very quickly. When the transistor is fully on or fully off, power loss is minimal because either there is no voltage drop across the transistor or there is no current passing through it.

A DC-DC converter (Fig. 4(a)) transforms input DC power at one voltage (V_d) to the output port at another voltage (V_{out}). In a boost converter, $V_{out} > V_d$ whereas in a buck converter, $V_{out} < V_{in}$. A boost converter is shown in Fig. 4(a) where the transistor switch is pulse width modulated (PWM) so that the control is the on/off duty cycle. The circuit's operation is as follows. When the transistor switch is on, current is stored in the inductor. The input is disconnected from the output circuit. The load is temporarily supplied by the charge stored in the capacitor. When the switch

[1] The response time depends on the sizes of the control/supply orifices used. Also, it should be pointed out that pressure compensation and load sensing pumps use integrated hydro mechanical feedback on the servo piston which have better response time (0.05 to 0.09s). However, the flexibility of the control is limited.

(a) Eaton family

model	pump type	Displacement [cm^3]	W x H x L [mm^3]	volume [$10^{-3}m^3$]	weight
L25	Gear	43	134 x 157 x 163	3.4	9 kg (est)
70360	Piston, MDC	40.6	233 x 146 x 168	5.7	14.1 kg
72400	Piston, EDC	40.6	274 x 222 x 170	10.3	27 kg

(b) Sauer-Danfoss family

model	pump type	Displacement [cm^3]	W x H x L [mm^3]	volume [$10^{-3}m^3$]	weight
GNP3	Gear	44	150x 121 x 148	2.7	n/a
M44	piston, DDC	43.5	239 x 182 x 189	8.2	25 kg
M46	Piston, EDC	46	264 x 242 x 200	12.7	33 kg

Table 1. Comparison of two manufacturers' gear pump and variable displacement piston pumps with various control options. The trend is typical of other manufacturers. DDC refers to direct displacement control where swash plate is directly controlled via manual lever. MDC refers to manual displacement control where the manual lever controls a valve that controls the hydraulically actuated swash plate. EDC is electronically controlled swash plate. EDC for the Sauer-valve is shown in Fig. 3(b).

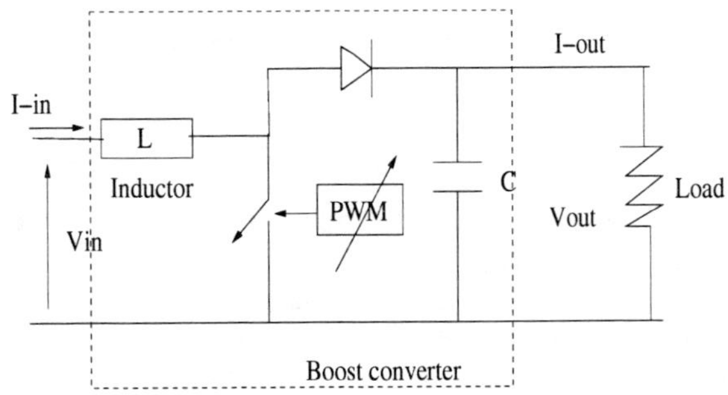

(a) DC-DC boost converter

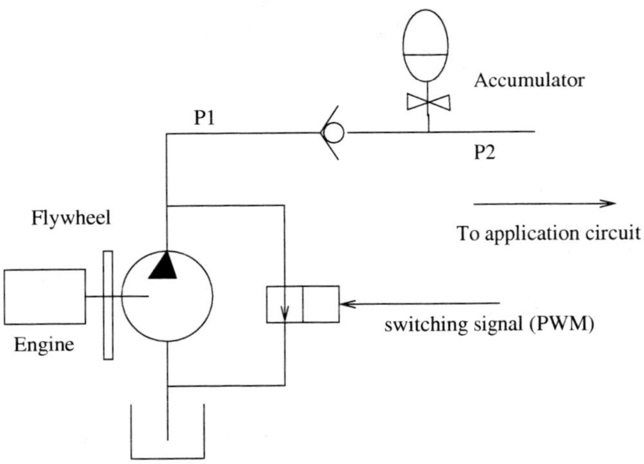

(b) PWM variable displacement pump

Figure 4. Electrical DC-DC boost converter and its hydraulic analog

is turned off, current from the input and the current stored in the inductor are used to feed the output load and to replenish the capacitor.

The operation of the hydro-mechanical analog in Figs. 4(b) is exactly similar, with the inductor replaced by the inertia of the fixed displacement pump and possibly an extra flywheel, the capacitor replaced by an accumulator (for storing fluid at high pressure), and the transistor switch replaced by a high speed on/off valve. The accumulator serves to filter out the flow ripple due to the on/off flow output. The flywheel serves to filter out the load on the engine. If the prime mover's speed is regulated, then the flywheel will not be necessary. Roughly speaking, in steady state operation, the actual pump flow or the effective pump displacement will be a fraction of the flow rate / displacement of the fixed displacement pump, as determined by the duty ratio of the on/off valve. Thus, flow rate modulation can be achieved by varying the duty ratio of the on/off valve, in either open loop or closed loop.

Fig. 5 shows that the concept is easily adaptable to situations where a single pump is driving multiple circuits. Each additional circuit is controlled by a PWM valve/ accumulator combination which serves the function of lossless variable transformer and low pass filter. Notice that hardware controlled variable displacement pumps cannot be easily adaptable to this situation.

Assuming that the on/off transition times are small compared to the times when the valve is in either the fully on or fully off states, the proposed device in Figs. 4(b) or 5 will be very efficient. It is because the throttling loss through the valve will be negligible when either the flow through the valve, or the pressure drop across it, is nearly zero.

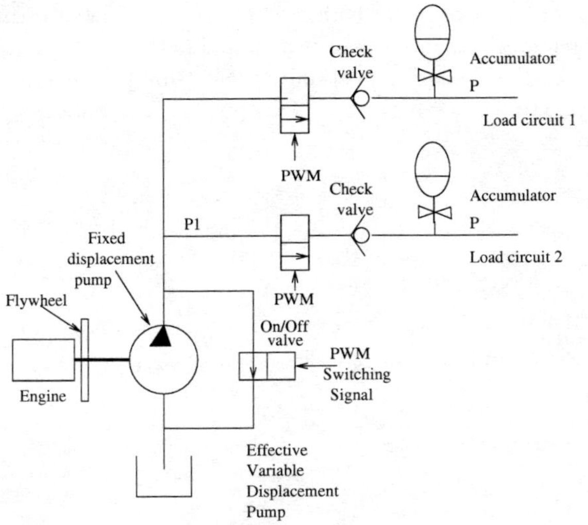

Figure 5. Adaptation to a single variable displacement pump driving multiple circuits

The hydraulic boost converter in Fig. 4(b) can achieve the desired weight and size saving because simple gear pumps, which are compact and light, can be used as the fixed displacement pump. As shown in table 1, the weight and volume of a fixed displacement gear pump are only 1/3 of those of an EDC pump. Even if a fixed displacement piston pump is used to take advantage of their lower leakage and noise properties, the extra swash plate control unit will no longer necessary, and a saving of 30-50% will likely be achieved.

Beside the topology in Fig. 5, the concept of common pressure rail (CPR) from which each hydraulic device obtains energy via a throttle-less hydraulic transformer [3, 4] is another possibility for situations when multiple circuits are fed by the same pump. The switched mode pump control concept proposed in this paper can also be extended easily to the efficient and compact implementation of hydraulic transformers as well. Notice that the hydraulic pump is itself a mechanical-to-hydraulic transformer.

3.1 Dynamics

We assume that the gas charged accumulator obeys the ideal gas law and operates under adiabatic conditions. Ignoring the on/off valve dynamics and assuming that the valve has a sufficiently large orifice so that the pressure drop across it is negligible when it is open, the dynamics of the circuit in Fig. 4(b) are given by:

$$J_f\dot{\omega} = -u(t)\frac{D}{2\pi}P + \Gamma(t) \tag{1}$$

$$\dot{P} = \gamma\frac{P}{V}\underbrace{\left[u(t)\frac{D}{2\pi}\omega - Q_{out}(t)\right]}_{-\dot{V}} \tag{2}$$

where $u = 0$ corresponds to the valve being opened, and $u = 1$ corresponds to the valve being closed; J_f is the inertia of the flywheel/pump, ω is the pump speed, D is the pump displacement, Γ is the engine torque, P is the accumulator/output pressure, and γ is the ratio of heat capacities at constant pressure and and at constant temperature. V is the compressed gas volume in the accumulator, given by:

$$PV^{\gamma} = P_0 V_0^{\gamma} \tag{3}$$

where P_0 and V_0 are the pre-charge pressure and the initial gas volume of the accumulator. Q_{out} is the load flow.

We assume that the valve input signal $u(t)$ will be pulse width modulated (PWM) so that for a duty ratio $s(t) \in [0,1]$, $u(t)$ is determined by comparing the $s(t)$ with a periodic saw-tooth function:

$$u(t) = \begin{cases} 1 & \text{if } s(t) \geq (t/T) mod 1 \\ 0 & \text{if } s(t) < (t/T) mod 1 \end{cases} \tag{4}$$

where T is the PWM period. If the PWM switching frequency $(1/T)$ is sufficiently fast [5], the "states space" average system [1], obtained by treating the state variables and inputs as constants during the time averaging procedure, is valid and is given by:

$$J_f\dot{\omega} = -s(t)\frac{D}{2\pi}P + \Gamma(t) \tag{5}$$

$$\dot{P} = \gamma\frac{P}{V}\underbrace{\left[s(t)\frac{D}{2\pi}\omega - Q_{out}(t)\right]}_{-\dot{V}} \tag{6}$$

In the steady state, we have the equilibrium condition,

$$Q_{out} = \frac{s(t)D}{2\pi}\omega; \qquad \Gamma = \frac{s(t)D}{2\pi}P \tag{7}$$

which is the usual ideal positive displacement pump relationship with $s(t)D$ playing the role of the effective displacement of the pump which can be controlled by software via the duty cycle $s(t)$.

3.2 Software enabled features

The PWM control input $s(t)$ can be used to modify the behavior of the pump via various types of feedback control algorithms, and thus enabling many possible useful features. We list some of these below.

Ideal pump: This can be achieved by designing an appropriate feedback control law so that the ideal pump relationship (7) is strongly stabilized at the desired duty cycle s.
Pressure control: this can be achieved by controlling the duty cycle so that the output pressure P follows some desired trajectory. From (5)-(6), we see that this is a state stabilization problem.
Flow control: this can be accomplished by placing a small orifice downstream to the PWM pump outlet, and then controlling $s(t)$ so that the pressure drop across the orifice corresponds to the desired flow rate.
Direct load control: if the load is modeled, then Q_{out} will be a function of P and the properties of the load. $s(t)$ can then be used to directly control the relevant variables.
Compensation for engine behavior: in current systems, the engine is regulated at constant speed ω. Using the PWM pump, variation in ω can be accounted for (e.g. by maintaining $s(t)\omega(t)$ constant).
Optimal engine operation: This is related to the above. Since variation in engine speed can be accounted for, the engine can be operated at the speed that is most efficient for the power demand according to the engine's own speed-torque curve.
and others ...

Notice that these features can be switched from one type to another by simply changing the control software. For this reason, we refer to this type of pump as *software enabled variable displacement pumps*. Notice also that the above features are only practical if the pump displacement can be varied sufficiently fast.

4 A Design Case study

In this section, we use a case study to consider the performance trade-offs due to various design parameters. The simple circuit in Fig. 4(b) is considered with the load being a simple orifice:

$$Q_{out} = C\sqrt{P} \tag{8}$$

The system is sized according to an experimental setup that is under construction. The displacement of the pump is $D = 4.542$cc. The pre-charge gas volume in the accumulator is $V_0 = 100$cc. The gas constant for N_2 is $\gamma = 1.4$. The orifice coefficient in (8) is $C = 1.75\times10^{-8}m^3/s/Pa^{0.5}$.

In order to compare with a conventional variable displacement pump, the prime mover is assumed to be operating at constant speed $\omega = 104.7 rad/s$ (1000 RPM). This gives a maximum flow rate of $7.57\times10^{-5}m^3/s$ (1.2 gpm). The dynamics are reduced to:

$$\dot{P} = \gamma\frac{P^{1+1/\gamma}}{P_0^{1/\gamma}V_0}\left[u(t)\frac{D}{2\pi}\omega - C\sqrt{P}\right] \tag{9}$$

where $u(t) = 0$ when the on/off valve is open, and $u(t) = 1$ when it is closed. The "states-space" averaged system is in turn reduced to:

$$\dot{P} = \gamma\frac{P^{1+1/\gamma}}{P_0^{1/\gamma}V_0}\left[s(t)\frac{D}{2\pi}\omega - C\sqrt{P}\right] \tag{10}$$

where $s(t) \in [0,1]$ is the duty ratio of the PWM input.

We consider the scenarios when we need to control the load flow through the orifice, from the initial flow of $Q_0 = 0.5\frac{D}{2\pi}\omega = 3.78\times10^{-5}m^3/s$ (0.6 gpm) to the final flow of $Q_f = 0.6\frac{D}{2\pi}\omega = 4.54\times10^{-5}m^3/s$ (0.72gpm), corresponding to the duty cycles of $s = 0.5$ and $s = 0.6$ respectively. Since the orifice is fixed, these correspond to the desired pressures of $P_0^* = 4.70\times10^6$Pa and $P_f^* = 6.77\times10^6$Pa.

4.1 Open loop response

For the open loop tests, the duty cycle is changed from $s = 0.5$ to $s = 0.6$ at $t = 5s$. The response when the accumulator pre-charge of $P_0 = 0.5\times10^6$Pa and the PWM period of $T = 0.1s$ is shown in Fig. 6. By $t = 5s$, the system has settled down at a steady oscillatory mode around the desired pressure P_0^*. After $t = 5s$, the average pressure rises to the desired pressure P_f^*. The response time for this transition is 1.741s as determined from the time required for the average value to reach 95% of the overall change. This may be satisfactory but for the large peak-to-peak pressure ripples, which are as large as 16.2% of the average pressure. The average and ripple pressure characteristics translate directly into those of the load flow according to the orifice equation (8).

Accumulator sizing The flow ripple can be improved by increasing the pre-charge pressure of the accumulator. Figure 7 shows the case when it has been increased 8 folds to $P_0 = 4\times 10^6$ Pa. While the flow ripple has been reduced to 3.7% of the mean pressure, the time response has been significantly increased to 7.56s. From (9), it can be seen that the term $\gamma\frac{P^{1+1/\gamma}}{P_0^{1/\gamma}V_0}$ plays the role of the volumetric stiffness (dP/dV) of the accumulator. Indeed, for the same operating pressure, increases in either the

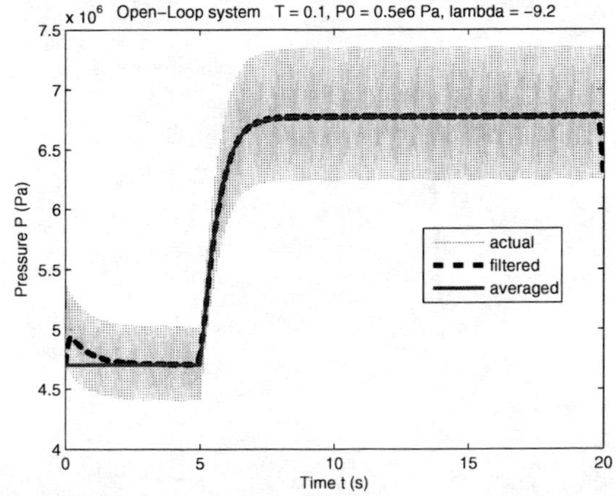

Figure 6. Open-loop pressure response: $T = 0.1$s, $P_0 = 0.5 \times 10^6$Pa. The response time is 1.741 and the peak-to-peak ripple is 16.2% of the average pressure at the final operating condition.

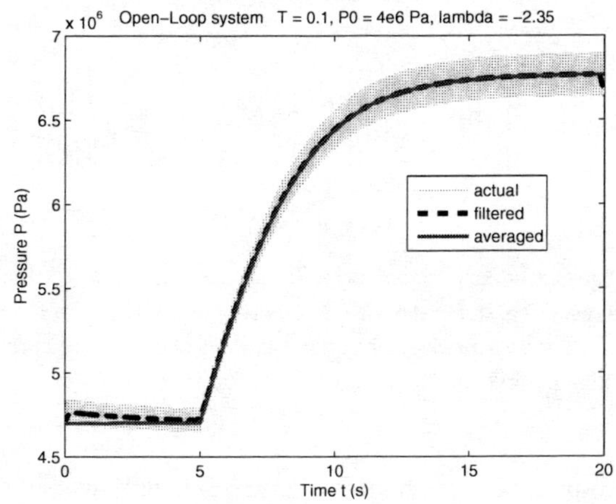

Figure 7. Open loop pressure response: $T = 0.1$s, $P_0 = 4 \times 10^6$Pa. The response time is 7.56s and the ripple is 3.7% of the average pressure at the final operating condition.

pre-charge pressure P_0 or the accumulator volume V_0 decreases its stiffness, which decreases ripple and slows down response. Notice that P_0 cannot be higher than the operating pressure for it to be functional.

PWM frequency As PWM period T decreases to 0.05s, the ripple size decreases to 1.8% without affecting the response time as illustrated in Fig. 8. The trade-off between ripples and open loop response times can be seen for each PWM period T in Fig. 11.

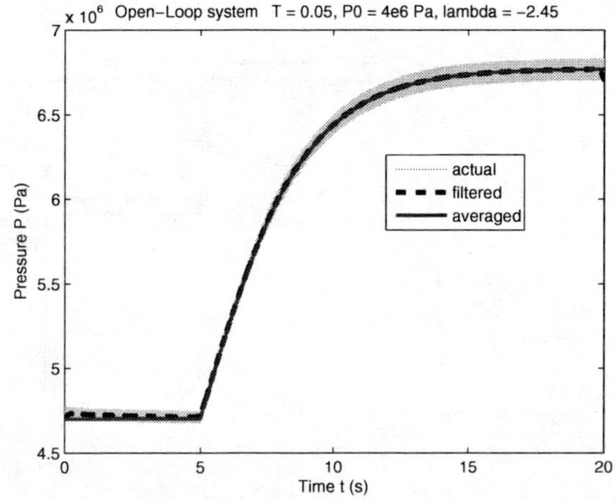

Figure 8. Open-loop pressure response: $T = 0.05$s, $P_0 = 4 \times 10^6$Pa. The response time is 7.55 and the ripple is 1.8% of the average pressure at the final operating condition.

4.2 Closed loop response

The open loop response shows that the ripple size can be reduced by increasing the PWM frequency (decreasing T). T is limited by the responsiveness of the on/off valve. Since throttling occurs when the valve is in transition (partially open), energy efficiency dictates that the transition times should be small compared to the times when the valve is either fully open or fully closed. For a given PWM frequency, feedback control can be used to improve the response time without adverse effects on the ripple size.

To investigate the usefulness of the feedback, the duty cycle $s(t) \in [0,1]$ is determined by feedback as follows:

$$s(t) = s_{ff} + K(P* - P(t)) \tag{11}$$

where s_{ff} is the feedforward term, P^* is the desired equilibrium pressure, P is the sensed instantaneous pressure at the accumulator outlet, and K is the feedback gain determined by placing the pole of the linearization of the "states space" averaged system (10) as negative as possible without saturation. The PWM control design procedure followed here is traditional and similar to the literature (e.g. [1,6,7]) in that the control law is derived based on a nonlinear model obtained using the "states space" averaging method. Typically, the control bandwidth is recommended to be an order of magnitude lower than the PWM frequency. For example, [6] uses a 20kHz PWM frequency on an electromechanical system and is successful in tracking a 5Hz trajectory. Authors in [7] successfully used a 26Hz PWM to track a 1.25Hz trajectory on a pneumatic system. Notice that the PWM frequency in these cases are at least 20 times higher than the frequency of the trajectory.

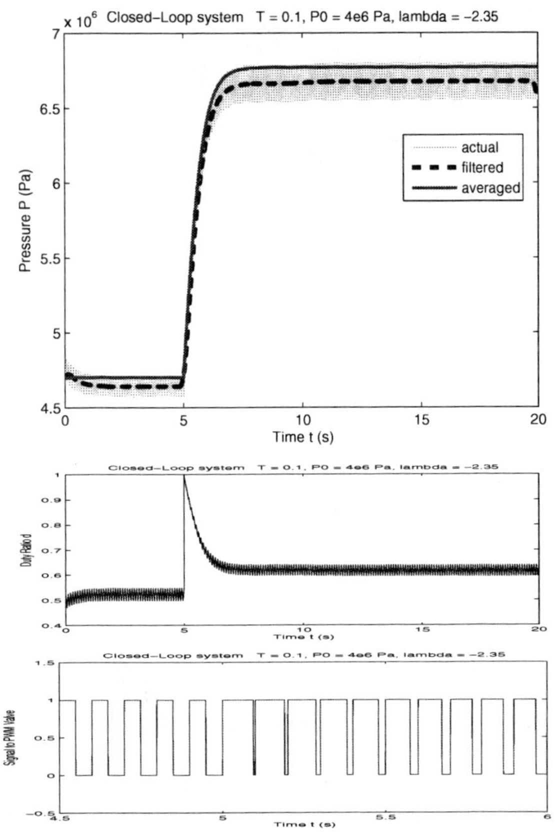

Figure 9. Closed-loop response for $T = 0.1$s, $P_0 = 4 \times 10^6$Pa. Top: pressure response, middle: duty ratio $s(t)$ and bottom: valve inputs $u(t)$. The response time is 1.56s and the peak-to-peak ripple is 3.7% of the average pressure at the final operating condition.

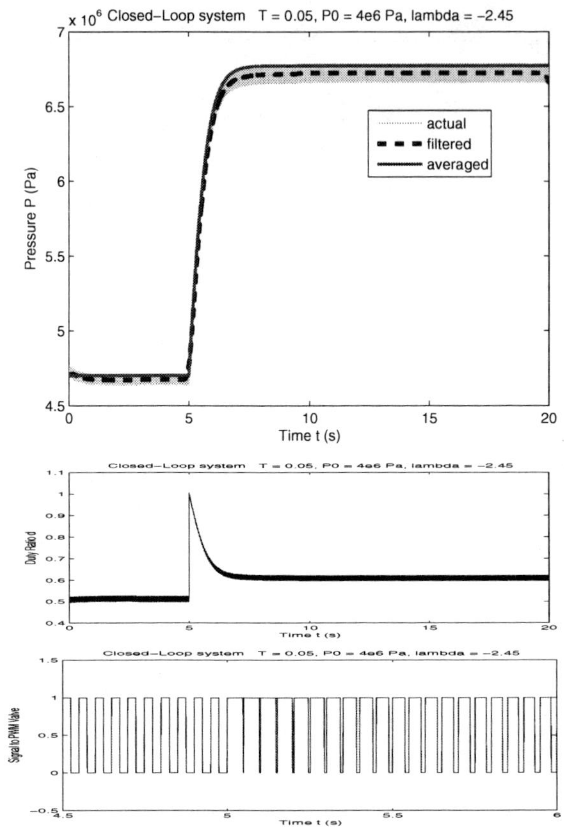

Figure 10. Closed-loop response for $T = 0.05$s, $P_0 = 4 \times 10^6$Pa. Top: pressure response, duty ratio $s(t)$ and valve inputs $u(t)$: The response time is 1.52s and the peak-to-peak ripple is 1.8% of the average pressure at the final operating condition.

Figures 9 and 10 show the closed loop responses when the accumulator pre-charge pressure is $P_0 = 4 \times 10^6$Pa and the PWM periods are $T = 0.1$s and $T = 0.05$s respectively. Compared with Figs. 7 and 8, the response times have been significantly reduced (from 7.56s to 1.56s, and 7.55s to 1.52s respectively) without increasing the ripple size. The large transient increase (right after $t = 5$s) in duty cycles $s(t)$ is responsible of the quickened response. If the PWM duty cycle is allowed to saturate, response times can be improved further. The trade-offs between ripple size and response times for different T's and pre-charge pressures P_0 for the closed loop case is shown in Fig. 11. Decrease in PWM period causes a slight but noticeable decrease in response time for all pre-charge pressures.

4.3 Frequency dependent averaging

In the closed loop responses (Figs. 9 and 10), the trajectories of the "states-space" averaged model (10) under the linear control law (11), as well as the actual average response (computed by a zero-phase Boxcar filter with a length approximately equal to $2T$), are plotted. Notice that the "state-space" averaged trajectories reach the desired pressures, but the actual average pressure trajectories tend to be lower. Hence, the "states space" average model over-estimates the average pressure. Since the control design is based on the average model, the control scheme generates pressures and flows that are below the the desired values. In contrast, for the open loop cases, the average models predict the actual average responses accurately (Figs. 6-8).

The discrepancies that exist in the closed loop cases are due to the implementation of the PWM control in (4) as predicted in [2, 5]. Because of the inevitable ripple in the response, a ripple function is also imposed on the duty cycle $s(t)$ in (11) and (4). The mean of this function is typically different from that obtained from (11) assuming that the mean pressure P is fedback. In [2], the steady state bias and qualitative differences are predicted, and a frequency dependent averaging that takes this effect into account is proposed. The discrepancy between the state-

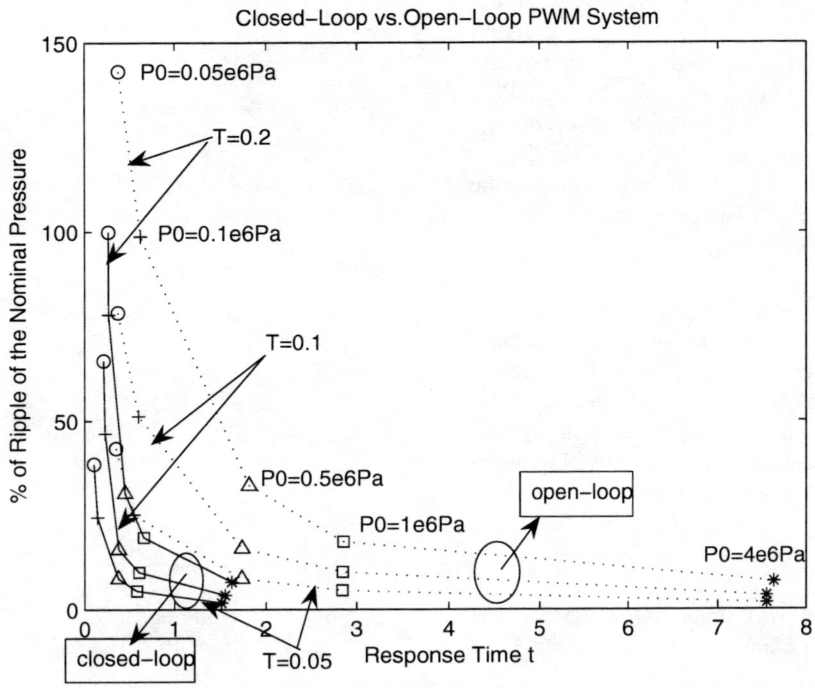

Figure 11. Open loop and closed loop trade-off between peak-to-peak ripple size and response time for each PWM period T and accumulator pre-charge pressure P_0.

space averaged system and the actual average system response becomes more significant as the PWM frequency decreases. Following [2], the state-spaces averaged model in (10) is modified to be:

$$\dot{P} = \gamma \frac{P^{1+1/\gamma}}{P_0^{1/\gamma} V_0} \left[\tau_s(t) \frac{D}{2\pi} \omega - C\sqrt{P} \right] \quad (12)$$

where $\tau_s(t)$ is the modified average duty cycle determined by solving the implicit equation:

$$\tau_s(t) = s_{ff} + K(P^* - (P(t) + \psi(t)))$$

where $\psi(t)$ is an estimate of the ripple function which is in turn a function of $\tau_s(t)$. The readers are referred to [2] for details. Figure 12 shows that the modified average model can much better predict the actual average behavior of the system.

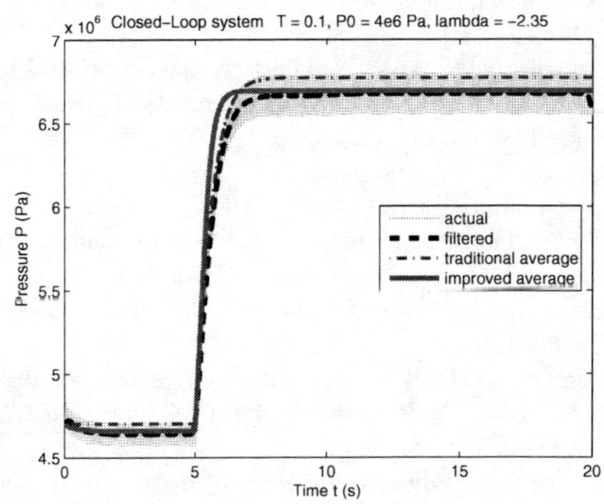

Figure 12. Actual, zero-phase filtered, "states-space" averaged, and frequency dependent averaged closed-loop pressure responses for $T = 0.1$s, $P_0 = 4 \times 10^6$Pa.

5 Conclusions

A fixed displacement pump, an accumulator, and an on/off switching valve under PWM control have been proposed to achieve the function of a variable displacement pump. The device is potentially more compact and controllable than a hardware controlled variable displacement pump so that a variety of software enable features can be implemented. The proposed device is the hydro-mechanical equivalent of a switched mode DC-DC boost converter in power electronics. Two important functional characteristics of the new device are ripple size and response time. It is shown that increases in the accumulator pre-charge pressure and volume can decrease ripple size at the expense of response times. This apparent trade-off can be alleviated by feedback control to achieve fast response time without adverse effect on the ripple size. Feedback control using PWM control must be implemented with care since the conventional "state-space" model may not be valid. The frequency dependent averaging technique proposed in [2] has been shown to be effective in predicting the average behavior for the new device. Increasing PWM frequency reduces ripple size and undesirable PWM effects.

These results point toward several research directions that are currently being pursued: 1) design of an on/off switching valve that is capable of very high PWM frequency; 2) integrated design of the accumulator, on/off valve and pump that maximizes the ripple/responsiveness trade off of the system; 3) control strategies that take into account relatively low PWM frequencies that can be achievable by mechanical systems.

Acknowledgment

This material is based upon work supported by the National Science Foundation under grant number ENG/CMS-0409832.

REFERENCES

[1] N. Mohan, T. Undeland, and W. Robbons, *Power electronics: converters, applications and design.* John Wiley and Sons, 2nd ed., 1995.

[2] B. Lehman and R. Bass, "Switching frequency dependent averaged models for pwm dc-dc converters," *IEEE Transactions on Power Electronics*, vol. 11, no. 1, pp. 89–98, 1996.

[3] *Innas Hydraulic Transformer - http://www.innas.com/IHT.html.* 2001.

[4] L. Gu, M. Qiu, B. Jin, and J. Cao, "New hydraulic systems made up of hydraulic power bus and switch-mode hydraulic power supplies," *Chinese Journal of Mechanical Engineering*, vol. 39, no. 1, pp. 84–88, 2003.

[5] B. Lehman and R. Bass, "Recent advances in averaging theory for pwm dc-dc converters," in *Proceedings of the IEEE CDC*, pp. 4467–4471, 1996.

[6] D. G. Taylor, "Pulse width modulated control of electromechanical systems," *IEEE Transactions on Automatic Control*, vol. 37, no. 4, pp. 524–528, 1992.

[7] E. J. Barth, J. Zhang, and M. Goldfarb, "Sliding mode approach to pwm-controlled pneumatic systems," in *Proceedings of the 2022 ACC*, pp. 2362–2367, 2002.

Proceedings of IMECE2005
2005 ASME International Mechanical Engineering Congress and Exposition
November 5-11, 2005, Orlando, Florida USA

IMECE2005-81711

COMPARISON OF HYDRAULIC PUMP FAULTS DIAGNOSIS METHODS: WAVELET VS. SPECTRAL ANALYSES

Yingjie Gao
Department of Mechatronic Engineering
Yanshan University
Qinhuangdao, Hebei, 066004, China
yjgao@ysu.edu.cn

Qin Zhang
Department of Agricultural and Biological Engineering
University of Illinois at Urbana-Champaign
Urbana, Illinois, 61801, USA
qinzhang@uiuc.edu

Xiangdong Kong
Department of Mechatronic Engineering
Yanshan University
Qinhuangdao, Hebei, 066004, China
xdkong@ysu.edu.cn

ABSTRACT

This paper introduces two faults diagnosis methods, a conventional spectral analysis method and a wavelet transform method, for hydraulic pump applications. The fundamental technologies of both methods, as well as their performance in detecting a few common hydraulic pump defects, are described in this paper. The performance of both diagnoses methods were evaluated based on experimental results. In order to eliminate the effects of border distortion arising from applying wavelet transform to finite-length signals, the pump outlet pressure in this case, a preprocess on the obtained signals is carried to clean up the errors prior to faults diagnosis analysis. Validation results obtained from both methods in analyzing the same data sets indicated that the wavelet transform based method showed a more sensitive and robust detecting capability than that obtained from a spectrum analyses approach.

Keywords: Wavelet transform, FFT spectrum analysis, hydraulic pump, multi-resolution analysis (MRA), fault detection.

INTRODUCTION

Fault diagnostics provide an effective tool of detecting and identifying equipment faults, which can help equipment users to achieve more consistent operating performance, reduced operational and maintenance costs, and improved operation safety. Hydraulic systems are used in a wide variety of applications ranging from precision control on machine tools to task implementation on construction equipment. The reliable and proper operation of a hydraulic pump is critical for any hydraulic systems. Therefore, a sensitive and accurate faults detection and identification methods for hydraulic pump health monitoring and diagnosis have great interests to the industry.

Effective program for hydraulic pump health monitoring would include equipment inspection, fluid analysis and monitoring by processing operational signals, such as vibration analysis of the shell, temperature, flow rate and pressure [1-4]. Efforts have been made to implement on-line monitoring and fault detection techniques using directly monitored operating parameters. Jardine et al. [2] used a proportional hazards modeling statistical approach to monitor a hydraulic motor health condition of a mine haul truck loader by regularly analyzing oil samples. Although oil analysis has been proven as a good method for timing oil changes and even fault detections, it has two drawbacks of the difficulty of obtaining a representative of oil sample and the temporal requirements for sampling, testing, and evaluating results. Treuhaft et al. [3] presented real-time monitoring of pistons and slippers wear condition of an axial piston pump by utilizing radioactive tracer technology. Through producing characteristic radionuclides (isotopes) in the slippers and pistons, the worn off in an operating pump can be estimated by monitoring the gamma-ray activity in the circulating fluid continuously under various operating conditions. Nevertheless, the limitation of this method is obvious as it is impossible to produce pistons and slippers with isotopes characteristic on industry pumps. Flow data as well as temperature signals are not fit to detect the malfunction of pump due to the characteristic of tightly associated with operation condition. Therefore, reliable means of analyzing more readily sensed signals is desirable [4].

Many different diagnostic tools have been developed for faults diagnosis and performance monitoring. Wolfman, et al. [5] proposed a multi-model approach for centrifugal pump fault detection and diagnosis. The model consists of three sub-models of pump, pipe, and mechanical subsystems. Individual neural-fuzzy sub-model was used to represent relative normal state for each subsystem. Evaluation residuals then were designed to implement the fault detection. Parametric models have been developed and tested for on-line diagnostics of hydraulic pumps and performance prediction [6-9]. Although a few proven theoretical methods have been developed for on-line hydraulic pump faults diagnosis, these methods rely on the accuracy of parametric models chosen [4]. Because the complexity and irrelative among pump operating parameters, it requires some sensitive analysis methods to capture some slight changes on pump operation characteristics to achieve reliable and sensitive pump faults diagnostics. Hancock and Zhang [4] analyzed pump outlet pressure to detect defects of piston pumps. They created a fuzzy-neural network classifier to sort processed feature signals using FFT and wavelet packet analyses. Zhou et al. [10] presented an information fusion application to identify pump malfunctions in terms of spectrum analysis and limit checking results on pump outlet pressure and oil temperature at return line after a relief valve. Both studies [4, 10] showed the outlet pressure signal could be a valuable source for monitoring the performance of hydraulic pumps.

Fourier analysis breaks down a signal into constituent sinusoids of different frequencies. For many signals, Fourier analysis is extremely useful because the signal's frequency content is of great importance [11]. From this decomposition, a power spectrum can be calculated to determine which frequencies are most prevalent in the signal. Similarly, wavelet analysis, a waveform signal analysis method performed by breaking up an evaluating signal into shifted and scaled versions of a standard wavelet, can identify feature signals in multiple decomposed band window of the original signal [12]. Where FFT decomposes a signal into scaled versions of a sine wave, wavelet analysis can decompose a signal into both scaled and shifted versions of almost any waveform preserving the time parameter of the signal. Both methods are sensitive to changing of the evaluating signals in interested frequency bands. Therefore, these two methods are promising faults detection techniques for their capability on extracting feature patterns for faults diagnosis.

This paper compares the effectiveness of spectral analyses and wavelet transform methods on diagnosing hydraulic pump faults. The signal analysis processes and their faults detecting capability of both methods are described in this paper. Both methods were validated experimentally. In order to eliminate the effects of measurement errors arising from noises and system vibrations, a preprocess on raw data is carried to clean up these errors prior to faults diagnosis analysis.

METHODOLOGIES OF DATA PROCESSING

Spectrum Analysis

Fourier transforms decompose a set of pressure signal into its frequency content by comparing various scales of a sine wave. It results a frequency versus amplitude relationship. Here, the discrete Fourier transform in Eq. (1) can be calculated using the Fast Fourier transform (FFT) algorithm.

$$F_n = \sum_{k=0}^{N-1} f(k) e^{-(2\pi)nk/N} \tag{1}$$

Where, f is the sampled pressure signal, N is the length of the sampled pressure signal vector. The results of transform are the Fourier coefficients F_n at frequency n Hz.

From Fourier transform, the power spectrum of the pressure signal can be calculated and used to determine which frequencies are most prevalent in the signal.

Wavelet and Wavelet Packet Analysis

Wavelet analysis is a waveform signal analysis method performed by breaking up an evaluating signal into shifted and scaled versions of a standard wavelet. There are two important differences between wavelet analysis and Fourier transforms. Where FFT decomposes a signal into scaled versions of a sine wave, wavelet analysis decomposes a signal into scaled and shifted versions of any waveform. Selecting a waveform to be similar to a feature signal provides more sensitivity to detecting changes [12]. Secondly, wavelet analysis retains the time dimension of the data. The chosen wavelet is compared to local sections of the signal through translation of the wavelet along the signal at different dilations or scales of the mother wavelet. Wavelet coefficients are preserving the time parameter of the signal. This distinguishing feature of wavelet analysis allows the time of a specific event to be identified. For this research, discrete wavelet transformation (DWT) is used. The DWT uses the power-of-two logarithmic scaling of both the dilation and translation steps, known as a dyadic grid arrangement. The discrete wavelet transform of the finite sequence of sampled pressure signal f using a dyadic mother wavelet is shown in Eq. (2).

$$W_{m,n}(f) = \sum_{k=0}^{N-1} f(k) h_m(k - 2^m n) \tag{2}$$

Where, m, n range over an integer space, $h_m(k - 2^{-m} n)$ is the synthesis wavelets, the discrete equivalents to mother wavelet $\psi_{m,n}(t) = 2^{-m/2}\psi(2^{-m} - n)$.

The results of the DWT are many wavelet coefficients W, which are a function of scale parameter m and shift parameter n. By applying m_0 level discrete wavelet decomposition, the finite sequence of sampled pressure signal f can be represented by the sum of its smooth approximation as follows:

$$\begin{aligned}
f &= A_{m_0} + \sum_{m=1}^{m0} D_m \\
A_{m_0} &= \sum_{n=-\infty}^{+\infty} a_{m_0} g_{m_0}(k - 2^{m_0} n) \\
D_m &= \sum_{n=-\infty}^{+\infty} d_m h_m(k - 2^m n) \\
a_{m_0} &= \sum_{k=0}^{N-1} f(k) g_{m_0}(k - 2^{m_0} n) \\
d_m &= \sum_{k=0}^{N-1} f(k) h_m(k - 2^m n)
\end{aligned} \tag{3}$$

where, A_{m_0} is the coarser approximation of f in level m_0, D_m leads to the differences among each dilation, a_m and d_m are approximation and detail coefficients at level m, respectively. $g_m(k-2^{-m}n)$ are called the synthesis scaling functions [13].

In harmonic analysis, such a decomposition procedure is referred to as a 'two-channel' sub-band filtering scheme. The inlet sequence is convolved with two filters of one low-pass and one high-pass. The outlet results in two sequences of sub-samples. The resulting sub-bands signal are then reassembled to perform wavelet analysis.

An efficient way to implement this scheme is fast wavelet transform algorithm. From the wavelet scaling sequence, this algorithm conducts two pair of filters, low-pass filter and high-pass filter for signal decomposition (L_D, H_D) and filters pair for signal reconstruction (L_R, H_R) as expressed below:

$$
\begin{aligned}
L_R &= \frac{g(k-n)}{\sqrt{\sum_{n=1}^{2m}(g(k-n))^2}} \\
H_R(j) &= (-1)^j L_R(2m+1-j) \\
L_D(j) &= L_R(2m+1-j) \\
H_D(j) &= H_R(2m+1-j) \\
j &= 1,2,\cdots,2m
\end{aligned}
\tag{4}
$$

where, m is the order of wavelet scaling sequence.

Then, the approximation and detail coefficients of sampled signal can be obtained by convolving the signal sequence with the low-pass filter for approximation, and with the high-pass filter for detail, followed by dyadic decimation [11]. The fast wavelet transform based signal decomposition could e illustrated by Fig. 1.

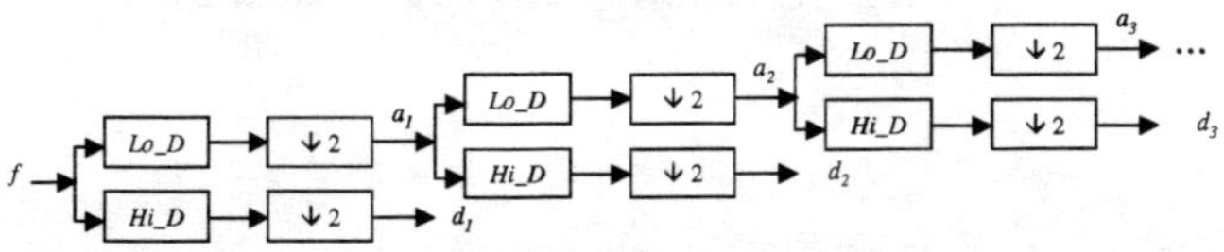

Fig. 1 Schematic wavelet decomposition scheme

Unlike wavelet analysis, which decomposes only the approximation, wavelet packet analysis decomposes both approximations and details for further analysis. Hence, wavelet packet analysis is capable of identifying feature signals in multiple decomposed bands. Schematically, the scheme of a three-level decomposition wavelet packet analysis to reassemble the original signal and the sub-band by each sub-sampled signal can be represented as shown in Fig. 2.

From wavelet packet analysis of the pressure signal, the wavelet coefficients corresponding to interested sub-band are calculated and used to determine if defects of the monitored pump occur.

Data Preprocessing

The border distortion phenomenon exists at the boundary of the examining data when applying wavelet transform to

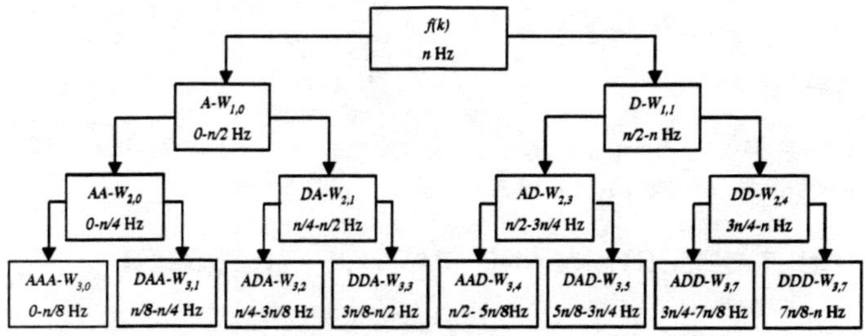

Fig. 2 Wavelet packet decomposition, coefficients and frequency sub-band

finite signal sequence. For eliminating the distortion a twin-sliding window is applied in order to improve the accuracy of fault detection. A bigger window, namely 'bound-window', is used to scan the sampled data and the wavelet packet decomposition is carried out on the data the window covered, while an observing-window, smaller than the bound-window in width, is used to extract the intercept wavelet coefficients. Then, the extracted 'true-wavelet coefficients' are used to conduct the malfunction symptom. The twin-sliding window approach is illustrated in Fig. 3.

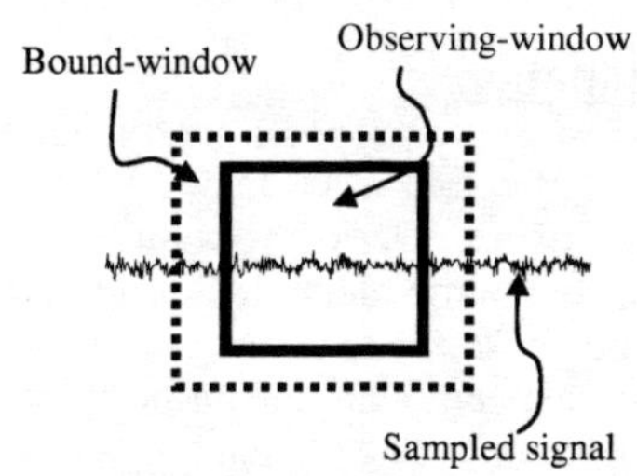

Fig. 3 Dealing with border distortion

Residuals Generation

In order to apply the diagnosis method on-line, a method is developed based on the residual between normal signal and the fault signal. While the residual exceeds a threshold, it indicates that the performance of the pump has a malfunction.

By applying Fourier transform, the summation of power of each fault signal from interested frequency sub-band is compared with normal signal to produce a relative residual as demonstrated in Eq. (5).

$$
FR = \frac{\sum FP - \sum NP}{\sum NP} \tag{5}
$$

where, *FR* is Fourier residual, *FP*, *NP* are fault and normal power of signals.

Similarly, when applying an n level wavelet packet transform to the original pressure signal, the 2^n wavelet coefficients can be obtained for the signal. The energy of each wavelet coefficient vector, the summation of square of each element in a coefficient vector, is calculated to construct a wavelet feature parameter. Then the fault wavelet feature parameters are compared with the normal feature parameters to determine the wavelet residuals of wavelet analysis.

$$WR_i = \frac{\sum FW_{n,i}^2 - \sum NW_{n,i}^2}{\sum NW_{n,i}^2} \quad (6)$$

where, *WR* indicates the residual of the wavelet analysis, *FW*, *NW* stand for the fault and normal wavelet coefficients.

FAULT DETECTION AND DIAGNOSIS RESULTS

Signal Selecting

The common faults of an axial piston pump include swash plate wear, loose of the ball-socket joints, bearing failure, and fatigue failure of the central return spring [10]. Those faults will be reflected in the pump discharge pressure and are normally buried in the pressure pulsating signals. Furthermore, other fault scenarios, such as cavitations, pipe resonance and leakage, will also be reflected in the discharge pressure signals. Thus, the pressure signal carries sufficient information but a lot of them are buried in background noise signals because of its property of 'inner measurement'. When compared with vibration signal, the pressure signal is considered more suitable for the analysis and detection of faults of the axial pump. This paper considers two kinds of common faults of the axial piston pump, swash plate wear and loose ball-socket joints as case study examples.

Experimental Validation

The validation test of the proposed fault diagnosis methods are conducted on a laboratory scale hydraulic system test-rig. The test-rig consists of an electro-hydraulic servo valve, two testing pumps (a normal pump and a defected pump), and other auxiliary devices and sensors. The pumps used in this test are 10 ml/rev fixed displacement axial piston pumps. Both pumps were operated at 1,470 rpm. The system pressure was 6.5E6 Pa set by a relief valve. The pressure sensors are installed on the discharge port of the pumps for collecting the pressure signals. When one pump was being tested, the other pump was shut off to avoid any possible inference to the discharge pressure of the testing pump. The obtained pressure signal was analyzed on-line using a MATLAB program developed for this research based on the method discussed earlier in this article.

For a pump with seven pistons, the rational flow pulsation frequency is $\omega_1 = 14\pi n/60$ (rad/s), the disturbance frequency due to piston ball-socket excitation is $\omega_2 = 2\pi n/60$ (rad/s), and the disturbance frequency due to swash plate excitation (ω_3) depends on the worn condition of the swash plate. Under normal operating conditions, the pump speed (n) was set at 1470 rpm, and the swash plate was worn within the high-pressure area. This resulted in the following parameters for this case: $\omega_1 = 171.5$ (Hz), and $\omega_2 = \omega_3 = 24.5$ (Hz). In this case, ω_1 is the highest frequency, and ω_2, ω_3 are each $\omega_1/7$. Therefore, the proper critical frequency should be around ω_2. The outlet signals of the normal condition and two malfunction scenarios are shown in Fig. 4.

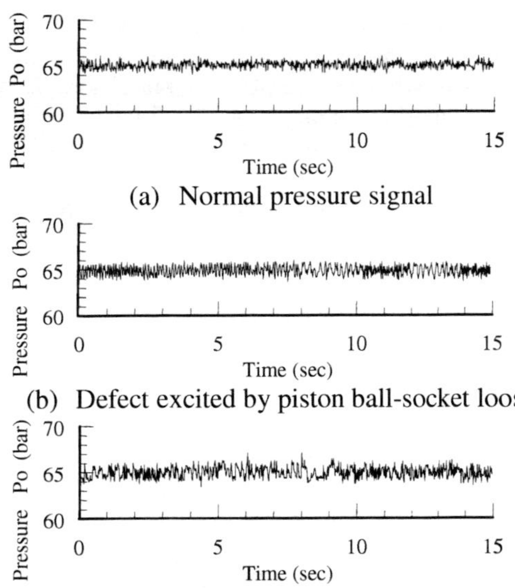

(a) Normal pressure signal

(b) Defect excited by piston ball-socket loose

(c) Defect excited by swash plate worn

Fig. 4 Test pressure signals

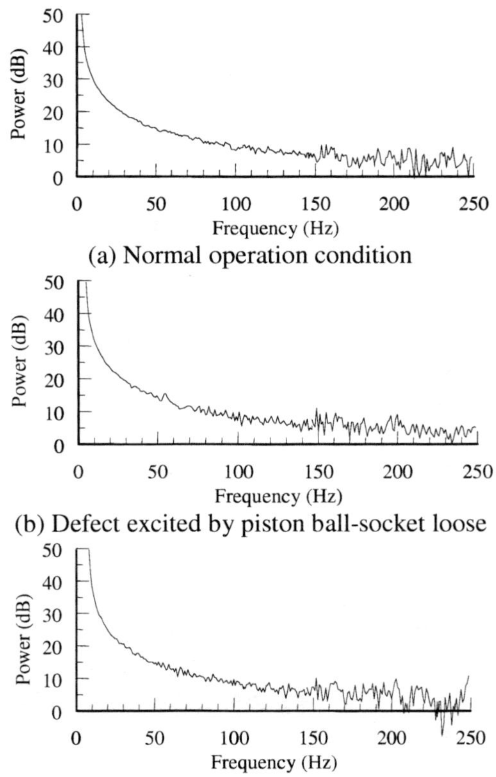

(a) Normal operation condition

(b) Defect excited by piston ball-socket loose

(c) Defect excited by swash plate worn

Figure 5 Fourier transform of test pressure signals

Comparing the original signals shown in Fig. 4, the results indicate that the pulsations of the outlet pressures from all the pumps are slightly higher than that from the normal pump, but the differences are insignificant. These results verified that the original pulsation pressure signals were not capable of providing sufficient information to support pump health diagnosis. Figure 5 shows the Fourier transform result for the test results. From these plots, the feature frequencies are difficult to be identified. The power spectrum changes are very small around 20-55 Hz region by comparing the plots in Fig. 5.

It is not easy to detect the malfunctions from these Fourier transform results.

By applying wavelet packets transform to the original pressure signals, it is noticed that the third level detail coefficient $W_{3,1}$ changes the most while other coefficients' changes are small. Figure 6 shows the wavelet coefficients $W_{3,1}$ of the original pressure signals.

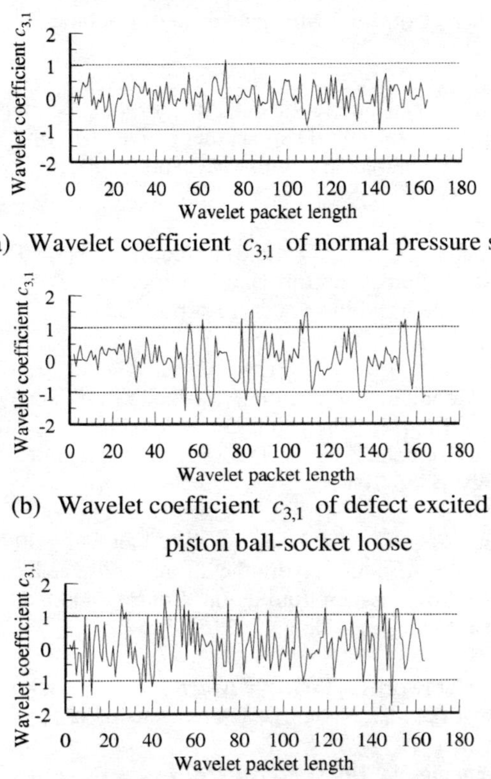

(a) Wavelet coefficient $c_{3,1}$ of normal pressure signal

(b) Wavelet coefficient $c_{3,1}$ of defect excited by piston ball-socket loose

(c) Wavelet coefficient $c_{3,1}$ of defect excited by swash plate worn

Fig. 6 Wavelet packets applying to the test pressure signals

Table 1 Test results: Fourier residuals

Signal	f_{normal}	f_{fault1}	f_{fault2}
Spectra analysis *FR*	0	0.089	2.738

Table 2 Test results: wavelet residuals

Signal		f_{normal}	f_{fault1}	f_{fault2}
Wavelet packet analysis *WR*	$WR_{3,0}$	0	0.003	0.002
	$WR_{3,1}$	0	0.747	0.956
	$WR_{3,2}$	0	0.534	0.215
	$WR_{3,3}$	0	0.816	0.544
	$WR_{3,4}$	0	0.264	0.748
	$WR_{3,5}$	0	0.207	0.542
	$WR_{3,6}$	0	0.528	0.215
	$WR_{3,7}$	0	0.481	0.388

While the threshold for Fourier residual *FR* is set to 1, and the threshold for wavelet residual *WR* is set to 0.5, it is noted that both of the defects (the defect excited by piston ball-socket loose and swash plate worn) can be detected by using wavelet based method as shown in Table 1 and Table 2. The wavelet residuals at sub-band (23-46 Hz) $WR_{3,1}$ and sub-band (69-93 Hz) $WR_{3,2}$ are most sensitive to the change of pump performance. Therefore, $WR_{3,1}$ and $WR_{3,2}$ can be applied to perform on-line fault detection of the pump.

Comparison with the wavelet approach, Fourier transform based method can detect the pump malfunction excited by swash plate worn. But this method missed the defect excited by piston ball-socket loose.

Comparing the results shown in Fig. 4 – Fig. 6, Table 1 and Table 2, the raw signals, the discharge pressures, showed little difference between the normal pump and the defected pumps. This indicates that the surface data (the pump outlet pressure), is not able to provide sufficient information to support pump health diagnosis. After applying wavelet decomposition on the raw pressure signals from three pumps, wavelet residuals from both defected pumps (within 23 to 46 Hz and 69 to 93 Hz) resulted in wilder variations than that of the normal pump. The experimental tests showed that the third level wavelet packet decomposition could extract the feature signals from pump discharge pressure signal for diagnosing the piston pump health conditions. The variation of wavelet residuals WR1 (within 23 to 46 Hz) validates the assumption that the defects of a piston pump will be reflected on some frequency bands of pump discharge flow pulsation. More importantly, the patterns of the coefficient changes were different for different types of pump defects. This fact verified that the wavelet analysis method can improve the capability of diagnosing the health conditions of the piston pumps by decomposing the original pulsation pressure signals, and that the patterns and the amplitudes of wavelet coefficients obtained from different decomposed sub-band windows are relevant to the types of pump defects.

CONCLUSIONS

This study investigated the methods for real time faults diagnosis of the axial piston pump by comparing wavelet based method and Fourier transform based method. The outlet pressure signals of three pumps are employed to carry out the evaluation of the diagnosis methods. Wavelet method decomposes the outlet pressure signal into eight sub-band signals at the third level. The residuals between the sub-band coefficients of the normal signal and those of the fault signal are used to monitor the performance of the pump. By applying Fourier transform to the original pressure signals, the summations of the power spectrum within a sub-band is calculated to produce the residuals. Based on results obtained from using both methods in analyzing the same sets of test data, the following conclusions can be drawn:

- The use of discharge pressure provided the direct information for the diagnosis of the health of monitored pump. This can improve the diagnostic accuracy by overcoming the limitations caused by noise and disturbances acting on the indirectly measured signals.
- The decomposition of the original pressure signals resulted in sub-band informative signals. Reassembling these sub-bands signals and comparing them with those of a normal signal resulted in distinguishable changes from normal and defective pumps.

- The differences in the patterns and amplitudes of the resulting wavelet coefficients within different band windows provided distinguishable features to identify the types of pump defects.
- Wavelet packet transform based fault diagnosis method showed a more sensitive and robust detecting results for the tested pump faults than that obtained from a spectrum analyses approach.

ACKNOWLEDGMENTS

This research was funded in partially by Hebei Education Department Doctoral Funds (China) and USDA Hatch Funds (ILLU-10-352 AE). UIUC College of Engineering Yee scholarship provides a scholarship fund to support Dr. Yingjie Gao conducting research at the University of Illinois at Urbana-Champaign. Any opinions, findings, and conclusions expressed in this publication are those of the authors and do not necessarily reflect the views of the University of Illinois and USDA.

REFERENCES

[1] Fey, C. G., Totten, G. E., and Sun, Y. H., 1999, "Analysis of common failure modes of axial piston pumps," ASTM Special Technical Publication, 1339, pp.299-317.

[2] Jardine, A. K. S., Banjevic, D., Wiseman, M., Buck, S., and Joseph, T., 2001, "Optimizing a Mine Haul Truck Wheel Motors' Condition Monitoring Program: Use of Proportional Hazards Modeling," Journal of Quality Maintenance Engineering, 7, pp.286-301.

[3] Treuhaft, M. B.; Eberle, D. C.; Timmons, S. A., and Wendel, G. R., 2003, "Wear Measurement of a Large Hydraulic Fluid Power Pump Using Radioactive Tracer Wear Technology," ASME- The Fluid Power and Systems Technology Division, Vol. 10, pp.55-64.

[4] Hancock, K. M., and Zhang, Q., 2005, "Maintenance and Fault Diagnosis Tools For Hydraulic Pumps," Proceedings of the 50th National Conference on Fluid Power, Las Vegas, Nevada, March 16-18, pp.361-369.

[5] Wolfram, A., Fussel, D., Brune, T., and Isermann, R., 2001, "Component-Based Multi-Model Approach For Fault Detection And Diagnosis Of A Centrifugal Pump," Proceedings of The American Control Conference, Arlington, June 25-27, pp.4443-4448.

[6] Kaliafetis, P., and Costopoulos, Th., 1995, "Modeling and Simulation of an Axial Piston Variable Displacement Pump with Pressure Control," Mechanism and Machine Theory, 30, pp.599-612.

[7] Manring, N., and Johnson, R., 1996, ''Modeling and Designing a Variable-Displacement Open-Loop Pump,'' Journal of Dynamic Systems Measurement And Control-Transactions of the ASME, 118, pp.267-271.

[8] Karkoub, M., Gad, O., and Rabie,M., 1999, "Predicting axial piston pump performance using neural networks," Mechanism and Machine Theory, 34, pp.1211-1226.

[9] Gerlter, J., Costin, M., Fang, X., Kowalczuk, Z., Kunwer, M., and Monajemy, R., 1995, "Model-based Diagnosis for Automotive Engines Algorithm Development and Testing on a Production Vehicle," IEEE Transactions on Control System Technology, 3, pp.61-68.

[10] Zhou, R., Lin, T., Han, J., and Yan D., 2002, "Fault Diagnosis OF Airplane Hydraulic Pump," Proceedings of the 4th World Congress on Intelligent Control and Automation, June 10-14, Shanghai, China, pp.3150-3152.

[11] The Mathworks, 2005, "*Wavelet Toolbox User's Guide*," The Mathworks Inc., Natick, Massachusetts, USA, pp. 1-9.

[12] Daubechies I., 1992, "*Ten Lectures on Wavelets*," Capital City Press, Montpelier, Vermont, USA, pp.130-146.

[13] Rioul, O., and Duhamel, P., 1992, "Fast Algorithms for Discrete and Continuous Wavelet transform," IEEE Transactions on Information Theory, 38, pp.569-586.

Proceedings of IMECE2005
2005 ASME International Mechanical Engineering Congress and Exposition
November 5-11, 2005, Orlando, Florida USA

IMECE2005-82335

FLUID EFFECTS IN SMALL GAPS – A CHALLENGE FOR SENSOR APPLICATION

Dr.-Ing. C Stammen
RWTH Aachen University
Institute for Fluid Power Drives and Controls
D-52074 Aachen, Germany
Phone +49 241 80-27533, Fax +49 241 80-22194
E-mail: christian.stammen@ifas.rwth-aachen.de

ABSTRACT

This paper is motivated by the difficulty of pressure distribution measurement in the small gap between valve spool and housing or sleeve. The research project "Design of Balancing Grooves for Hydraulic Valves using Analysis of Flow and Pressure Distribution" at IFAS aims at providing a design tool for balancing grooves. It is based on the idea that the application of modern simulation techniques together with experimental results enhances knowledge about necessity and design of balancing grooves. Main influence on the pressure distribution is the gap height profile. Although valve spool and sleeve can be examined separately, it cannot be concluded to the geometrical information about the gap between these parts as accurate as needed to verify the simulation model.

The paper describes the development of miniature pressure sensors made of manganin wire. Measurements showing the current results are presented. As an outlook, the integration of gap height measurement into the pressure sensors is proposed.

INTRODUCTION

Improvement of hydraulic valves needs enhancements in actuators, sophisticated control and advanced spool design to reduce the effect of flow forces. Spool dynamics are influenced by mass inertia, friction, stiction and flow forces. For high dynamic performance and to shift operating range to higher pressures and flow rates, lateral forces effecting stiction and friction also have to be taken into account. The lateral forces are on the one hand caused by inhomogeneous pressure distribution around the gaps between the valve spool and its housing. On the other hand, asymmetries in flow result in lateral forces next to the metering edges. Due to these lateral forces the spool may come in contact with the sleeve and the resulting friction forces may lead to hydraulic locking of the spool, when they overcome the maximum actuator force. Especially for directly operated valves usually using a proportional solenoid to drive the spool there are limitations in the maximum force.

The problem of hydraulic lock is well known in spool valve design. The first research on the hydraulic locking phenomena was carried out by Sweeney[1], followed and completed by Blackburn [2]. To prevent this phenomenon, circumferential balancing grooves are used. These balancing grooves equalise the pressure around the spool. At the persisting gap areas between the grooves, the pressure distribution may still be inhomogeneous, but the differences are on a lower level and the lateral forces decrease. Increasing the number of grooves leads to lower friction, but to higher leakage.

For gaps with perfect cylindrical geometry, there would be no need for balancing grooves. Although machine tools were vastly improved, it can be assumed that remaining differences between ideal and real geometry cause the asymmetries in pressure distribution.

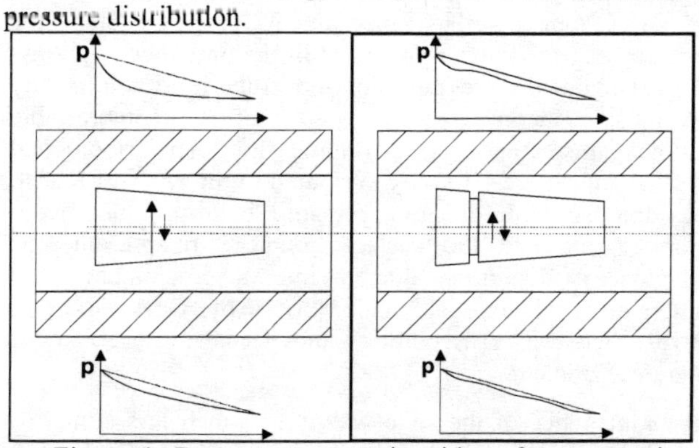

Figure 1: Pressure on opposite sides of an eccentric gap, with(left) and without(right) balancing groove [2]

Former studies concentrated on the effect of balancing grooves in tapered clearances. When the gap diverges in the direction of flow, there are forces pressing the spool towards the wall as shown in **Fig. 1** (left). In converging clearances, there is a stabilising effect of tapering. As valves move in both directions and also the direction of pressure difference may change, tapering cannot solve the problem of hydraulic locking in valves.

Nevertheless, numerical approaches by solving the 1- or 2-dimensional Reynolds equation for tapered clearances by Borghi [3] or later Milani [4] are also applicable on mainly cylindrical gaps with geometric failures. The computation of the pressure distribution in gaps with non-ideal geometry helps to understand the phenomenon of hydraulic lock and gives the possibility to compare the effects of different groove designs. For a single balancing groove, Fig. 1 (right) gives an idea about the equalising effect.

At present, there are no design rules available to find the optimum number and geometry for balancing grooves. The research project "Design of Balancing Grooves for Hydraulic Valves using Analysis of Flow and Pressure Distribution" at the Institute for Fluid Power Drives and Controls (IFAS) aims at providing a design tool for balancing grooves.

SIMULATION

Using the 1- or 2-dimensional Reynolds equation one cannot conclude to the value of the occurring friction force and the effect of contact pressure between the rough surfaces of spool and sleeve. The research work is based on the idea that the application of modern simulation techniques used for the calculation of losses in hydrostatic machines together with experimental results enhances knowledge about the necessity and the design of balancing grooves depending on geometry and working conditions. Therefore a modified 2-dimensional Reynolds equation for rough gaps taken from Kleist [5] is used for simulation of both pressure distribution and related friction forces. Kleist [5] developed a simulation program that enables calculation of the eccentric piston motion and the pressure build-up across the length of the gap or around its circumference for simulated dimensioning of the tribological contacts in a displacement unit. He then conducted experiments to investigate the eccentric motion and compared the measured values with the calculated results. A simulation tool for a complete axial piston pump (PUMA) emerged from the research work performed by Sanchen [6]. The simulation program differentiates between mechanical, hydraulic and tribological models. The mechanical part contains the kinematics and the kinetic processes in the unit. The hydraulic model calculates the pressure build-up in the fluid volumes by balancing the incoming volume flow of the compressible medium against the outgoing volume flow. The tribological model, which can be primarily attributed to the work of Kleist, allows for the hydrodynamic pressure build-up, the mixed friction regime and the surface contacts. In present, the tribological model is integrated into a co-simulation of commercially available simulation tools by Deeken [7] to achieve a better user friendliness and decrease the effort for further development.

As the kinematics of the spool valve are much less complex, the mechanical part of PUMA is simplified. The hydraulic part is re-used to compute the effects inside the balancing grooves, collecting the in- and outgoing flow over the connected gaps. **Fig. 2** depicts the combination of tribological computations of the 2-dimensional Reynolds equation with the grooves modelled as hydraulic capacities.

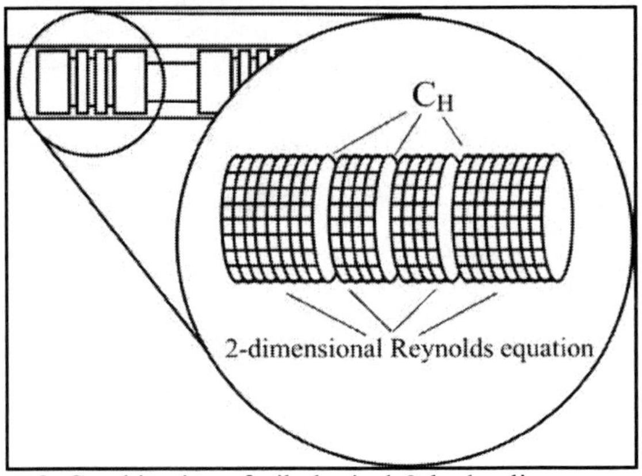

Figure 2: Combination of tribological & hydraulic computation

PRESSURE MEASUREMENT IN NARROW GAPS

Examining the pressure distribution around the valve spool is difficult because of the small tolerances between spool and sleeve. When measuring the pressure in a narrow gap it is necessary not to influence the gap geometry. Every connecting hole for pressure sensors leads to a considerable geometry deviation. Therefore, it is necessary to search for sensor principles which are easy to integrate without effects on gap geometry.

Before customizing the PUMA program code, the modified 2-dimensional Reynolds equation for rough gaps is implemented in MATLAB to gain first experience with the modelling of the gap geometry and the coupling of several gaps with the grooves separating them. The MATLAB script was also used to compute the effect of a pressure sensor connecting hole with a diameter of 1 mm. A annular gap with a diameter of 10 mm, a length of 10 mm and 6 µm nominal gap height was modelled, the pressure dropping from 2 bar to 1 bar across the gap. For computation, the gap surface was divided into 100 x 100 rectangular elements. The hole is simulated by an increased gap height of 25 µm in an 10 x 10 area. The maximum difference in pressure distribution is up to 6% of the pressure drop at the gap as seen in **Fig. 3**.

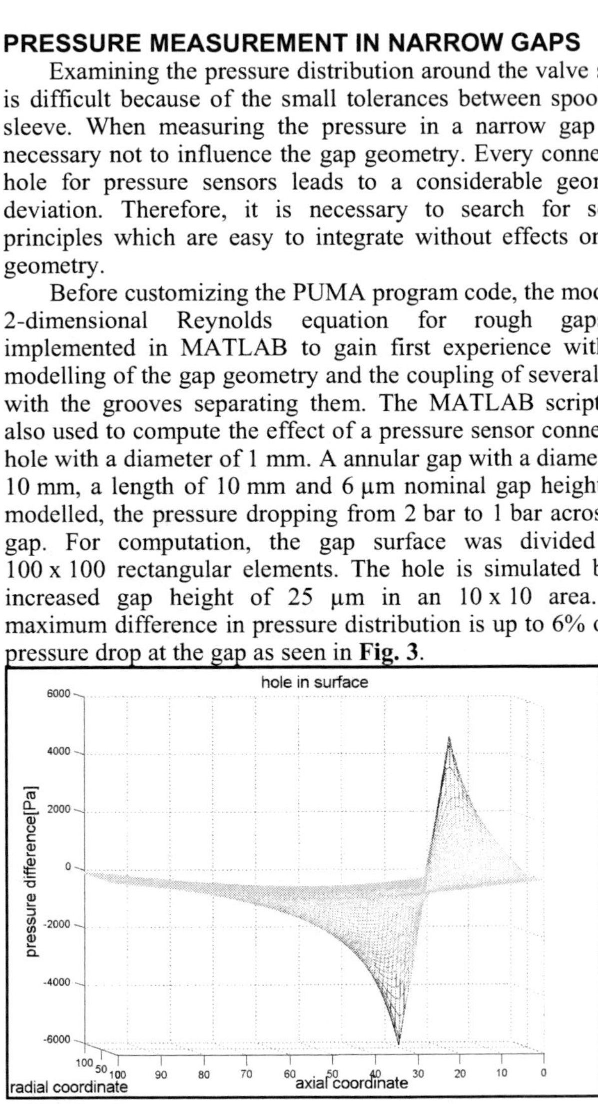

Figure 3: Pressure differences inside cylindrical gaps with / without hole of 10x10 grid elements

Additionally, a densely stacked arrangement of sensors is necessary. The sensor size limits the number of measurement points. To connect the sensors to the gap, small holes in the sleeve are needed. The pressure buildup in the gap with local height h is strongly influenced as the hole diameter $D >> h$. Taking these restrictions into account, most conventional pressure sensors are unsuitable.

SENSOR FOR PRESSURE MEASUREMENT IN GAPS

Restrictions

To match the restrictions for pressure measurement in narrow gaps, miniature pressure sensors are developed.

When measuring the pressure in a gap it is necessary not to influence the geometry of the gap itself. To show the occurring difficulties, **Figure 4** shows a sensor array using miniature pressure transducers. To connect the sensors to the gap, small holes in the sleeve are needed. Even with a tiny hole diameter, the pressure buildup in the gap is influenced as Fig. 4 shows. In addition the dynamic response of the pressure measurement decreases. The size of the small sensor still limits the number of measurement points, even when placed far from the gap.

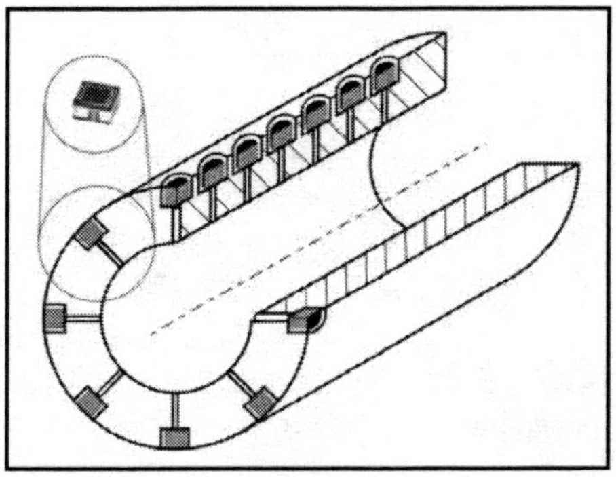

Figure 4: Arrangement using conventional miniature pressure sensors

One approach to reduce the number of sensors, when the geometries can be supposed to be ideal or their effect is negligible, is moving the part the sensor is attached to in order to sample the conditions in different positions. As described above, there is a strong variation of the gap height in axial and circumferencial direction due to the manufacturing process and uncertainties in both sleeve and spool. This causes a change in gap height when moving the two parts relatively to each other leading to a completely different pressure distribution. It can be concluded that in this case the measured pressures at certain positions cannot be combined to the real resulting force.

Miniature Manganin Pressure Sensor

To avoid the problems in the use of conventional pressure sensors, a miniature pressure sensor made of manganin wire is developed. Manganin is a copper-manganese-nickel alloy with a low strain sensitivity, but a relatively high sensitivity to hydrostatic pressure. Resistance changes slightly as a function of applied pressure and is linear up to high pressures. Manganin wires have therefore been used to build high-range fluid pressure cells. This work is based on the experiences with contact pressure under piston rod sealings by Wernecke [8], further developed to investigate pressure distribution and laminar flow in gaps of hydrostatic machines [5].

In further applications, the wires were placed in a circumferential groove and covered with resin to measure the average pressure at different axial positions of sensor and spool. To provide a more local measurement, the manganin wire is now coiled. With the used 30 µm wires and 40 mm active length, coils of less than 1 mm diameter and length are possible. The resin can be levelled to match the sleeve contour, hence the gap geometry is not effected.

Figure 5 depicts on the left side the different states of integration. After first successful tests with a loop (1), further examinations used a coiled and resin-covered manganin wire (2). There was almost no difference in the measurement results, so the development proceeded towards the complete integration of a coiled wire inside of a 1.5 mm hole (4). On the right side of **Fig. 5** 2 manganin wire pressure sensor prototypes on top (5) or inside (4) of a M5 screw are shown.

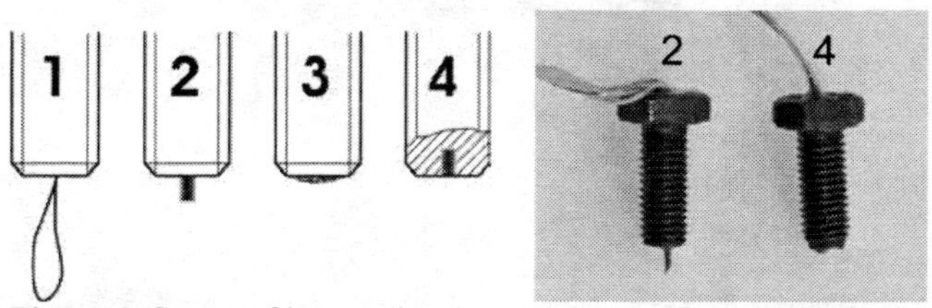

Figure 5: States of integration / prototypes

A Wheatstone bridge consisting of two fixed resistors, a variable resistor for trimming and the manganin coil is used. As manganin is strongly effected by temperature differences, a compensation would be advantageous. To keep the circuit simple, the pressure sensors will be calibrated after the test stand reaches its operating temperature.

The Wheatstone bridge is supplied by a battery to prevent noise. To amplify the output voltage, a three-stage combination of operational amplifiers OP27G with a total amplification of about 200.000 to 300.000 is used. The current development concentrates on the choice of the appropriate resin and the optimum coil design.

Pressure Measurement Results

On a test stand the manganin pressure sensor and a conventional pressure transducer were tested with pressures changing from 0 to 200 bar. **Figure 7** shows the output signal of the miniature pressure sensor which is very much affected by noise. Low-pass filtering and static offset compensation leads to a smoother signal matching the reference pressure (**Fig. 8**). The remaining differences are similar to a 1 Hz sine wave with an amplitude of ca. 10 bar. The cause of this disturbance is still under examination to enable more dynamic measurements. For the static measurement of the pressure distribution averaging is a suitable method to compensate this effect.

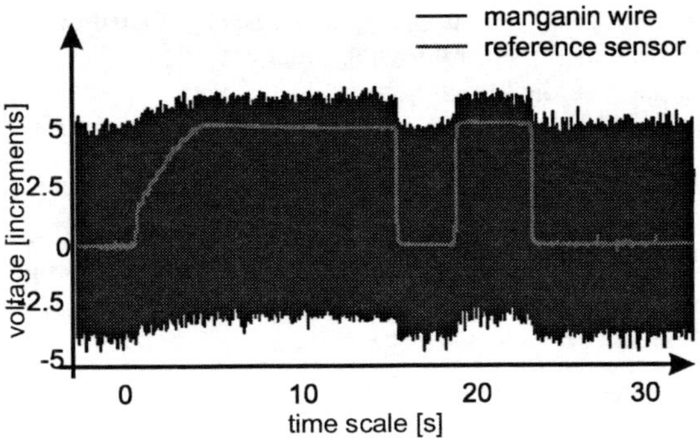

Figure 7: Temperature compensated arrangement using miniature manganin wire sensors

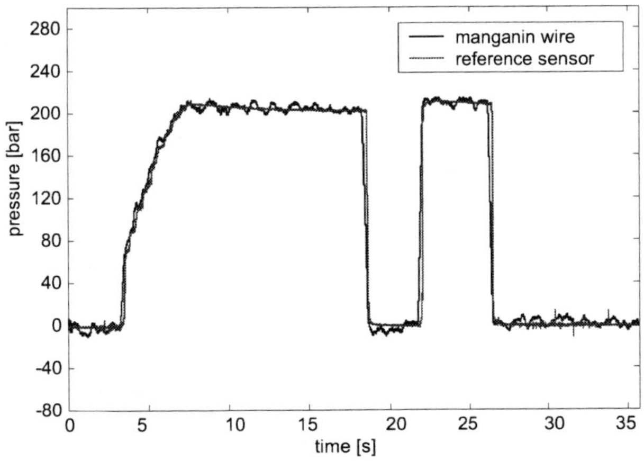

Figure 8: Low-pass-filtered signal

GAP HEIGHT MEASUREMENT

Oil Film Resistance

Figure 9 shows the principle of oil film resistance measurement. Electrodes are placed around the sleeve and the spool is connected to ground. Each electrode is connected via a fix resistor to the supply voltage.

As a major advantage of this gap height sensor principle, it is easily integrable into the manganin wire pressure sensor described above. The manganin wire is coiled around a copper wire which can later work as electrode for gap height measurement when the epoxy casting resin is machined off and the wire is levelled to the sleeves surface. Then, pressure and gap height can be examined at the same position on the gap surface.

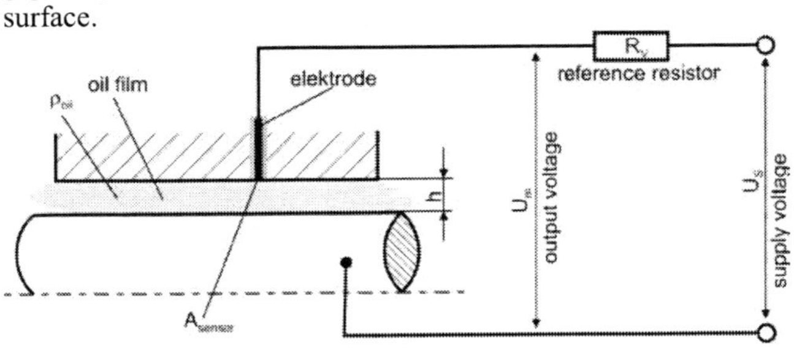

Figure 9: Gap height sensor

The resistance of the oil film and the fix resistor work as a potential divider. From the measured voltage the film resistance and the gap height can be obtained.

The resistance R_{oil} of the oil film inside the gap is proportional to the gap height:

$$R_{oil} = \rho_{oil} \ * h / A_{sensor}$$

The high specific resistance of mineral oil ($\rho_{oil} = 11 \cdot 10^9\ \Omega \cdot \mathrm{mm}$) and the limited measurement area A_{sensor} lead to a limited measurement range and a high resistance of the circuit. **Figure 10** depicts change in the output voltage versus ration of oil film resistance to reference resistance. In order to keep the resistance of the fixed reference resistor within the range of reasonable values up to 10 MΩ, a low oil film resistance R_{oil} is required.

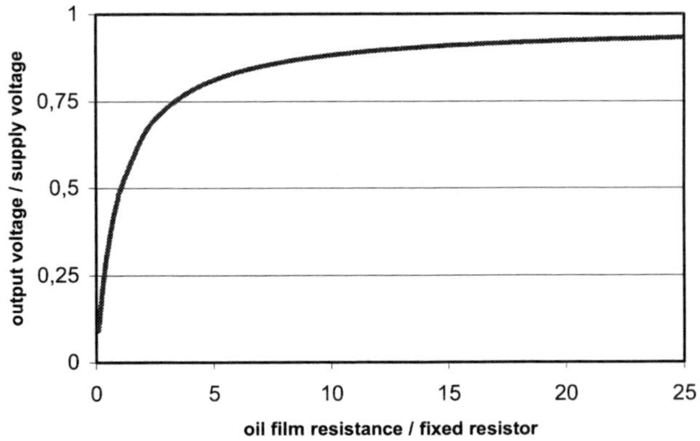

Figure 10: gap height sensor

Sensor prototypes are examined on a test stand. Part of the work is the benchmarking of different fluids with low specific resistance or additives to lower the specific resistance of mineral oil.

CONCLUSIONS

This paper presents the current state of a newly set-up project to examine the hydraulic locking phenomenon both theoretically with a modified 2-dimensional Reynolds equation and experimentally using miniaturised and improved sensor principles for pressure and gap height measurements. The miniature pressure sensors have been successfully tested, next steps are the verification of the gap height measurement and the integration into the sleeve.

Experimental data of gap height and pressure provides a deeper understanding of their effects on hydraulic lock. A successful conclusion from the resulting contact pressure, lateral forces and torques to the experimental results can finally provide a tool for balancing groove design. Schlemmer is giving summary of the total project in [9].

ACKNOWLEDGMENTS

The author would like to thank for the sponsorship of the research project "Design of Balancing Grooves for Hydraulic Valves using Analysis of Flow and Pressure Distribution" by the AiF (German Federation of Industrial Co-operative Research Associations) using funds provided by the German

Ministry of Economics and Technology (BMWi) in the period from 01/03/2004 to 28/02/2006 (AiF No. 13978).

REFERENCES

[1] Sweeney, D.C.: Preliminary Investigation of hydraulic lock, Engineering Vol 172, 1951
[2] Blackburn, J.F.; Reethof, G. & Shearer, J.L.: Fluid Power Control, MIT-Press John Wiley, New York 1960
[3] Borghi, M..: Hydraulic locking in spool-type valves: tapered clearance analysis, Proc. Inst. Mech. Engrs., 2000 Vol. 215 Part I
[4] Milani, M..: Designing hydraulic locking balancing grooves, Proc. Inst. Mech. Engrs., 2001 Vol. 215 Part I
[5] Kleist, A.: Berechnung von Dicht- und Lagerfugen in hydrostatischen Maschinen, PhD thesis RWTH Aachen 2002
[6] Sanchen, G.: Auslegung von Axialkolbenpumpen in Schrägscheibenbauweise mit Hilfe der numerischen Simulation, PhD thesis RWTH Aachen 2003
[7] Deeken, M.: Simulation of the tribological contacts in an axial piston machine, Proceedings of IMECE 2004, 2004, Anaheim, California USA
[8] Wernecke, P.W.: Untersuchung der physikalischen Vorgänge in Spalten von Hydraulikdichtungen, PhD thesis RWTH Aachen 1983.
Locking Forces in Directional Spool Valves
[9] Schlemmer, K: Analysis of Side Thrust Reduction by Peripheral Pressure Balancing Grooves, o+p Ölhydraulik & Pneumatik 07/2005, Vereinigte Fachverlage Mainz

Proceedings of IMECE2005
2005 ASME International Mechanical Engineering Congress and Exposition
November 5-11, 2005, Orlando, Florida USA

IMECE2005-79019

SWITCHMODE HYDRAULIC POWER SUPPLY THEORY*

Cao Jianwei/ Zhejiang University/ The State Key Laboratory of Fluid Power Transmission and Control/38 Zheda Load, Hangzhou, Zhejiang Province, 310027/ China/ 87953028, cjwcolin@zju.edu.cn

Gu Linyi/ Zhejiang University/ The State Key Laboratory of Fluid Power Transmission and Control/38 Zheda Load, Hangzhou, Zhejiang Province, 310027/ China/ 87953028, lygu@zju.edu.cn

Wang Feng/ Zhejiang University/ The State Key Laboratory of Fluid Power Transmission and Control/38 Zheda Load, Hangzhou, Zhejiang Province, 310027/ China/ 87953028,dieter@zju.edu.cn

Qiu Minxiu/ Zhejiang University/ The State Key Laboratory of Fluid Power Transmission and Control/38 Zheda Load, Hangzhou, Zhejiang Province, 310027/ China/ 87953028, cmee_qiu@emb.zju.edu.cn

KEY WORDS

Hydraulic, Pressure, Conversion, Switchmode

ABSTRACT

Switchmode hydraulic power supply is a new kind of energy-saving pressure converting system, which is originally proposed by the authors. It is mainly applied in multiple-actuator hydraulic systems, and installed between hydraulic pump and actuators (one switchmode hydraulic power supply for one actuator). It can provide pressure or flow rate that is adapted to the consumption of each actuator in the system by boosting or bucking the pressure, with low power loss, and conveniently, through high-speed switch valves, just like a hydraulic pressure transformer.

There are two basic types of switchmode hydraulic power supply: pressure boost and pressure buck. Their structures and working principles are introduced. The dynamic characteristics of two typical types of switchmode hydraulic power supply, the pressure boost type and the pressure buck type, were analyzed through simulations and experiments. The performances were evaluated, and improvements on the efficiency of switchmode hydraulic power supply were proposed.

INTRODUCTION

A practical advantage of hydraulic systems is their multiple-actuator characteristic, i.e., multiple actuators can be driven simultaneously by a single hydraulic power supply. Currently, there are three kinds of multiple-actuator hydraulic systems in terms of the power supply used, the throttle control systems, the load sensing systems and the secondary regulation systems. To compare the performance of different systems, one can take a three-actuator system as an example. The energy losses of these three kinds of multiple-actuator hydraulic systems are shown in Figure 1 (a)-(c), where flow rate is shown on the x-axis and pressure is shown on the y-axis. The three blank rectangle parts in Figure 1 marked by 1, 2, and 3 represent the effective power

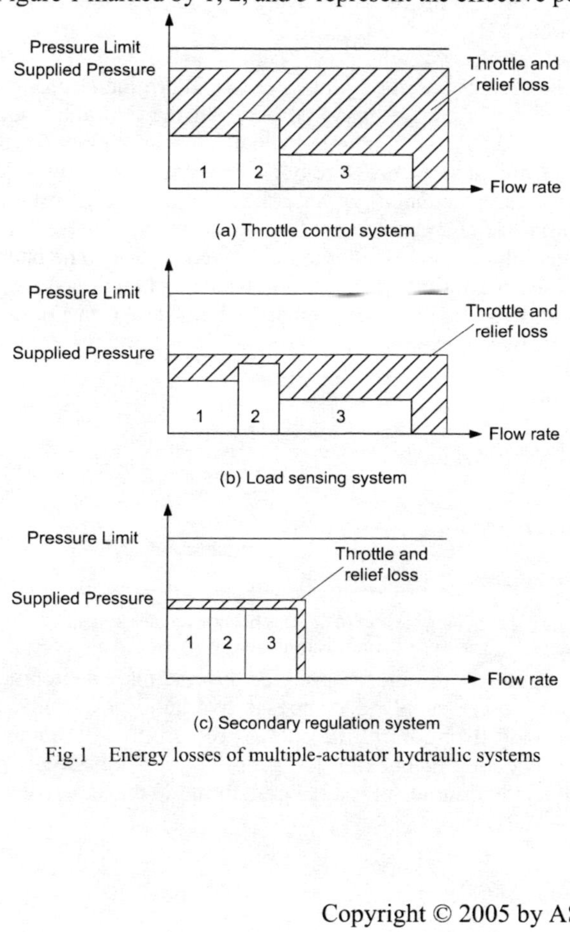

Fig.1 Energy losses of multiple-actuator hydraulic systems

* Chinese National Natural Science Foundation project (No. 50205024)

(pressure*flow rate) consumed by three actuators. The dashed lines illustrate the output flow rate and pressure of hydraulic power supply, while the shadowed parts show the power losses.

In the throttle control system, as shown in Figure 1a, the supplied hydraulic pressure is fixed on a large value even when the actuators' loads are small. This leads to large throttle and relief losses. The load sensing system is thus developed in the 1970's, in which the supplied pressure can be real-time controlled to fit the maximum load through a variable-displacement pump or a three-way pressure compensator. The throttle losses from reducing valve or flow rate control valve can be reduced considerably [2,3]. The load sensing system is therefore widely used in the construction machinery field. However, for actuators with non-maximum loads, e.g. block 1 and 3 in Figure 1b, throttle losses are still large resulting in a pronounced energy waste.

In the secondary regulation system developed in the 1980's, the actuators', i.e. hydraulic motors' displacement can be real-time controlled in order to make their output torques match their loads, as shown in Figure 1c (The pressures and flow rates of actuators are changed because of their variable displacements). Throttle losses from reducing valves or flow rate control valves can be minimized and part of the energy wasted in braking process can be recycled [4]. Therefore, the secondary regulation system is applied more and more widely in industry. Unfortunately, the secondary regulation system is not capable of dealing with pressure boost or pressure buck with low energy loss. As a result, the output pressure of the hydraulic power supply must be maintained on a higher level to ensure proper actions. The size of variable displacement hydraulic motor is limited by the supplied pressure, and its cost is considerable. Furthermore, some throttle losses through flow-rate restriction components (to avoid zero load) still exist in the main circuit.

Switchmode hydraulic power supply, as new energy-saving hydraulic system based on a new theory, is originally proposed in this paper. It can be installed between the hydraulic pump and actuators (one switchmode hydraulic power supply for one actuator), and achieve pressure boost or pressure buck with low energy loss conveniently, just like a hydraulic pressure transformer, thus overcomes the disadvantages of the fore mentioned three traditional hydraulic systems [5-9]. The energy losses for a multiple-actuator hydraulic system equipped with a switchmode hydraulic power supply are illustrated in Figure 2, using the same example as previous. It is obvious that the energy losses are significantly reduced.

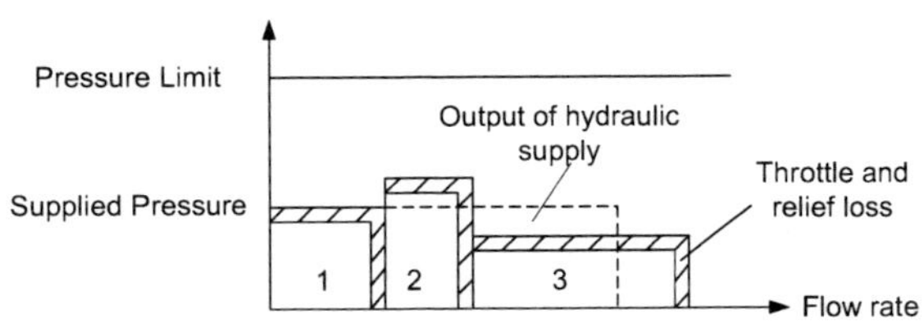

Fig.2 Energy loss of multiple-actuator hydraulic system made of switchmode hydraulic power supplies

In contrast to those traditional systems, switchmode hydraulic power supply decomposes hydraulic power to each actuator, and then control the output pressure to fit their loads through boosting or bucking the supplied pressure. Other than traditional hydraulic pressure transformer, the switchmode hydraulic power supply is a new hydraulic transformer, which has advantages in its low cost and its good performance. Generally traditional hydraulic transformer consists of a variable displacement hydraulic motor and a fixed displacement hydraulic pump. The shafts of the hydraulic motor and the hydraulic pump are connected mechanically. The pressure transform is realized through changing the displacement of the variable displacement hydraulic motor, which is usually done by servo valve [10,11]. The complex structure and the high cost have limited its applications in many areas [12]. The new hydraulic transformer—switchmode hydraulic power supply, which consists of high-speed valves, hydraulic motor and hydraulic accumulator, is simpler in structure and much lower in cost than the traditional one. Using valve control technique rather than volume control technique, the system has advantage in its rapidity. Therefore, switchmode hydraulic power supply is a new energy-saving hydraulic system suitable for different kinds of industrial applications. The basic structure and working principle are introduced in the following section.

NOMENCLATURE

B_m= viscous damping of hydraulic inductor's motor
β_e= fluid bulk-modulus
C= hydraulic capacitance
C_{im}= leakage of hydraulic inductor's motor
D= close time ratio of switch valve in one period
D_m= volumetric-displacement of hydraulic inductor's motor
J_m= moment of inertia for the flywheel and the rotational part of hydraulic inductor's motor
p_b= tank pressure
p_0= pre-charge pressure of accumulator's bladder
p_1= input pressure
p_2= output pressure
p_3= high frequency switching volume's pressure
$q_{V,1}$= flow rate through hydraulic inductor
$q_{V,2}$= output flow rate
$q_{V,3}$= flow rate through high-speed switch valve
$q_{V,4}$= flow rate absorbed by hydraulic capacitor
$q_{V,5}$= flow rate through high-speed check valve
T= period of PWM command
V_0= initial volume of accumulator's bladder
V_2= volume of accumulator's bladder
V_3= high frequency switching volume's volume
ω_m= rotate velocity of hydraulic inductor's motor

STRUCTURE AND PRINCIPLE

Switchmode hydraulic power supplies need three key elements: a volume in which the pressure switches at a high frequency, a flow rate storing component (high frequency hydraulic inductor) and a pressure filter (high frequency hydraulic capacitor).

The high frequency pressure switching volume can be made of a high-speed switch valve and a high-speed check valve. The frequency of high-speed valves should be several times higher than the pressure switching frequency, and the volume should be as small as possible to shorten the pressure fluctuation time.

High frequency hydraulic inductor should drive a high inertia load. Its flow rate should keep constant while its pressure decrease switches from positive to negative repeatedly at high frequency. The most likely hydraulic inductor is a hydraulic motor that drives high inertia load (flywheel). The only disadvantage is that the distributor will be damageable under its high frequency switching pressure and the switch frequency of switchmode hydraulic power supplies will be restricted by the entrapped oil problem of hydraulic motor, which creates a dead zone in switching process.

High frequency hydraulic capacitor has the ability of absorbing high frequency pressure fluctuation, and is able to take in or send out large instantaneous flow rate, which is several times higher than the system's output. In order to keep the output flow rate constant, it should be able to absorb flow rate rapidly while the input flow rate in main circuit is several times higher than its output, and also spit out flow rate rapidly while the input flow rate decrease to zero. The switch frequency of input flow rate equals to the switch frequency of switchmode hydraulic power supplies. The most likely hydraulic capacitor is bladder accumulators. The only disadvantage is that the output pressure range is restricted by the initial pressure in its bladder, and the hysteresis of bladder's distortion according to pressure fluctuation, which creates a dead zone in switching process, restricts the switch frequency of switchmode hydraulic power supplies. Two or three accumulators should be connected in parallel in case that their poppet valves are closed under hydrodynamic force while they spit out flow rate.

The function of switchmode hydraulic power supply is divided into two types: pressure boost switchmode hydraulic power supply and pressure buck switchmode hydraulic power supply.

PRESSURE BOOST POWER SUPPLY

Pressure boost principle: Pressure boost switchmode hydraulic power supply consists of hydraulic inductor, high-speed switch valve, check valve and hydraulic capacitor. Its structure sketch is shown as figure 3.

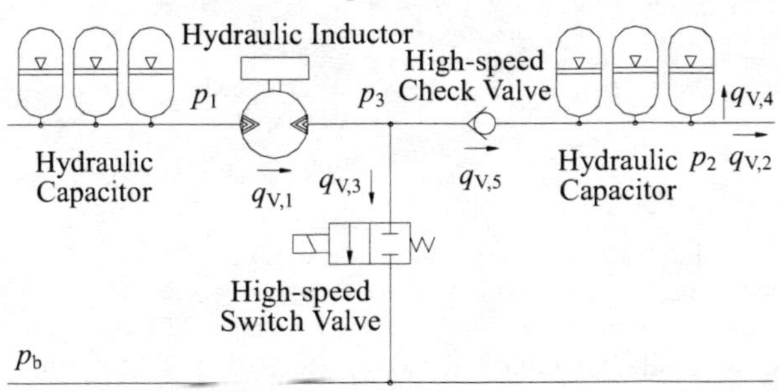

Fig.3 Pressure boost switchmode hydraulic power supply

In pressure boost switchmode hydraulic power supplies, the output pressure p_2 cannot be changed instantaneously because of its connection with a hydraulic capacitor, and the flow rate through hydraulic inductor $q_{V,1}$ also cannot be changed instantaneously.

The fan-out pressure of hydraulic inductor (high frequency switching volume's pressure) p_3 switches from maximum output pressure to zero frequently according to the control of high-speed switch valve.

The flow rate through high-speed switch valve $q_{V,3}$ also switches from $q_{V,1}$ to zero frequently according to the control of high-speed switch valve.

The sum of the output flow rate $q_{V,2}$ and the flow rate absorbed by the hydraulic capacitor $q_{V,4}$ equals to $q_{V,1}$ when the switch valve is shut down but equals to zero when it opens.

Providing the high-speed switch valve a Pulse-Width Modulation (PWM) signal, whose period is T, and close time ratio in one period is D, the input pressure p_1 can be boosted to $p_2=p_1/D$ to fit its actuator at the expense of the decrease of output flow rate.

The process of pressure boost switchmode hydraulic power supply is divided into two stages: energy storing stage and flow rate complement stage.

In energy storing stage, the switch valve is open under the control of PWM signal, so that the fan-out port of hydraulic inductor (The high frequency pressure switching volume) connects with tank directly. Therefore,

$$p_3 = 0 \tag{1}$$

The high-speed check valve is closed under reverse pressure as long as the output pressure is not zero. Therefore,

$$q_{v,5} = 0 \tag{2}$$

These would lead to the following results: The input pressure p_1 accelerates high frequency hydraulic inductor (a high-speed motor drives a flywheel) and stores energy,

$$J_m \frac{d\omega_m}{dt} + B_m \omega_m = D_m (p_1 - p_3) \tag{3}$$

$$q_{v,1} = D_m \omega_m + C_{im} (p_1 - p_3) \tag{4}$$

The output flow-rate $q_{V,2}$, which is decided by its actuator, is complemented by the hydraulic capacitor (accumulator) at the output side. Therefore,

$$q_{v,4} = q_{v,5} - q_{v,2} \tag{5}$$

The output pressure p_2 would decrease while the accumulator release flow rate.

$$\frac{dV_2}{dt} = -q_{v,4} \tag{6}$$

$$p_2 = p_0 \left(\frac{V_0}{V_2} \right)^{1.4} \tag{7}$$

This stage would continue $(1\text{-}D)T$ seconds. Ignoring the leakage, friction and dead zone time of high frequency hydraulic capacitor and inductor, and assuming the high-speed switch valve and check valve is ideal (the response time delay and throttle loss all equal zero),

$$\frac{dq_{v,1}}{dt} = \frac{D_m^2}{J_m} p_1 \tag{8}$$

$$\frac{dp_2}{dt} = -1.4 \frac{p_0}{V_0} \left(\frac{V_0}{V_2} \right)^{2.4} q_{v,2} \tag{9}$$

In flow-rate complement stage, the switch valve is closed suddenly under the control of PWM signal. Since the flow rate through the hydraulic inductor $q_{v,1}$ cannot be changed instantaneously, a sudden brake of high inertia load occurs and leads to the sudden increase of the fan-out pressure of hydraulic inductor (high frequency switching volume's pressure) p_3. Therefore,

$$\frac{dp_3}{dt} = \frac{\beta_e}{V_3} q_{v,1} \tag{10}$$

This transitional period can only continue a very short time. Take the HSV series high-speed valve for example, it is a two-position two-way switch valve. The maximum switch frequency is more than 200Hz. Assume the high-speed switch valve works at the frequency of 10Hz and the close time ratio is

50%, the time of flow-rate complement stage would be 0.05s. Suppose the flow rate through hydraulic inductor $q_{v,1}$ is 37.5 L/min, the bulk modulus of the hydraulic oil β_e is 17000bar, the high frequency switching volume V_3 is 5.65ml (a hydraulic pipe with its diameter 6mm and its length 200mm), according to eq.(10), the time when p_3 increases from 0 bar to 200bar would be 0.00012s. It is really a very short time in the flow-rate complement stage. when p_3 increases to be higher than its output pressure p_2, the check valve is opened. Therefore,

$$p_3 = p_2 \tag{11}$$

$$q_{v,5} = q_{v,1} \tag{12}$$

In this stage, eq.(3)~(7) are still correct. These would lead to three results: First, the output pressure p_2 is still lower than its input pressure p_1. The high frequency hydraulic inductor continues accelerating and storing energy. Its flow rate $q_{v,5}$ provides the output flow rate $q_{v,2}$, and the excess flow rate $q_{v,4}$ is stored in the hydraulic capacitor (accumulator) at the output side. As the flow rate through the hydraulic inductor $q_{v,5}$ increases continuously, the flow rate absorbed and complemented by the hydraulic capacitor $q_{v,4}$ increased accordingly, which leads to the increase of output pressure p_2. Second, the output pressure p_2 is higher than its input pressure p_1, and the high frequency hydraulic inductor is decelerated and releases the stored energy, but the released energy is still less than the stored energy in energy storing stage. Its flow rate $q_{v,5}$ still increases and leads to the increase of output pressure p_2. Third, the released energy equals to the stored energy in energy storing stage. Its flow rate $q_{v,5}$ and output pressure p_2 reach their balance point.

This stage would continue DT seconds. Also ignoring the leakage, friction and dead zone time of high frequency hydraulic capacitor and inductor, and assuming the high-speed switch valve and check valve is ideal (the response time delay and throttle loss all equal zero),

$$\frac{dq_{v,1}}{dt} = \frac{D_m}{J_m}(p_1 - p_2) \tag{13}$$

$$\frac{dp_2}{dt} = 1.4\frac{p_0}{V_0}\left(\frac{V_0}{V_2}\right)^{2.4}(q_{v,1} - q_{v,2}) \tag{14}$$

The pressure boost switchmode hydraulic power supply would repeat the above process while the next period of PWM signal begins.

When the increase of input flow rate and output pressure in the first stage equal to their decrease in the second stage, the system arrives its balance point. Still ignoring the leakage, friction and dead zone time of high frequency hydraulic capacitor and inductor, and assuming the high-speed switch valve and check valve is ideal, the output pressure of switchmode hydraulic power supply p_2 is direct proportional to its input pressure, and is inverse proportional to the close time ratio. The input flow rate $q_{v,1}$ is direct proportional to its output flow rate $q_{v,2}$, and is inverse proportional to the close time ratio:

$$q_{v,1} = \frac{q_{v,2}}{D} \tag{15}$$

$$p_2 = \frac{p_1}{D} \tag{16}$$

The fluctuation of its output pressure p_2 is direct proportional to its output flow rate $q_{v,2}$, period T, and the open time ratio (1-D), but is inverse proportional to the hydraulic capacitance C.

$$p_{2,\max} - p_{2,\min} = \frac{q_{v,2}}{C}(1-D)T \tag{17}$$

From eq.(15) and (16), the converting efficiency of the switchmode hydraulic power supply could be 100% under ideal condition, i.e., the output power equals to the input power. The pressure could be boosted without any power loss.

Through PWM control, the pressure boost power supply could provide the proper power (**not flow rate!**) that is adaptive to its consumption to realize an optimal energy-saving effect.

PRESSURE BUCK POWER SUPPLY

Pressure buck principle: changing the connection of the hydraulic inductor, high-speed switch valve and check valve of pressure boost power supply, a pressure buck switchmode hydraulic power supply is constructed. Its structure sketch is shown as figure 4.

Owing to this new hydraulic control technique, the discharged oil that flows back into the tank can be duplicated used and the energy conversion efficiency increases accordingly. The duplicated usage of the oil leads to the deceleration of the hydraulic inductor (hydraulic motor) and so the flywheel slows down. This is an energy consumption process and the energy consumed in this stage needs to be stored in the last stage. So the working process can be divided into two stages: energy storage stage and energy consumption stage.

Fig.4 Pressure buck switchmode hydraulic power supply

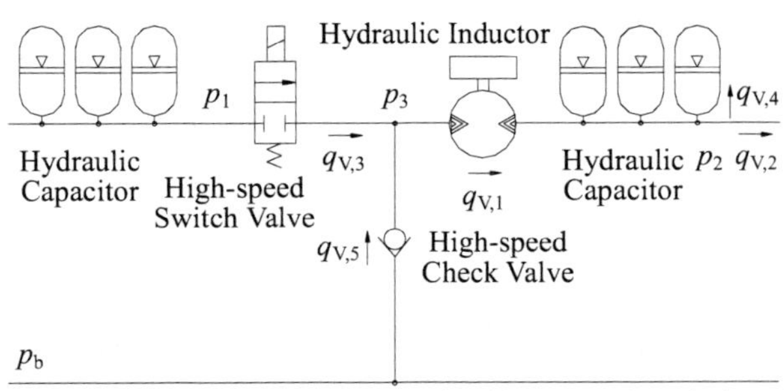

In the energy storage stage, the high-speed switch valve opens under the control of PWM signal, while the high-speed check valve closes under the high pressure. The oil flows into the hydraulic inductor directly, $P_3 = P_1$. Under this high pressure, the hydraulic inductor is accelerated and the flywheel speeds up, so the energy is stored.

While in the energy consumption stage, the high-speed switch valve closes suddenly under the control of PWM signal. Due to the large inertia of the flywheel, the rotate speed of the hydraulic inductor will not change instantly and the flow through it will not change instantly either. However, this continuous flow could lead to the sudden drop of P_3. When P_3 drops until it is lower than P_b, the high-speed check valve opens, $P_3 = P_b$. The discharged oil that flows back into the tank flows into the hydraulic inductor through the high-speed check valve. The hydraulic inductor is then decelerated and the flywheel slows down, the energy stored in the last stage releases. In this stage, the discharged oil does not flow back into the tank while recycles in the hydraulic circuit.

When next PWM signal periods begins, the system will repeat the above working process.

Pressure buck switchmode hydraulic power supply obtains a discontinuous flow rate through the high-speed switch valve firstly. Take advantage of the instantaneous low pressure in its high frequency switching volume, the flow rate is complement through the check valve to realize the increase of its output flow rate conveniently. Then the pressure is filtered and decreased via a LC filter made of a hydraulic inductor and a hydraulic capacitor.

Similar to pressure boost power supplies, the output pressure p_2, average input flow rate $\bar{q}_{V,3}$, and fluctuation of output pressure of switchmode hydraulic power supply under ideal condition are formulated as following:

$$\bar{q}_{v,3} = (1-D)q_{v,2} \tag{18}$$

$$p_2 = (1-D)p_1 \tag{19}$$

$$p_{2,\max} - p_{2,\min} = \frac{q_{v,1} - q_{v,2}}{C} DT \tag{20}$$

DYNAMIC CHARACTERISTICS OF SWITCHMODE HYDRAULIC POWER SUPPLIES

Theoretically speaking, as long as the characteristics of the hydraulic components are ideal enough, pressure boost and pressure buck could be realized without any power losses through the converting circuits shown in figure 3 and figure 4. However at present, the efficiency of the switchmode hydraulic power supplies are not satisfying because of the compressibility of hydraulic oil in high frequency pressure switching volume, throttle loss and On-off time delay of high-speed switch valve and check valve, additional hydraulic resistance and dead zone time of hydraulic inductor and hydraulic capacitor (bladder's elasticity and hysteresis of accumulator, viscous damping, leakage and entrapped oil in the high inertia hydraulic motor). To develop high frequency valves with large flow rates, high frequency hydraulic inductors and capacitors with low additional resistance, extra-high-speed hydraulic motors with large flow rates and low entrapped oil, and high frequency material for accumulator's bladder, increase its switching frequency (so that the size of the power supplies and the high frequency pressure switching volume can be decreased) is essential for improving the efficiency and reliability of switchmode hydraulic power supplies.

Unlike in traditional hydraulic systems, two different kinds of power loss would take place in the regulation process of switchmode hydraulic power supplies: dynamic power loss and instantaneous power loss.

Dynamic power loss is mainly caused by the throttle loss of high-speed switch valve and check valve, and additional hydraulic resistance in hydraulic inductor and hydraulic capacitor. The corresponding efficiency is defined as dynamic efficiency, which is represented as η_d, Dynamic power loss can be examined through separate characteristic experiments for each hydraulic component. Raising the maximum flow rate of high-speed switch valve and check valve, developing hydraulic inductors hydraulic capacitors with low hydraulic resistance are the only ways to enhance the dynamic efficiency.

Instantaneous power loss is mainly caused by the On-off time delay of high-speed switch valve and check valve, dead zone time of hydraulic inductor and capacitor, and the compressibility of hydraulic oil in high frequency pressure switching volume. The corresponding efficiency is defined as instantaneous efficiency, which is represented as η_s. Instantaneous efficiency can only be examined through the integrated characteristic experiment of the switchmode hydraulic power supplies because of its relation with switching frequency. The only way to reduce instantaneous power loss is to raise the response of hydraulic components, reduce their dead zone times and the high frequency pressure switching volume.

PRESSURE BOOST POWER SUPPLIES

For the pressure boost switchmode hydraulic power supply shown as figure 3, it is assumed that the entrance pressure is p_1=10MPa, but needs to be boosted to p_2=25MPa, and the fan-out flow rate should be constant $q_{V,2}$=60L/min. From eq.(1) and (2), the close time ratio of high-speed switch valve should be D=40%, and the input flow rate would be $q_{V,1}$=150L/min if its efficiency were 100%.

Assume the switch frequency is 5Hz, i.e., the period of PWM signal is T=0.2s.

To limit the fluctuation of output pressure in 6%, i.e., 1.5MPa, three parallel bladder accumulators (whose volumes are 3L) are applied as hydraulic capacitor, and are pre-charged at 12.5MPa. According to eq.(3), the capacitance is 650 [(L/min)/MPa]·s. The dynamic power loss of this hydraulic capacitor is about 8%.

Similarly, to limit the fluctuation of entrance flow rate to 10%, i.e., 15L/min, a bend-axial piston hydraulic motor (whose volumetric-displacement is 150ml/r) driving a flywheel (whose size is ϕ320×200 and its moment of inertia is J=1.6kg·m^2) is applied as a hydraulic inductor. Its maximum shaft-speed is less than 1000rpm. The inductance is 0.048[MPa/(L/min)]·s. and dynamic power loss caused by this hydraulic inductor is about 12%.

To reduce throttle loss, two cartridge valves, whose frequencies are 25Hz, are applied as the high-speed switch valve and check valve. Their power loss is about 1%.

Therefore, the dynamic efficiency of the pressure boost switchmode hydraulic power supply η_d is about 80%. Research on hydraulic capacitors and hydraulic inductors with lower additional resistance and higher efficiency is imperative for switchmode hydraulic power supplies. On the other hand, the slow response of hydraulic capacitor and hydraulic inductor also restricts the increase of the switch frequency and the decrease of their size significantly. Simulation shows that if the switch frequency were increased to 50Hz, the diameter and thickness of flywheel could be halved, and the volume of accumulator could be reduced by 1/5 according to estimates. The structure of switchmode hydraulic power supply could therefore be more compact and practical.

Besides the dead zone time of high frequency hydraulic inductors and capacitors, instantaneous power loss is mostly caused by the high frequency pressure switching volume in hydraulic system, which connects the output port of hydraulic inductor and the entrance of the two cartridge valves. The compressibility of oil in this volume causes the instantaneous pressure and flow rate in the switching process, which decrease the efficiency of the switchmode hydraulic power supplies significantly. This volume could be reduced to about 320ml if the valve block is optimally designed.

The dynamics of the pressure boost switchmode hydraulic power supply is shown as figure 5. Figure 5 shows that the gross efficiency of pressure boost switchmode hydraulic power supply is 63%. Eliminating the dynamic power loss, the instantaneous efficiency is less than 80%. The instantaneous power loss is mostly caused by the large size of hydraulic components such as hydraulic inductor and valves. The pressure transition time is too long in high-speed switching, causing high instantaneous pressure and flow rate. Only when the maximum motor shaft-speed in hydraulic inductor is increased, will the size of hydraulic components and high frequency pressure switching volume be decreased considerably. The instantaneous pressure and flow rate in the switching process could be reduced to increase the instantaneous efficiency.

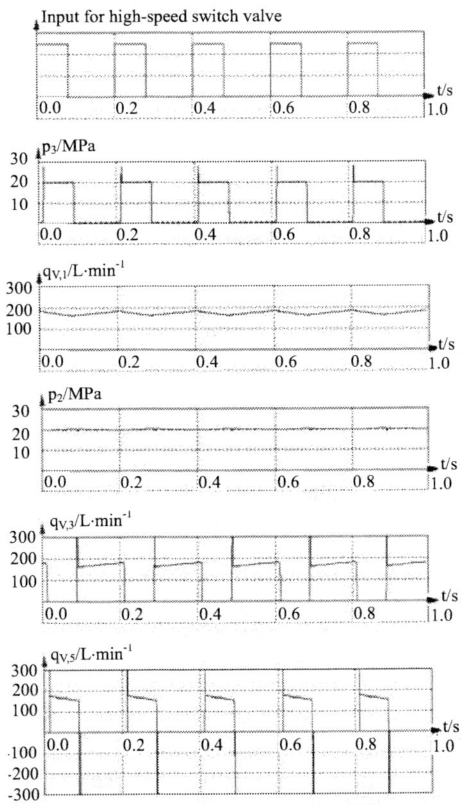

Fig.5 Dynamics of pressure boost power supply

While the maximum shaft-speed could be increased to 3000rpm, the volumetric-displacement, the high-speed switch volume and the instantaneous power loss could be reduced respectively 1/3, 40% and 50%. Research on high frequency hydraulic motors with higher maximum shaft-speeds and cartridge valves with lower throttle loss and smaller size is imperative for switchmode hydraulic power supplies.

PRESSURE BUCK POWER SUPPLY

The dynamic process of pressure buck switchmode hydraulic power supply is shown as figure 6. The pressure needs to be bucked to P_2=6MPa, and the fan-out flow rate should be constant $q_{V,2}$=240L/min. Therefore, the close time of the switch valve should be D=40%, and the average entrance flow rate would be $\bar{q}_{V,3}$=240 L/min if its efficiency were 100%. Accumulators are pre-charged at 3.0MPa. Figure 6 shows that the gross efficiency of pressure with flow rate increase of the switchmode hydraulic power supply is only 58%. Eliminating the dynamic power loss, the instantaneous efficiency is less than 75%.

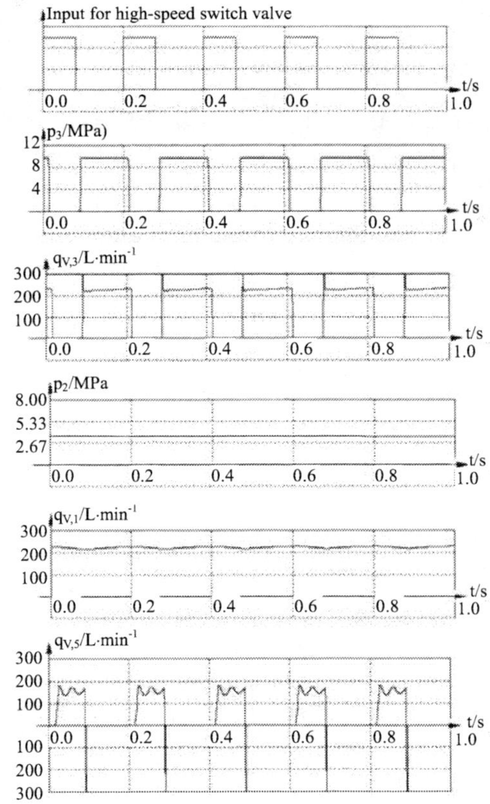

Fig.6 Dynamics of pressure buck power supply

The most successful application of the pressure buck power supply is the switch-mode deep-sea electric power generation system. It is a new electric power generation technique, in which electricity is generated from seawater pressure on the seafloor directly. The system consists of a high-speed switch valve, a high-speed check valve, a seawater hydraulic motor, a high inertia flywheel, an electric DC generator, a rectifier filter and a load, which is shown as Fig. 7.

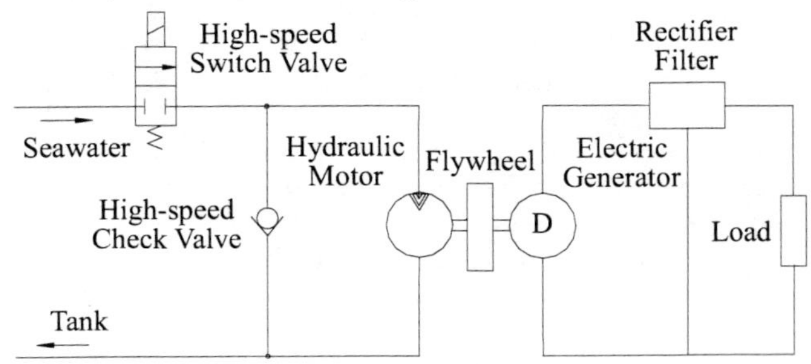

Fig.7 Switch-mode Deep-sea Electric Power Generation System

The experiment of switch-mode deep-sea electric power generation system has been carried out on the scientific investigation ship "DY 1" on June 12th, 2004. The system is launched to the seafloor at the depth of 2,400 m with a large pressure-resistant tank whose capacity is 0.2 m^3. A seawater hydraulic motor with its delivery capacity 2.5 ml/r and its rated rotate speed 1,450 r/min drives an electric DC generator with its rated output voltage 24 V and its maximum output power 1.5 kW and a flywheel with its inertia 1.5 kg·m^2. The high-speed switch valve is driven by a solenoid at a frequency of 8 Hz. The output power of the system is shown as Fig. 8. The system can supply electric power continually for about 48 minutes with the output power of 1,200W. So all the electric power system can supply is 0.96 kW·h and the energy conversion efficiency is about 70%.

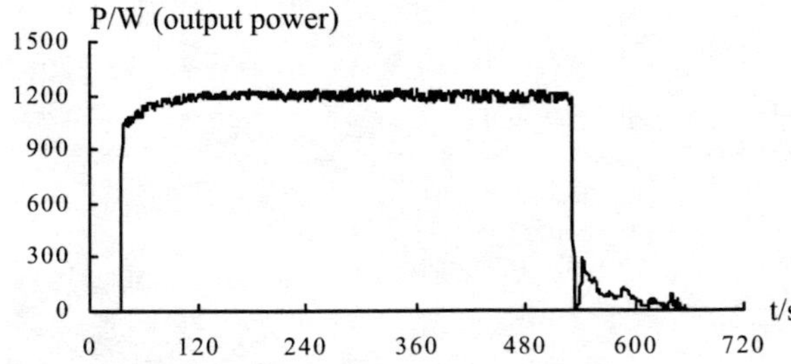

Fig.8 Output Power of Deep-sea Electric Power Generation System

CONCLUSION

(1) Switchmode hydraulic power supplies are new kind of energy-saving hydraulic systems, which could boost or buck hydraulic pressure with low power loss conveniently, and overcome the disadvantages of traditional hydraulic systems. Pressure boost is realized through the sudden brake of high inertia load, while pressure buck is realized through the LC filtering of a discontinuous flow rate and flow rate complement under low pressure.

(2) High frequency pressure switching volume, flow rate storing component (high frequency hydraulic inductor) and a pressure filter (high frequency hydraulic capacitor) are three key elements of switchmode hydraulic power supplies.

(3) Besides the dynamic power loss caused by the components' dynamic characteristics, there is another kind of power loss in the regulation process of switchmode hydraulic power supplies: instantaneous power loss, which is caused by the components' high-frequency characteristic.

REFERENCES

[1] Keith Billings.H, Eng.C., 1989, Handbook of Switchmode Power Supplies, Mc Graw-Hill Publishing Company, New York.

[2] Aoki.Y, Uehara.K, Karakama.T, et al., 1994, "Load Sensing Fluid Power System," Society of Automotive Engineerings Transactions, Section 2, Journal of Commercial Vehicles, **103(941714)**, pp.139-153

[3] W.back, B.Zahe, 1991, "Electrohydraulic Loadsensing," Society of Automotive Engineerings Transactions, Section 2, Part 1, Journal of Commercial Vehicles, **100(911814)**, pp.303-311

[4] W.back, Ch.Koegl, 1993, "Secondary control (Secondary Regulation)," Rexroth International Symposium on Fluid Power, JHPS, Tokyo

[5] Gu Linyi, Qiu Minxiu, Jin Bo, Cao Jianwei, 2003, "New Hydraulic Systems made up of Hydraulic Power Bus and Switchmode Hydraulic Power Supplies," Chinese Journal of Mechanical Engineering, **39(1)**, pp84-88

[6] Gu Linyi, Wang Feng, Chen Ying, Zhang Yan, 2004, "Research on Deep Sea Power Supply Technique From Seawater Pressure based on Switchmode Hydraulic Power Supply," Chinese Journal of Mechanical Engineering, **40(5)**, pp141-144

[7] Qin Huawei, Gu Linyi, Chen Ying, 2004, "Velocity Control System for High Inertia Loads based on Switchmode Hydraulic Power Supply," Chinese Journal of Mechanical Engineering, **40(9)**, pp106-110

[8] Ling Feng, Gu Linyi, 2003, "Compounded Switchmode Hydraulic Power Supply," Proceedings of the 4th International Symposium on Fluid Power Transmission and Control, Wuhan, China, pp591-594

[9] Zhang Xiaofeng, Gu Linyi, 2003, "Pressure Buck Switchmode Hydraulic Power Supply with Return Pressure," Proceedings of the 4th International Symposium on Fluid Power Transmission and Control, Wuhan, China, pp583-586

[10] Achten P A J, Fu Zhao, Georges EM, et al., 1997, "Transforming future hydraulics: a new design of a hydraulic transformer," Proceedings of the 5th International Conference on Fluid Power, SICFP'97, Linkoping, Sweden, 1997

[11] Achten P A J, Titus van den Brink, Johan van den Oever, 2002, "Dedicated design of the hydraulic transformer," Proceedings of 3rd International Fluid Power Conference. Aachen, Germany, 2002

[12] Herbert H K, Camarillo C. "Hydraulic Transformer," US Patent 3627451, 1971

Proceedings of IMECE2005
2005 ASME International Mechanical Engineering Congress and Exposition
November 5-11, 2005, Orlando, Florida USA

IMECE2005-79329

MODEL SIMPLIFICATION FOR NONLINEAR HYDRAULIC CIRCUITS

M. Alirand, IMAGINE SA, Roanne, France

N. Orand, IMAGINE Inc., Southfield, MI, USA

M. Lebrun, Université Claude Bernard, Lyon, France

ABSTRACT

The aim of this paper is to have a view of different levels of model simplifications for a simple hydraulic circuit. The bond graph is used to synthesize the simplifications and the different assumptions behind. Different models were built in order to keep continuity in model simplification. These models have the two causalities allowed for a particular physical phenomenon. It is believed that software will give access to different levels of model simplifications and the user will choose between these different levels the proper one that will answer to his question. However this proper model will only be achieved step by step using continuity in model simplification. The paper also shows that systems may include very different levels of power (pilot and main stages or thermal effects) and in such a case at least one C-, I- or R-element should be kept per power level.

INTRODUCTION

High fidelity physical models are generally necessary to represent complete systems like an automatic gearbox, injection circuits, mobile and industrial hydraulic circuits, cooling systems and this can only be achieved by detailed analysis. It is a truism to say that the process of synthesizing a numerical model consists of abstracting one or more aspects of real world phenomena into an ideal physical model. To do so, and what ever the physical domains are, engineers must represent the system behaviour using discrete inertial, dissipative and compliant elements. One can recognize here the elements introduced by the bond graph technique. As well, a multi-energy domain system (hydraulic, mechanical, thermal and electro-mechanical domains like in most automotive and aerospace applications) usually contains a significant number of components which can be represented with very complex models using software available today. As a consequence the synthesis process is a knowledge intensive and labour-intensive process.

Mathematical models are generally expressed in the form of state systems and are used to perform tasks such as, for example, sensitivity analysis like for manufacturing variability influences, design optimisation, controller design or simply to gain insight into the dynamic behaviour of the system. It is clear that optimal design requires a high number of system simulations. Since the considered software includes such features, it was essential to allow reduction of models to a more manageable size. Effective design of complicated systems requires simple models. As a consequence, the most difficult aspect of synthesizing a model is to figure how many physical effects (inertial, compliant and dissipative elements) are needed to adequately represent the real-world phenomena under study. This leads to the notion of the necessary and sufficient complexity for the final goal. This aspect had led some authors to the definition of proper models [1]. Proper models are defined as minimum complexity models with physically meaningful states and parameters (geometric information and material constants) required for fitting the modelling objectives. Proper models contain the minimum information needed to show the relationship between the design parameters and the dominant system dynamics and preserve the physical meanings of their structure and parameters. The physical meaning is essential from the engineering point of view.

As mentioned in [2] the methods to obtain simple (or proper) models are essentially divided in three main techniques: 1. the method based on the mathematical manipulation of the system's input-output relations, 2. the methods based on physical interpretation of the system's dynamic behaviour and 3. the combination of 1 and 2 [3]. The first category is essentially developed by the control community and uses modal truncation or balanced truncation [4, 5, 6, 7]. The transfer function H(s) and its reduced counterpart $\widetilde{H}(s)$ are used as a metric for approximation: if $\left\|\widetilde{H}(s)-H(s)\right\|<\varepsilon$ for some given allowable error e and allowed frequency domain, the reduced model is accepted as accurate. Since these methods are mainly based on transfer functions they are mostly used for "linear" systems and in general the physical sense of the state variables or the parameters is lost However for multi-domain systems including Hydraulic for which the dynamics can vary enormously during the simulation, these techniques are not well suited.

The second category uses procedures involving power exchanges between parts. As previously mentioned, a metric should be used as well as a frequency range of interest. This technique was first introduced by Rosenberg [8] and adapted and improved by Louca [1] by introducing something known as the activity.

$$A = \int |P(t)| dt$$

where P(t) is the power. The idea behind the absolute value is to store whatever the exchange is. Think about a spring, the activity stores during the extension and re-stores during the compression. The activity index is just the activity of an element divided by the total amount of activity in the system. The underlying intuition of this power criteria is the consideration that components associated with low power makes low contribution to the system dynamic behaviour. Activity index of inertial, dissipative and compliant elements is a standard feature of the software used.

Pressure components are always used in hydraulic circuits and thus have been widely analysed for their stability [9]. They can be viewed as multi domain components since they involve hydraulic and mechanical and sometimes electro mechanical devices. Merritt [10] has first presented the analysis of pressure components with a generic approach. The author has given some complementary analyses of the stability and the dynamic behaviour of pressure components [9]. As well, some model simplifications on such a component have been introduced in [11]. The idea here is to reuse the results of [11] but viewed with another approach highlighting the continuity in model simplifications and their different theoretical contributions. In the first part of the paper, the complex model of a pressure relief valve will be introduced. It will be used as a base line of the simplifications. From this model different levels of model reduction will be introduced and this step by step in order to properly understand the assumption and contribution of each. It is possible from the AMESim point of view to keep almost the same original sketch for most of the simplifications, however the final step is necessary a complete manipulation of the equations to obtain a reduced model. The second part is a discussion on model reduction and the interest for making automatic the simplification process.

MODEL SIMPLIFICATION TECHNIQUES APPLIED TO PRESSURE COMPONENTS

In [11], a complete validated gearbox system has been presented and simplified in a way to obtain a very simple model aiming for real-time application. A pressure relief valve model has been used to synthesize the different simplifications for the hydraulic part. The goal here is to reuse the same simplification techniques as in [11] but viewed using bond graphs and a more theoretical point of view. The model of the pressure relief valve coming from [11] is recalled in Figure 1 with its bond graph.

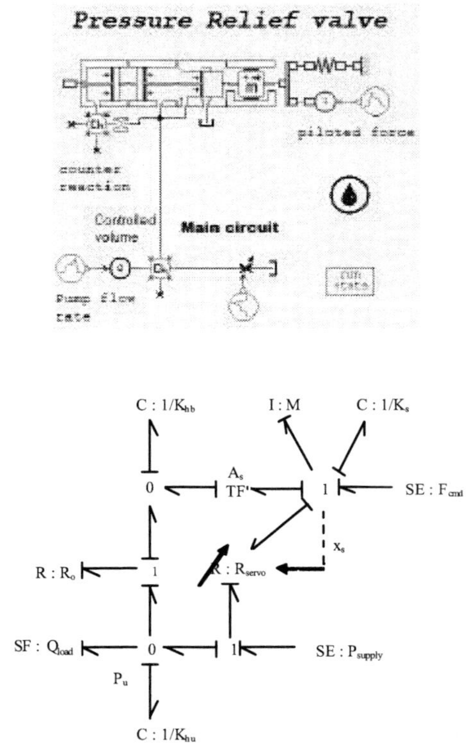

Figure 1. Pressure relief valve and its bond graph

For simplicity, the jet forces have been represented using a variable R-element. However in the linear domain the jet forces are represented by a combination of a C-element and a transformer [9]. Since the structure remains identical from the linear to the nonlinear domain, jet forces are clearly not R-elements. They are sometime represented as such for simplicity (a force function of a flow rate).

Since activity index mainly depend on the inputs, it is thus important to show their effects on the system. The model gives the response shown in Figure 2 for variation of supply flow rate (time = 0.15 s) and variation of force input (time = 0.1 s).

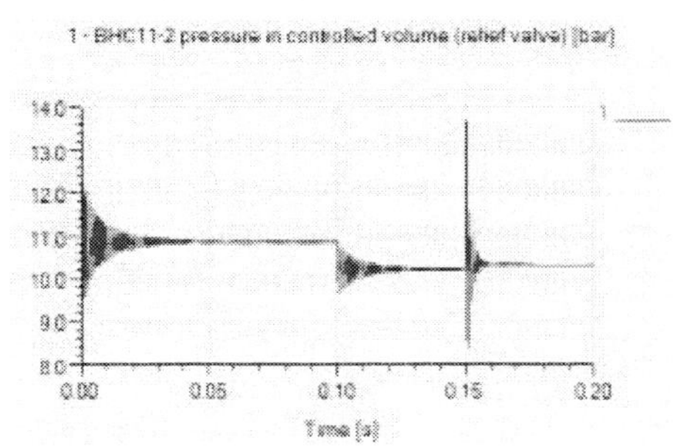

Figure 2. Regulated pressure for the two inputs

In order to simplify a model, the engineer must ask to the question: can some physical effects be omitted without introducing excessive error? In order to simplify the models the idea is to look at the activity index of each element involved in the model.

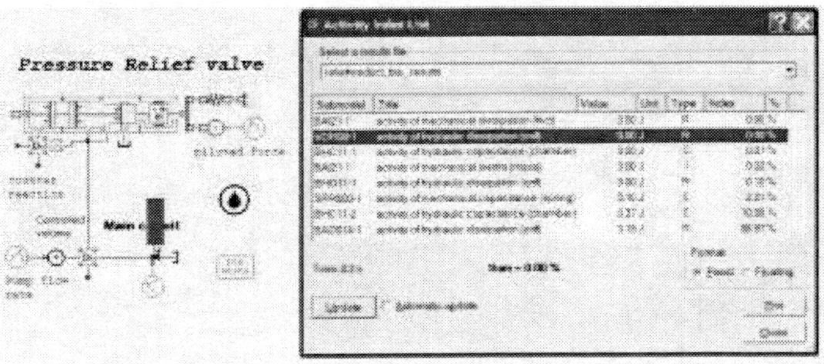

Figure 3. Activity index of each element

The orifice which corresponds to the variable load in the main circuit of the pressure relief valve model can be omitted as it remains closed during the entire period. The second contribution is the viscous friction applied on the spool valve which is also zero. The next element corresponds to the hydraulic chamber considered as the counter reaction chamber. The suppression of this component will constitute the first level of simplification. The mass of the spool appears after and thus should be eliminated as the second level of simplification. The next contribution corresponds to the fixed orifice that "feedback" the pressure information and will lead to the third level of simplification. The next step of simplification should involve the spring stiffness. However this should be analysed with the contribution of the hydraulic stiffness of the controlled volume. The two different possibilities for model simplifications have been introduced and discussed later.

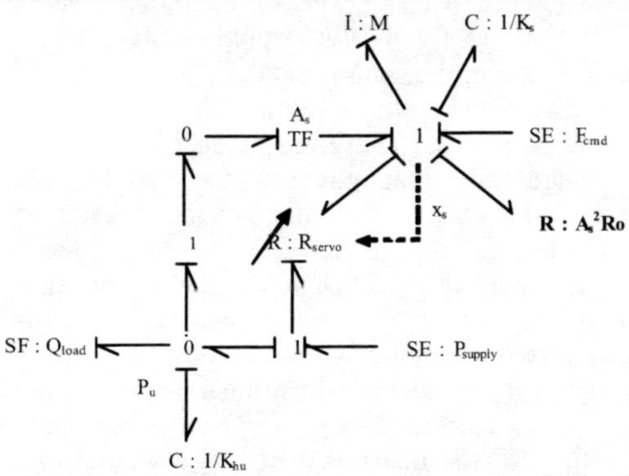

Figure 4. Bond graph of the first model reduction

As previously done in [9], the first step in a way to simplify the model is to remove the counter reaction chamber. In [9], the fixed orifice has been brought into the mechanical domain and as such, the orifice becomes a viscous damper. Attention should be brought of the fact that the flow regime is laminar most of the time and not turbulent, the equations being clearly different. The first level of simplification can be the one shown in Figure 4. As can be seen the junction structure has not been simplified in order to correctly see the simplification and the term which has changed is highlighted.

From the activity index of Figure 3, the next simplification will see the suppression of the spool mass. Since it will be removed, another element should give access to the velocity of the spool. To do so, the force calculation in the viscous damper is reversed. Instead of computing the force from the velocity, the velocity is computed from the total force (Figure 5 in blue).

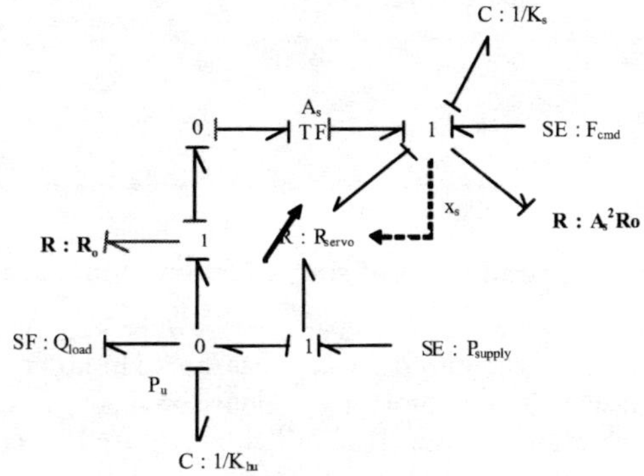

Figure 5. The two second levels of model simplification

Even if this simplification seems attractive it is possible to keep the orifice model since the flow regime can have an influence. In non linear domain such as Hydraulic it may become important to keep the non linearities in order to properly fit the original model. In [9], for low frequency it has been proven that the flow rate crossing the fixed orifice is directly linked to the spool velocity multiplied by the spool area. Instead of bringing the orifice into the mechanical domain as for the first level of simplification, it is possible to use the flow rate inside the fixed orifice and divided it by the spool area (Figure 5 in red). In this case and from a bond graph point of view the transformer of the hydraulic domain to the mechanical domain has a reversed causality. Instead of using the spool velocity to compute an induced flow rate almost similar to the flow rate crossing the fixed orifice, this flow rate is used to compute the spool velocity. From a bond graph stand point the two R elements corresponding to the orifice and the equivalent viscous damper are in the same causality: the flux is computed from an effort. That is why these two levels of simplification can be considered as identical. However the model using the Bernoulli equation is logically the better one since laminar and turbulent regimes are taken into account.

From Figure 3, the next step is to remove the fixed orifice effects. Here another AMESim feature can be used: the state count. The state count shows the effort done by the integrator to solve a particular state variable for the given tolerance. It is expected that if the solver has some difficulties on a particular state variable this means that it normally corresponds to a "high" frequency variable. It is also a way to get some information about model reduction if only low frequency effect have to be kept. Actually both the activity index and the state count should enable the user to simplify a model, however the state count only have access to state variables and thus power storage elements but no access to dissipative elements. From

Figure 6, the state variable that leads to difficulties seems to be the pressure (state variable of the chamber) and not the displacement (state variable of the spring).

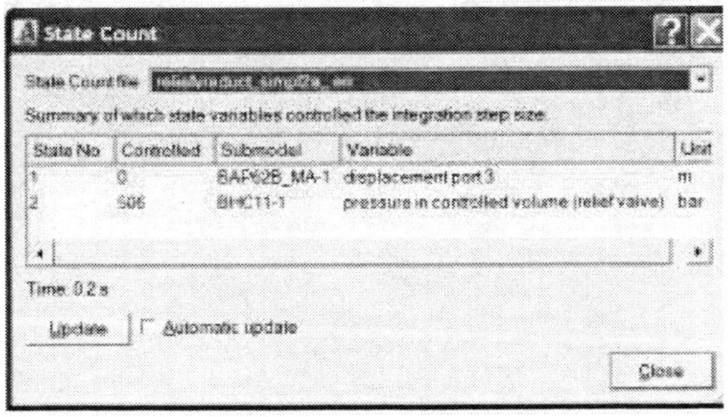

Figure 6. Second levels of simplification – State count

The issue is to know which C-element has to be kept or if it is possible to bring one into derivative causality? In order to well understand what will be done, the bond graph theory is of great help in such a case. If the fixed orifice is removed an element should give access to the spool motion (at least displacement). That is why in the bond graph Figure 7 the C-element of the spring is in derivative causality. It is always preferable to reverse a constant stiffness than a variable stiffness (hydraulic compressibility is clearly variable) to calculate a displacement.

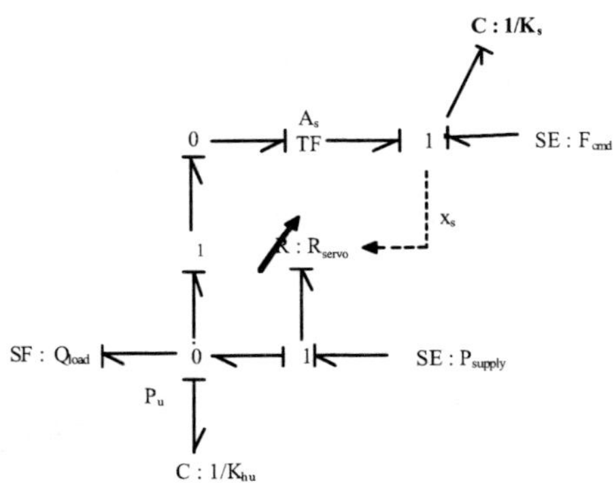

Figure 7. Bond graph of the third level of simplification

Since the bond graph of the new system includes a C-element in derivative causality the model naturally includes implicit loops. This model is difficult to manage since he implicit variable prevents this model to be used in a real time context.

The logical last simplification will thus use an equivalent spring of the two (mechanical and hydraulic) to remove the implicit loop. The main problem is to know which stiffness should be kept. Either the spring is taken into the hydraulic domain or the chamber is brought into the mechanical domain (see Figure 8).

From the activity index Figure 3, the spring stiffness should be removed and from the state count Figure 6 the chamber stiffness should be removed. A quick look at the two stiffness reveals that the spring is softer than the hydraulic chamber. This was expected from an engineering point of view since in most cases the pilot stage contains a soft spring compared to the hydraulic one. The hydraulic chamber should be gathered with the spring but the hydraulic chamber is so stiff that the equivalent stiffness is the only one of the spring. This aspect was given by the state count feature but not by the activity index feature.

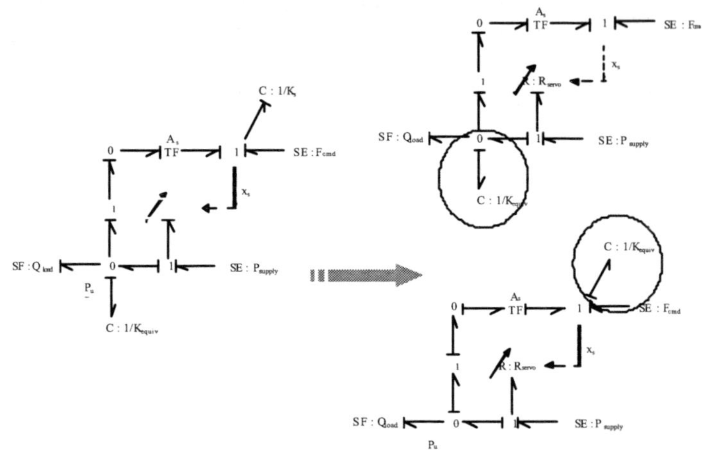

Figure 8. Bond graph of the fourth level of simplification

An interesting aspect can be noticed in the bond graph Figure 8. If the hydraulic chamber is kept there is no mean to obtain the velocity of the spool. Actually there is no more C- nor I- nor R-elements connected to the spool. This will also give the engineer information on the reliability of this new model simplification. Nevertheless, a new model was built that does not take into account the controlled volume [11]. In this particular case the model runs but gives wrong time responses. The flow rate going inside the piston model (the transformer), leading to spool velocity and thus spool displacement does not seem to reflect the proper spool displacement. There is a clear difference between the main circuit and the pilot stage from the point of view of power. Therefore, even a small error in the flow rate calculation that goes to the spool leads to an enormous mistake in the spool displacement resulting in the wrong flow rate crossing the valve. This type of "loop" makes fail the calculation. The aspect of power level between the main and pilot stages is of great importance. The activity index facility may give a wrong answer if the powers of two different "circuits" are very different. Unfortunately hydraulic piloted components have necessarily very different power levels. It is also interesting to note that this aspect is also true for thermal hydraulic applications. The power range into thermal domain is often very large compared to the power into the fluid domain (as well as their dynamics). As a consequence, both the two parts of the circuit (pilot and main stages) should be simplified independently: at least one power element should remain for each of the two parts in order to give access to the right behaviour. This conclusion can also be drawn from the bond graphs Figure 8 since one is none compatible with a solution.

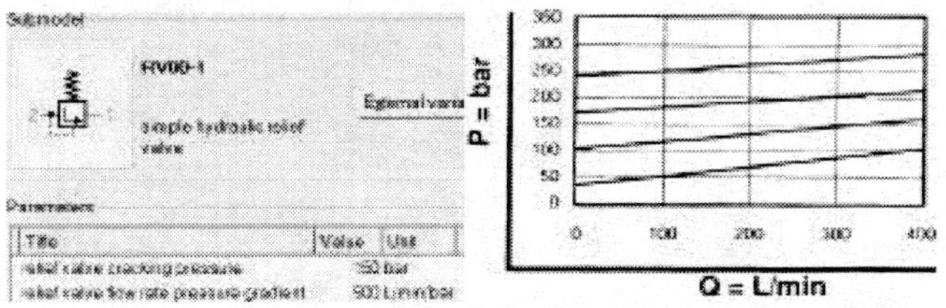

Figure 9. Pressure relief parameters

An interesting aspect now is the link between all these simplifications and the simplest model included in the hydraulic library. One should note that AMESim includes two different hydraulic libraries: one is dedicated to hydraulic component design (HCD library [12]) and the other is dedicated to circuits (Hydraulic library). The parameters of this hydraulic model are shown Figure 9 with pressure-flow rate characteristics given by a pressure relief valve manufacturer. This also enables the understanding of the required parameters.

The flow rate-pressure gradient gives access to an R-element since it is a relation between the flow rate and the pressure. In order to obtain an equivalent model of Figure 1, the AMESim simpler model should be connected to the controlled volume chamber making a RC system. Since this very simple model is generally connected in place of all the HCD icons, the model is thus almost similar to the third and fourth levels of simplification. However, further simplifications were achieved and the bond graph of the simpler model shown in Figure 10 illustrates them.

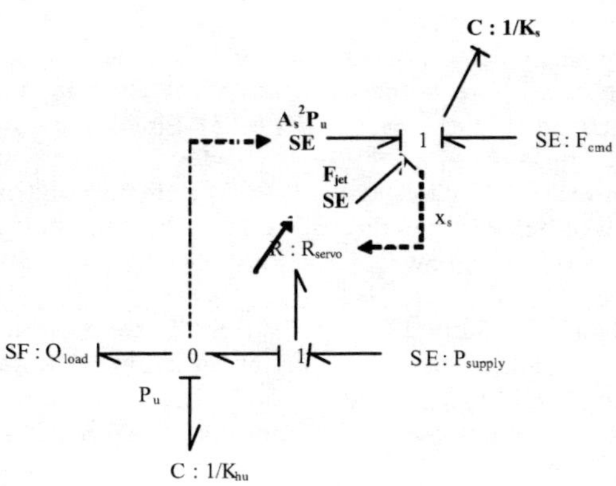

Figure 10. Bond graph of the simpler model

For the simpler model of Figure 10, there are no energy couplings between the main circuit and the pilot stage. This should be link to [13] where the junction structure has been simplified. In this particular case the induced flow rate due to motion of the pilot stage has no influence on the main stage and thus the procedure in [13] should apply. The pressure feedback is realized by using a source of force. The slope specified in the simpler model is linked to the spring and the jet forces effects [9]. The position of the spool can be calculated by the equation

$$x_s = \frac{A_s P_u - F_{cmd}}{K_s + 2C_q \pi D_s P_u \cos(\theta_{jet})}$$

and may use the Bernoulli equation. However it is even simpler than this since the flow rate is directly calculated from the slope which is viewed as a gain. There is no more Bernoulli equation. The previous non linear equation is combined with the Bernoulli equation in order to calculate the flow rate function of the pressure

$$Q = C_q \pi D_s \frac{A_s P_u - F_{cmd}}{K_s + 2C_q \pi D_s P_u \cos(\theta_{jet})} \sqrt{\frac{2}{\rho} P_u}$$

and then simplified by linear approximations. Such a simplification can only be achieved by symbolical manipulation of the previous equation. This has nothing to do with power exchange and is thus not related to bond graphs even if it can be used to give a view.

To conclude this part, one can say that it is a truism to say that the best model is always a compromised between the accuracy and the CPU time efficiency. The model simplification should reach this best level. From this point of view, [14] gives a possible idea of this best level. It was previously stated that at least one power element should remain per power level. For very simple models, it is usual that the power element of the low power part is algebraically reduced (and often linearised) in a way to remove the power exchange with the high power part, the dominant dynamics being into the high power part. This stays as the main assumption for the activity index use.

DISCUSSION ON MODEL SIMPLIFICATION

It is clear that competition enforces a reduction in design cycle time. Thus, the cost and time associated with building and testing prototypes are less and less tolerated. The clear alternative is today the use of modelling and computer simulation to evaluate design decisions. Current complaints of engineers towards modelling and simulation tools are first, that software require sophisticated users and secondly that models are often large and complex making interpretation as difficult as the real system. The first problem comes from generally speaking, novice engineers who are often not multi domain experts and thus lack the ability to fully utilize the model as well as the simulation tool. The second problem can be viewed as a consequence of the first or may come from the fact that it is easy to make complex a system when everything is not well understood and this unfortunately even for expert engineers. Modern software packages allow the building of "error free" models relatively quickly, but provide almost no guidance in selecting the appropriate physics to be included. As a result, engineers usually do not know what is important and so include everything. When the behaviour of a system is well understood it becomes easier to eliminate some parts that are not necessary to represent the suited objectives. It is the necessary way to synthesize a system and gain insight. It is difficult to believe that the proper model will be reach at the first try. However the authors believe that the development of a semi automatic procedure for model simplification is essential for the engineer. In the design process, it is generally necessary to have a complex validated model to build the initial confidence in simulation technique before proceeding to its reduction, the next step being the simplification in order to reach a proper model. Behind these aspects of simplifications, the idea of the determination of the major design parameters and their relationship to product performance is essential. In this process, simplifications may be viewed as a way to understand the nature of the system that is to gain system insight [11]. When using an automated modelling tool with symbolical manipulations, models are generated automatically with potentially specific parameters and system structure being added or eliminated from the model. This can pose a comprehensive problem for the user who has no control over the decisions made by the computer. The purpose of the tool is also to enhance the modeller's ability to understand the system. A model is usually a collection of interconnected ideal elements that represent the ideal dynamics of the system. The development of an appropriate model requires engineering intuition, judgement and experience that are gained through extensive exposure to system modelling. This is the reason the authors believe that the engineer should be kept as the master of decisions in the model simplifications.

CONCLUSION

As a conclusion there are a lot of possibilities to build a simplified model and software should allow continuity in model simplifications. This means that software should give access to all causalities for all physical effects. Care should be taken about storage elements since they lead to implicit loops if they are not reduced (i.e. gathered with others [15]). This is relatively important from the engineering point of view since the engineer should keep in mind the different assumptions behind all the simplification levels. The activity index as well as the state count facility is of great help to simplify a model since the first gives an aspect of power exchanges and the other the difficulty the integrator has to satisfy its error criterion. The bond graph technique enables the user to see which assumption is kept and which one is removed and is a very good way to synthesize the different levels of model simplifications but difficult to use in an industrial way.

ACKNOWLEDGMENTS

We would like to acknowledge the French Ministère de la Recherche for supporting this work.

REFERENCES

[1] Louca L.S., Stein J.L., Hulbert G.M., Spragme J., "Proper model generation: an energy-based methodology", 3rd Int. Conference on Bond Graph Modelling and Simulation, Phoenix, AZ, January 1997, pp 1-6

[2] Ye Y., Youcef-Toumi K., "Modeling in the physical domain: an optimization-based approach", 2002 ASME Int. Mechanical Engineering Congress & Exposition, New Orleans, Louisiana, November 2002, pp 1-8

[3] Ripert F.J., Chevrel P., Lefebvre D., "Physical simplification of MIMO linear models from elementary components manipulations", 2nd IFAC Symposium on System, Structure and Control, Oaxaca, Mexico, December 2004, pp 1-6

[4] Dauphin-Tanguy G., Borne P., Lebrun M., "Order reduction of multi-time scale systems using bond graphs, the reciprocal system and the singular perturbation method", Journal of Franklin Institute, vol 319, n°1/2, January/February 1985, pp 157-171

[5] Liu Z., Wagner J., "Non linear model reduction for dynamic and automotive system descriptions", Trans. ASME, Journal of Dynamic Systems, Measurements and Control, vol. 124, December 2002, pp 637-647

[6] Moore B.C., "Principal component analysis in linear systems: controllability, observability and model reduction", IEEE Trans. On Automatic Control, vol. AC-26, n° 1, February 1981, pp 17-32

[7] Varga A., "Numerical methods and software tools for model reduction", Proc. of the 1st MATHMOD Conf., Wien, Vol 2, 1994, pp 226-230

[8] Rosenberg R.C., Zhou T., "Power-based model insight", ASME Winter Annual Meeting, Chicago, IL, DSC-Vol. 8, December 1988, pp 61-67

[9] Alirand M., Favennec G., Lebrun M., "Pressure component stability analysis: A revisited approach", Int. Journal of Fluid Power, Vol 3, n°1, April 2002

[10] Merritt H.E., "Hydraulic Control System", John Wiley & Sons Publisher, 1967

[11] Alirand M., Jansson A., Debs W., "A first idea of continuity in model simplification for hydraulic circuit – An automatic gearbox real-time application", 9th Scandinavian Int. Conference on Fluid Power, SICFP'05, Linköping, Sweden, June 2005, pp 1-16

[12] Lebrun M., Richards C.W, "How to create good models without writing a single line of code", 5th Scandinavian Int. Conference on Fluid Power, Linköping, Sweden, May 1997, pp 19-32

[13] Rideout G., Stein J.L., Louca L.S, "System partitioning and physical-domain model reduction through assessment of bond graph", IMAACA 2004, Bond Graph Techniques for Modelling Dynamics Systems, Genoa, Italy, April 2004, pp 1-20

[14] Sendur P., Stein J.L., Peng H., Louca L.S., "An algorithm for the selection of physical system model order based on desired state accuracy and computational efficiency", 2003 ASME Int. Mechanical Engineering Congress & Exposition, Washington D.C., November 2003, pp 1-12

[15] Louca L.S., Stein J.L., "Physical interpretation of reduced bond graphs", 2nd IMACS Int. Multiconference: CESA'98, Hammamet, Tunisia, April 1998, pp 1-6

Proceedings of IMECE2005
2005 ASME International Mechanical Engineering Congress and Exposition
November 5-11, 2005, Orlando, Florida USA

IMECE2005-81687

SELECTION OF OPERATING MODES OF A MULTI-FUNCTIONAL HYDRAULIC DEVICE

Amir Shenouda
Intelligent Machine Dynamics Laboratory
Fluid Power Motion Control Center
Department of Mechanical Engineering
Georgia Institute of Technology
Atlanta, Georgia 30332
Email: gtg531p@mail.gatech.edu

Wayne J. Book*
Intelligent Machine Dynamics Laboratory
Fluid Power Motion Control Center
Department of Mechanical Engineering
Georgia Institute of Technology
Atlanta, Georgia 30332
Email: wayne.book@me.gatech.edu

ABSTRACT

The field of earth moving equipment is experiencing a transformation due to the introduction of more electronic control capability and advanced control concepts. Conventional hydraulic control systems are controlled by proportional directional spool valve. The construction of the spool valve is such that a given position of the spool determines the flow in and the flow out restriction sizes. Thus, metering in and metering out are dependent or coupled. A certain restriction size on the inlet corresponds to a certain restriction size on the outlet. Therefore, we have one degree of freedom. It can provide for good motion control but it cannot achieve energy saving potential at the same time.

In this paper, the concept of 'independent meter in / meter out' will be emphasized. Decoupling of meter in from meter out provides for more controllability and potential for energy saving in overrunning load cases when compared with a conventional spool valve controlled hydraulic system. A four-valve configuration controlling a hydraulic cylinder will be stressed. The four-valve configuration can operate in several modes because each of the four valves is controlled separately from the others. Five of these metering modes will be pointed out; two conventional modes and three regenerative modes.

An industry system supplier has introduced a novel distributed independent metering in/meter out hydraulic system with actuator force feedback. This control system is going to presented in brief. The distributed, actuator mounted implementation of this system maximizes the efficiency of the regeneration flow by minimizing the line losses associated with the recirculating fluid

This control system can be used to control a multi-functional hydraulic device, e.g. excavators, tractor loader backhoes, etc...An experiment was done where a typical dig cycle was performed by a tractor loader backhoe equipped with this new technology. The data recorded from this experiment is used to perform an off-line optimization of metering modes selection for each function during the cycle.

this off-line optimization can be used to maximize machine productivity or machine efficiency or a weighted combination of both. A general objective function will be presented and results will be given for maximizing machine productivity. Finally conclusions and future work suggestions are presented.

INTRODUCTION

Machines and robots can have movable members that can be operated by hydraulic actuators like a piston and cylinder arrangement that is controlled by a hydraulic valve. Hydraulic systems are used in many industrial applications because of their high power and force to weight ratio. Conventionally, in earth moving equipment for instance, the valve was located close to the operator cabin and the valves were operated manually to pro-

*Address all correspondence to this author.

vide flow to the cylinders. However, the trend is to move away from the manual operation tradition and use electronics for control purposes. This makes the hydraulic plumbing much easier and provides the chance for implementing modern control algorithms that are realized by an embedded microprocessor. Hydraulic system control is notorious for being difficult due to the many nonlinearities involved and the time varying characteristics that occur with temperature changes and wear, which makes microprocessor control capabilities even more valuable.

The construction of the four way spool valve is such that a given position of the spool determines the 'flow in' **and** the 'flow out' restriction sizes. Thus, metering-in and metering-out are coupled or dependent. A known restriction size on the inlet corresponds to a known restriction size on the outlet. Therefore, we have one degree of freedom, and we can control the pressure in just one chamber in this case. Thus, it can provide for good motion control but it cannot achieve energy saving potential at the same time [1]. Bu and Yao showed that this one dimensional system is stable [2]. However, once the desired motion of the actuator is determined, the control input is uniquely determined. That means that it cannot provide precise motion control while controlling pressure in both actuator chambers. This further means that its potential for energy savings is low. Besides, different applications require different four way spool valve designs to accommodate the different needs of the various systems. The current trend in industry is to design a valve that can be adapted to the various systems in practice [3].

Independent metering promises to achieve higher system efficiency. Research has been done to achieve independent metering and thus to be able to have an extra control dimension. This has the potential of achieving precise motion control while controlling pressures in both actuator chambers for energy savings. The idea has been used in industry for several years. Aardema suggest using two conventional four way directional valves: one to control meter-in and the other to control meter-out [4]. But this is an expensive solution since it replaces one valve by two, and it does not achieve complete independence of meter-in and meter-out. Other more promising solutions replace the typical four way servo valve with four poppet type valves that can be controlled separately. Patents by Deere & Company [5], Caterpillar Inc [6], and Moog Inc [7] demonstrate the concept with minor variations. The "Smart Valve" [8] is another demonstration of the concept of a programmable valve that achieves independent metering control.

The Electro Hydraulic Poppet Valve (EHPV) patented by HUSCO International is also an example of this type of valve [9–14]. The control system for using EHPVs in a hydraulic earth moving equipment was introduced by Tabor and Pfaff [15–18]. This kind of valve can be adapted to the different hydraulic applications in the market with no hardware changes, which makes them "economically attractive" [19].

The new independent metering concept replaces the four

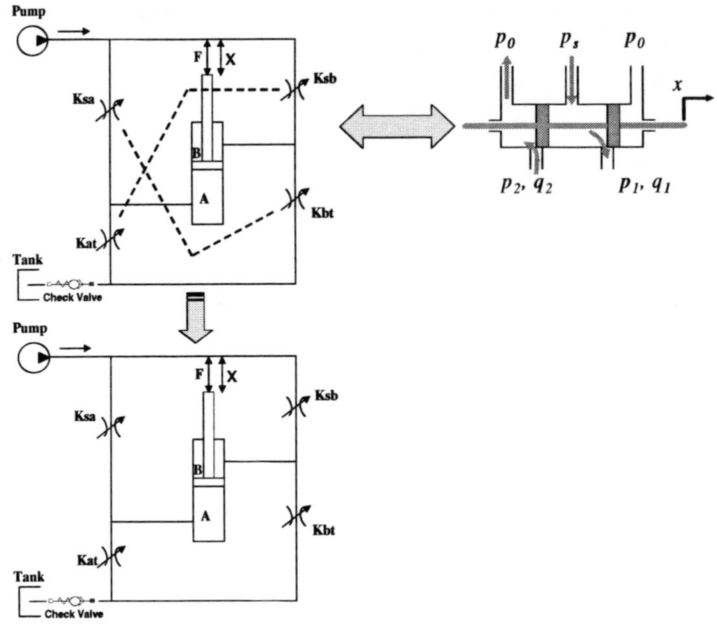

Figure 1. EVOLUTION FROM SPOOL VALVES TO INDEPENDENT METERING VALVE CONFIGURATION

way spool valve with four valves in one package. Through software changes, this package can be adapted to several hydraulic system applications [15, 20]. Four valves are needed to operate one cylinder. One valve connects one of the work ports of the cylinder to the supply line that contains pressurized oil. A second valve connects the other work port to the supply line. A third valve connects the first port to return line which leads to the tank. Finally a fourth valve is needed to connect the second port to the return line. Using this four valve configuration, while only two valves will be active for a given condition, allows for independent metering and also allows for energy regeneration [5, 9, 14–16], which has the potential of reducing the requirement on the pump and thus achieve highly valuable energy saving. **Regeneration flow can be defined as pumping the fluid from one chamber to the other to achieve motion control of the load with using no or minimum flow from the pump**. This regeneration flow concept is almost impossible to achieve using a regular directional control valve [19]. Fig. 1 shows the transformation from spool valves to independent metering valve configuration. The spool valve can be represented by four restriction that are not independent but mechanically coupled. The independent metering configuration breaks this mechanical linkage and uncouples the restrictions.

The four-valve configuration can operate in several modes because each of the four valves is controlled separately from the others. Five of these metering modes will be pointed out in the following section.

Hydraulic cylinder cavitation happens when the load is moving faster than the fluid is filling the inlet chamber and a zero or negative pressure appears. This is a problem because control laws do not apply in this case and controllability of the operation of the cylinder is extremely deteriorated. For example, in the case of overrunning load, where the load is being lowered with gravity assistant, losing controllability may lead to the load falling uncontrollably. Therefore, hydraulic systems have to take precautions against cavitation. In a case of overrunning load, spool valves are designed such that the outlet restriction is used to control the flow so as to prevent the load from falling at uncontrollable speeds.

METERING MODES

The bottom schematic in fig. 1 shows the concept of using a four-valve configuration to achieve independent metering. There are five metering modes that describe the operation of a hydraulic cylinder controlled by such a four-valve assembly. A description of the operation of these modes is presented next.

1. **Powered Extension Mode**: The operation of this mode is not significantly different from the operation of a cylinder controlled by a conventional proportional directional control valve. Flow is supplied at high pressure from the pump through valve K_{sa} into head chamber A of the cylinder causing the piston to move (extend). This piston motion forces the flow out of rod chamber B through valve K_{bt} to the tank. The difference, however, is that in this four-valve configuration K_{sa} and K_{bt} can be controlled separately, while in a conventional proportional valve they cannot be controlled independently as explained in the introduction.
2. **Powered Retraction Mode**: In this case, flow is supplied at high pressure from the pump through valve K_{sb} into rod chamber B of the cylinder causing the piston to move (retract). This piston motion forces the flow out of head chamber A through valve K_{at} to the tank.
3. **High Side Regeneration Extension Mode**: This is one of the regenerative modes that can be achieved by the four-valve configuration and could not be achieved by a conventional proportional directional control valve. Flow coming out of rod chamber B at high pressure does not go to tank through valve K_{bt}, but it circulates through valve K_{sb} and K_{sa} into head chamber A causing the piston to move (Extend). However, the flow coming out of chamber B is less than the flow needed in chamber A to achieve a certain speed because of the difference in areas ($A_a > A_b$). Flow out of either chamber is $Q = A \times \dot{x}$ and for a given velocity and with the difference in area ($Q_a > Q_b$). The remaining flow is supplied from the pump, but it is just enough to make up of the difference in flow while in powered extension mode all the flow is supplied from the pump. Thus, the high side regeneration extension mode has the potential to save energy.
4. **Low Side Regeneration Extension Mode**: Flow coming out of rod chamber B does not go to tank but it circulates through valve K_{bt} and K_{at} into head chamber A causing the piston to move (extend). However, the flow coming out of chamber B is less than the flow needed in chamber A to achieve a particular speed as explained above. The remaining flow can be supplied from the pump, but it is just enough to make up of the difference in flow and it is supplied at a low pressure.
5. **Low Side Regeneration Retraction Mode**: Flow coming out of head chamber A does not go to tank but it circulates through valve K_{at} and K_{bt} into rod chamber B causing the piston to move (retract). The flow coming out of chamber A is more than the flow needed in chamber B to achieve some speed as explained above. The extra flow goes to tank through the check valve. Thus, no pump flow is needed at all for this mode, which means a high potential for saving energy whenever this mode is used.

DISTRIBUTED VALVE ASSEMBLY CONFIGURATION

This section explains why independent metering using four valves is appealing and promises much more energy savings than the conventional spool valve technology.

Spool valves, a valve per cylinder, are stacked together and put under the operator cabin. Long pipes extend from that valve stack to the various cylinders. That constitute major line losses as flow has to go all the way to the cylinder and back to through the valve to tank. It is obvious from the analysis above that regenerative flow uses minimum pump flow if any. Moreover, a four-valve assembly controlling one cylinder can be mounted directly on the cylinder it is controlling. Figure 2 shows the whole backhoe mechanism with the valve assemblies mounted directly on the cylinders. If any of these valve assemblies is operating in the regenerative modes, flow would be circulating within the assembly and does not have to go through long pipes and thus must save energy that would have been otherwise lost as hydraulic losses in the pipes.

CONTROL SYSTEM

A schematic diagram of the hydraulic system used in this research is shown in figure 3 [15–18]. As shown in figure 3 a hydraulic actuator (cylinder) is connected between a supply pressure line to the pump, and a return line to the tank through four Electro-Hydraulic Poppet Valves (EHPVs), that meter the flow of oil into and out of the cylinder. These four EHPVs control each cylinder and hence each function. In a typical construction or agricultural equipment, there are multiple functions that need to be controlled simultaneously. Each function is controlled by a

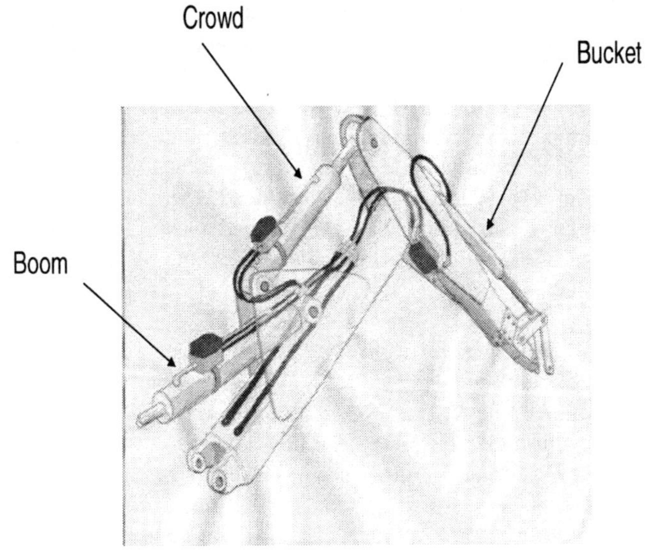

Figure 2. A BACKHOE MECHANICSM SHOWING THE FOURV-VALVE ASSEMBLY MOUNTED DIRECTLY ON THE CLINDERS

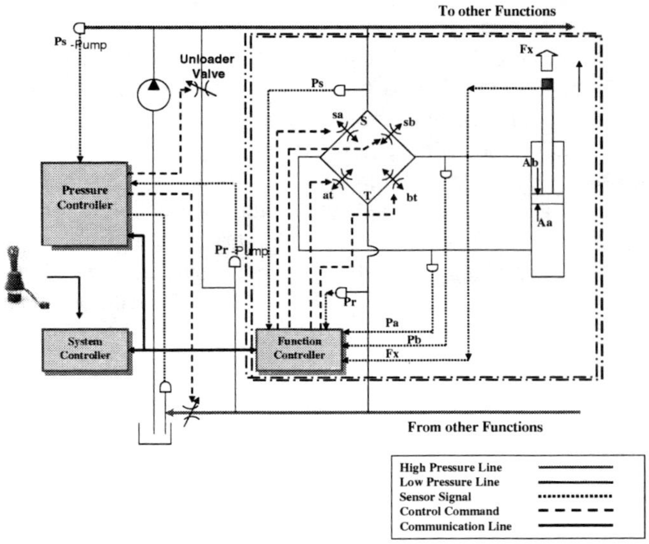

Figure 3. HYDRAULIC CONTROL SYSTEM [15–18]

set of four EHPVs. Only one cylinder and one set of valves are shown in figure 3. The pump shown in the figure is a fixed displacement pump, which requires an unloading valve to control the supply pressure. In this case the *Pressure Controller* controls the opening of the unloading valve and hence the supply line pressure. A variable displacement pump can be used instead and in this case the *Pressure Controller* directly controls the pump to supply the required set point pressure as will be explained later in this chapter.

As explained in section on metering modes, selected pairs of valves enable the operation of the hydraulic actuator in different metering modes. For example, in powered extension mode, valves 'sa' and 'bt' will be opened. In powered retraction mode, valves 'sb' and 'at' will be opened. In high side regeneration extension mode, valves 'sa' and 'sb' will be opened. In low side regeneration extension mode and low side regeneration retraction mode, valves 'at' and 'bt' will be opened ('sa' will also be opened in low side regeneration extension mode for makeup flow). Thus, two of the four valves are opened depending on the metering mode. The degree to which each valve is opened determine the valve conductance coefficient K. This coefficient is determined based on specific priorities [15–18].

The control system is a microprocessor controlled system that coordinates between the different blocks shown above. The ***System Controller*** organizes the whole operation of the system and communicates with the ***Function Controller*** and ***Pressure Controller*** blocks. The messages exchanged between these three blocks are communicated through a CAN bus protocol.

Six pressure sensors are shown in figure 3. Two pressure sensors measure the pressure in both the head and rod chambers. A third sensor measures the supply pressure right before the four-valve assembly. The fourth sensor measures the pressure right after the the four-valve assembly before the flow goes into the return line. These four sensors are input to the ***Function Controller***, which through the execution of a software program determines the signals to be sent to each of the four valves i.e. operates the valves by determining conductance coefficient of each valve. The other two sensors Ps-Pump and Pr-Pump are used by the pressure controller to determine the openings of the unloading valve and the return valve in the case of a fixed displacement pump, or to control the pump directly in case of a variable displacement pump.

Figure 4 is a functional block diagram for the architecture of the control system [15–18]. The figure shows in more details how the various control routines are distributed among ***System Controller***, ***Function Controller***, and ***Pressure Controller***.

The operator moves the joystick to some position that translates into moving more than one function (hydraulic cylinder) in the equipment. A certain joystick position is an input to the ***System Controller*** which maps into a ***velocity*** requirement on the hydraulic cylinders. This mapping function can be linear or nonlinear and can be implemented by an arithmetic expression or a lookup table.

These commanded velocities for the various cylinders require a particular amount of flow from the pump that can be available or not depending on how much the pump can supply. In other words, the simultaneous flow requirements of the various cylinders require a total flow from the pump. This total can exceed the total flow that the pump can supply at a given pressure. In this case, the control system should be able to divide the available flow from the pump among the different functions and the desired velocities of the corresponding cylinders will not be achieved. Thus, a ***Flow Sharing*** routine, which is part of the ***System Controller***, is necessary. It takes the **desired velocities** and the **available flow** as inputs and divides the flow between the different functions according to priorities such that the ratios of commanded velocities for different functions remain the same. Thus, the output of the ***Flow Sharing*** routine is a set of modified velocities that are achievable by the system according to the available flow [18].

For the ***Flow Sharing*** routine to divide up the flow to the different functions, it needs to know the current operating metering modes of the different functions. The metering mode and the commanded velocity determine the amount of flow needed by each function. That is why the ***Flow Sharing*** routine communicates with the ***Metering Mode Selection*** routine as shown in figure 4. The ***Metering Mode Selection*** routine is part of the ***Function Controller*** and it selects the metering mode that each function should be operating in.

The modified commanded velocities calculated by the ***Flow Sharing*** routine are then sent to the ***Valve Opening*** routine, which as part of the **Function Controller**, is responsible for operating the four EHPV valves to supply flow to the cylinder based on the operating metering mode. In other words, based on the achievable velocity calculated by the ***Flow Sharing*** routine, and the operating metering mode of each function, the ***Valve Opening*** routine determines which valves are opened and which valves are closed. It also determines the extent to which the valves are opened based on control priorities.

The force that the actuator needs to exert (to move a load) is determined by the sensed pressures at the workports and the knowledge of the actuator geometry. Workport pressures in addition to supply and return pressures are all sensed and fed back to the ***Pressure Controller***, which determines the highest pressure that the pump needs to supply. It then controls the unloading and return valves shown in figure 3 if the pump is fixed displacement or directly controls the pump if it is a variable displacement pump to raise (or lower) the pressure supplied by the pump depending on system requirements.

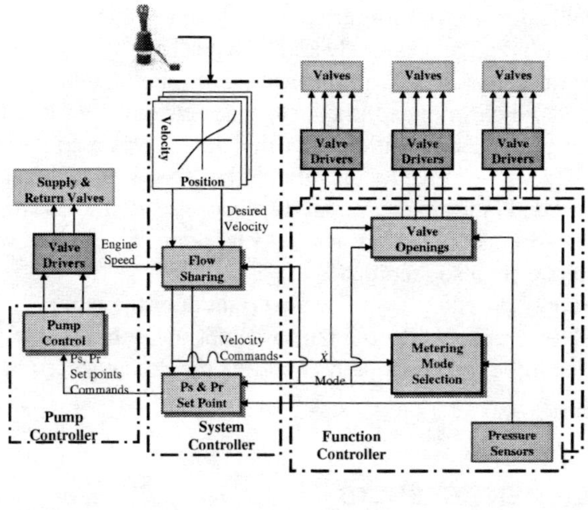

Figure 4. CONTROL SYSTEM DETAILED ARCHITECTURE [15–18]

Table 1. TYPICAL DIG CYCLE SIX PHASES - EXPERIMENT

Phase	Cycle Description
Phase I	Combined Crowd IN, Bucket CURL, and Boom RAISE
Phase II	Combined Boom RAISE and Crowd OUT raising loaded bucket out of trench
Phase III	Combined Boom RAISE and Swing OUT
Phase IV	Bucket DUMP
Phase V	Combined Swing IN and Boom LOWER just above trench
Phase VI	Boom LOWER into trench

A TYPICAL DIG CYCLE

Table 1 below shows the six typical phases of a typical dig cycle. An expert operator performed this typical dig cycle on the backhoe to ensure consistency and that the cycle does not take more time than what an operator in the field would take to perform it.

The operator started with the bucket inside the trench with the bucket loaded. Phases I and II just get the whole mechanism with the loaded bucket out of the trench. Then the operator swings the mechanism away from the trench while still raising the boom. The operator then dumps the load of the bucket and reverses phases I, II, and III to get back inside the trench and the whole cycle repeats. The cycle that was recorded took 14.24 seconds to complete. Data were recorded at 400 Hz. Experimental setup is discussed in the following section.

Experiment

The experiment was done in the facilities of the industry supplier to prove the operability of the control system. Three cylinders (crowd, bucket, and boom) of a typical mid-size tractor loader backhoe were equipped with the four-valve assemblies as schematically shown in figure 2, the swing cylinders were controlled with regular spool valves. The typical dig cycle described above was performed. Workport pressures for all cylinders were recorded which implies the typical loads encountered in a typical dig cycle. The position of each cylinder was recorded and velocities were derived from filtered position signals. Velocities recorded indicate the required speeds of operation during the cycle. Workport pressures and velocity data for the two swing cylinders were also recorded.

Supply pressure was measured right after the pump assembly. Return pressure to the tank was not measured, but it was assumed to be the cracking point of the check valve used in the system.

EXPERIMENT RESULTS

As an example of the results obtained from this experiment, supply and workport pressures for crowd function is shown in fig. 5. In the figure, it can be noticed that the supply pressure is equal to one of the workport pressures at parts of the cycle. This means that crowd function is driving the pump at this portion of the cycle. In other words, crowd determines the pressure required from the pump while the other functions require less pressure. When the workport pressures are higher than supply pressure then crowd is in a regeneration mode where it does not need high pressure from the pump but rather uses trapped pressures in the chambers to achieve the motion. Workport pressures along with cylinder geometries can be used to calculate the forces that each function is subjected to during the whole cycle. This is going to be used, along with velocities required of each cylinder during the cycle, to optimize the metering modes selection in a later section.

Fig. 6 shows the position and velocity of crowd throughout the cycle. From the figure, extension and retraction can be easily recognized by velocity direction. Positive direction means the cylinder is extending, while negative direction means it is retracting. For example between the 7th and 8th second in the cycle, the pressure in rod chamber (b) is higher than the supply pressure and pressure in head chamber (a). At the same time the position and velocity figure shows that crowd is in extension. Then it can be concluded that crowd was operating in High Side Regeneration Extension Mode.

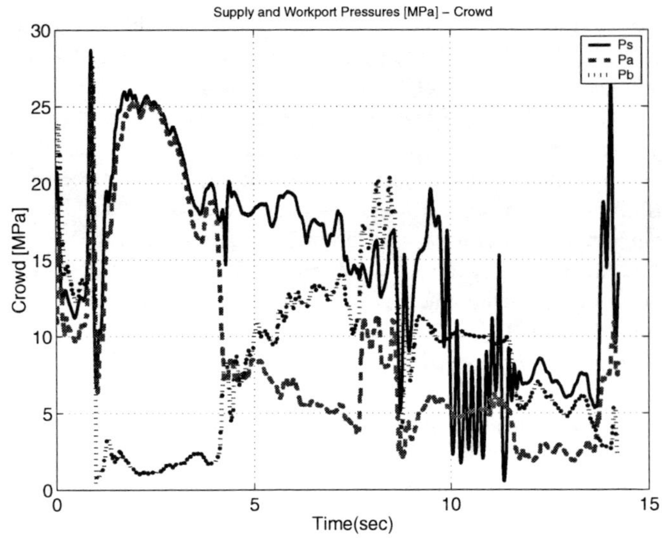

Figure 5. SUPPLY AND WORKPORT PRESSURES FOR CROWD FUNCTION THROUGHOUT THE DIG CYCLE

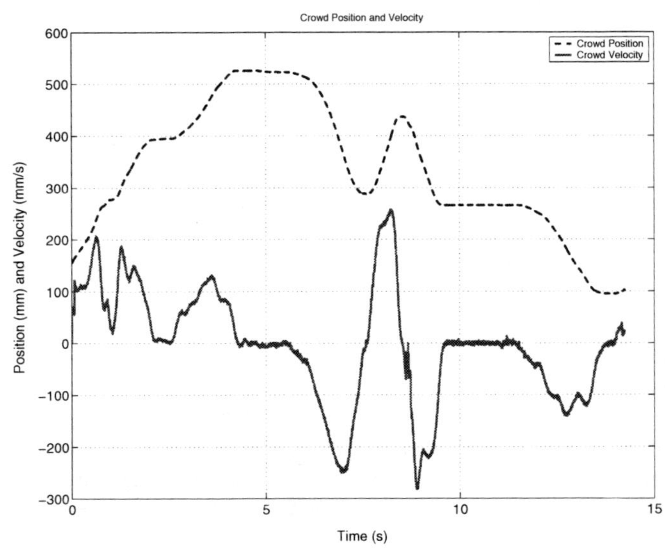

Figure 6. CROWD FUNCTION CYLINDER POSITION AND VELOCITY THROUGHOUT THE DIG CYCLE

METERING MODE SELECTION OFF-LINE OPTIMIZATION

The mid-size tractor loader backhoe used in the experiment has four functions: crowd, bucket, boom, and swing. Therefore, there are 125 metering mode combinations possible at any instant in the cycle. The optimization code explained in this section makes use of the experimental data obtained from the tractor loader backhoe typical dig cycle experiment such as commanded velocities, workport pressures, which represents the load the function experience, and some geometry of the four functions cylinders. At each instant in the cycle the code uses the experimental data, simulates the control algorithm implemented on the machine, tries all the possible combinations of metering modes and selects the best mode that each function should be operating in to maximize a fitness function.

Fitness Function

The fitness function can take into consideration the machine productivity and efficiency. Productivity of a tractor loader backhoe, as an example of mobile hydraulics equipment, is usually expressed in volume of dirt removed per hour. In order to quantify productivity in this optimization scheme, the ratio of achievable velocity to the commanded velocity is going to be used. Other quantities can be used to express productivity like the cylinder's output power. Minimizing cylinder friction or minimizing input power needed by the pump can be a measure of machine efficiency.

Different metering modes have different load and speed characteristics. A comparison between High Side Regeneration Extension Mode and Powered Extension mode is given next as an example to show such differences in load and speed characteristics.

Comparison Between High Side Regeneration Extension and Powered Extension Modes Figure 7 shows a cylinder in two different modes: High Side Regeneration Extension (HSRE) mode and Powered Extension (PE) mode. The maximum force that can be achieved in both modes can be easily derived to be:

$$Fmax_{PE} = P_s A_a \tag{1}$$

$$Fmax_{HSRE} = P_s(A_a - A_b) \tag{2}$$

Where P_s is the supply pressure. Similarly, the achievable speed can be derived in both modes to be:

$$\dot{x}_{PE} = \frac{Q_s}{A_a} \tag{3}$$

$$\dot{x}_{HSRE} = \frac{Q_s}{(A_a - A_b)} \tag{4}$$

Where Q_s is the supply flow

For the same Q_s and P_s it is clear from these equations that Powered Extension is capable of providing a larger maximum force than High Side Regeneration Extension, while High Side Regeneration Extension is capable of achieving higher maximum speeds than Powered Extension.

There are regions of performance overlap where both modes can drive a load at a desired speed with different supply pressure and flow. On the other hand, there are regions of performance where only one mode can achieve the desired performance and this mode has to be chosen. Similar comparison can be easily made between the different modes.

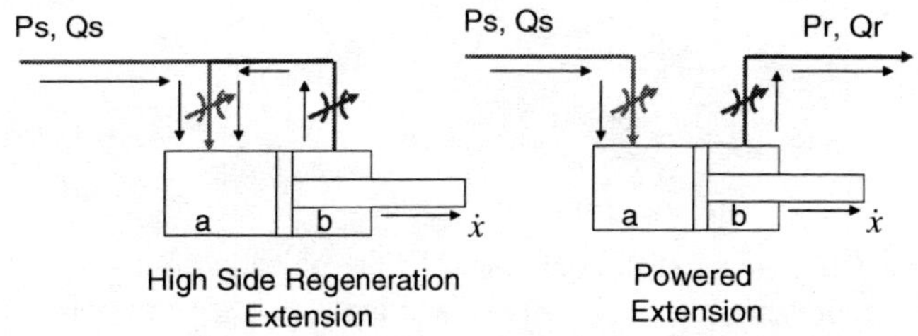

Figure 7. COMPARISON BETWEEN CHARACTERISTICS OF HIGH SIDE REGENERATION EXTENSION AND POWERED EXTENSION MODES

Fitness Function Equation

This is given by the equation:

$$Fitness = \alpha_1 fit_1 + \alpha_2 fit_2 + \alpha_3 fit_3 + \alpha_4 fit_4 \tag{5}$$

Where:

fitness is the fitness function to be maximized by the optimization code.

α_1, α_2, α_3, and α_4 are weighting factors for the four components of the objective function, fit_1, fit_2, fit_3, and fit_4 respectively, to allow the user to give relative importance to one over another (or to remove it as a whole from the fitness function by setting α_1, or α_2, or α_3, or α_4 to zero).

fit_1 is the first component of the fitness function, and it is a measure of machine productivity. It is the summation of the achievable speed divided by commanded speed for each of the four function crowd, bucket, boom, and swing. The achieved speed is constrained not to exceed the commanded speed. It can be represented by this equation:

$$fit_1 = \sum_{i=1}^{4} \frac{(\dot{x}_{Achieved})_i}{(\dot{x}_{Commanded})_i} \tag{6}$$

fit_2 is the reciprocal of friction in the individual cylinders. Thus maximizing fit_2 means minimizing friction in the cylinders. It can be represented as:

$$fit_2 = \sum_{i=1}^{4} \frac{1}{(F_f)_i} \tag{7}$$

fit_3 is the summation positive power output from the cylinders, which is the power that the cylinder exerts on the load. It can be represented by this equation:

$$fit_3 = \sum_{i=1}^{4} \left[(P_a A_a - P_b A_b)\, \dot{x}_{Achieved} \right]_i \Big|_{Positive} \quad (8)$$

fit_4 is the fourth component of the objective function. It represents the reciprocal of the input power to the pump. The optimization scheme tries to maximize a fitness function, so in this case we try to maximize the reciprocal of input power to minimize input power. The reciprocal of input power is calculated as the overall pump efficiency divided by hydraulic power output from the pump. It is given by the following equation:

$$fit_4 = \frac{\eta_{ov}}{P_{sp} Q_s} \quad (9)$$

Constraints

Some constraints need to be enforced on the code. For example, to avoid having a negative equivalent pressure or a negative term under the root. In practice, the control system shuts the four valves until pressure builds up to a value where the desired motion becomes possible.
Pressure is constrained so that it does not go below some prescribed pressure, which is the minimum that the pump would supply.
Pressure is constrained by a maximum value, which is the maximum pressure the pump can supply.

OPTIMIZATION OF METERING MODE SELECTION

Figure 11 in the appendix shows a flowchart for the optimization code. Two tables are also provided in the appendix section. Tab. 2 shows the nomenclature used in the figure.

Data was taken at a $400Hz$ frequency. However, the metering mode is not going to change that fast because it involves closing and opening valves and other physical changes that cannot happen that fast. Therefore, the optimization of metering mode selection used a much lower frequency. The results shown in this chapter were obtained using the data points at a frequency of $8Hz$.

All the possible mode combinations for the four functions are tried at each instant in the cycle. Commanded velocity (including direction) and the mode combination being tried are input to a `MATLAB` function called `PMO` to determine if the modes are possible for the functions. For example, powered extension mode cannot be selected for a function that has a negative commanded velocity and thus in retraction.

If the modes are possible, the metering modes along with some cylinder geometry information are sent to another function called `QsQrFlows` that performs the flow sharing calculations and determines the supply and return flows for each function and also determines the achievable velocity versus the commanded velocity.

The new achievable velocity, some geometries, workport pressures information, and maximum equivalent valve conductance value are input to another `MATLAB` function called `PSP` that determines the required pressure set point that the pump should supply for each function and then the maximum pressure set point is chosen to be the system pressure set point.

The pressure set point determined by `PSP`, cylinder geometries, and metering modes are input to another `MATLAB` function K_{Calc} that performs the most complicated set of calculations in this code. It determines the individual valve conductance values that would satisfy pressure constraints, and commanded velocity. It also comes up with an achievable equivalent valve conductance value and a possibly modified achievable velocity.

This information along with some geometries and the metering mode are input to a `MATLAB` function called `PaPb` that calculates the workport pressures that will result in if the metering mode combination being tried is the combination of modes in operation for the different functions. This is mainly used to calculate the instantaneous positive power out from the cylinders and is also used to estimate the friction in the cylinders.

Now the code has all the information to calculate the objective (utility) function, which is compared the objective function value from the previous metering mode combination. If the new objective function value is greater than the previous objective function, then the present metering mode combination is stored as the optimum combination along with all the information pertaining to this combination like the system pressure set point, individual valve conductance coefficients, pump efficiency,...etc. Thus, at the end of the 125 runs, the code determines which combination of metering modes optimizes the objective function.

This whole sequence is repeated again for the next instant in the cycle. The cycle takes 14.24 sec and at $8Hz$ frequency, the cycle amounts to 114 instants that this optimization sequence has to be repeated.

OPTIMIZATION RESULTS

This section presents the results for maximizing the objective function. The optimization code explained above was run with all $\alpha's = 1$. No preference in weighting was given to any of the four terms. Swing was controlled by spool valves so it had to operate in Powered Extension whenever it was moving in a particular direction and Powered Retraction when it was moving in the other direction, so swing modes are not presented. Only optimized metering modes selected for crowd, bucket, and boom are presented. Fig. 8 shows the optimized modes for crowd through out the cycle along with the speed of the cylinder, fig. 9 shows the optimized modes for bucket, and fig. 10 shows the optimized

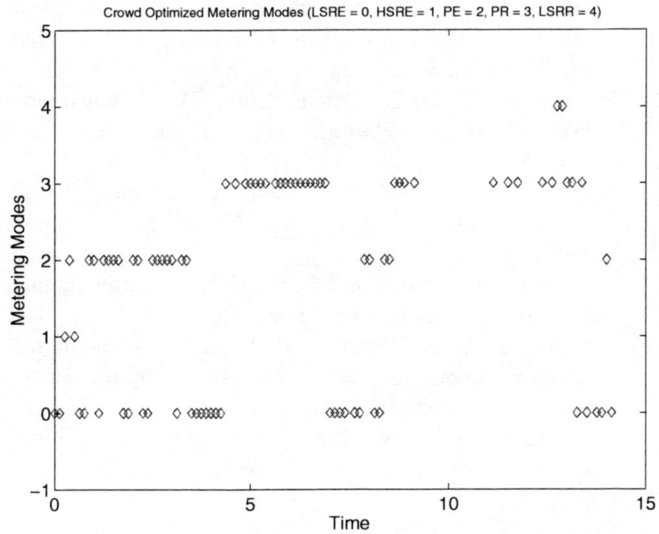

Figure 8. OPTIMIZED CROWD MODES DURING THE DIG CYCLE

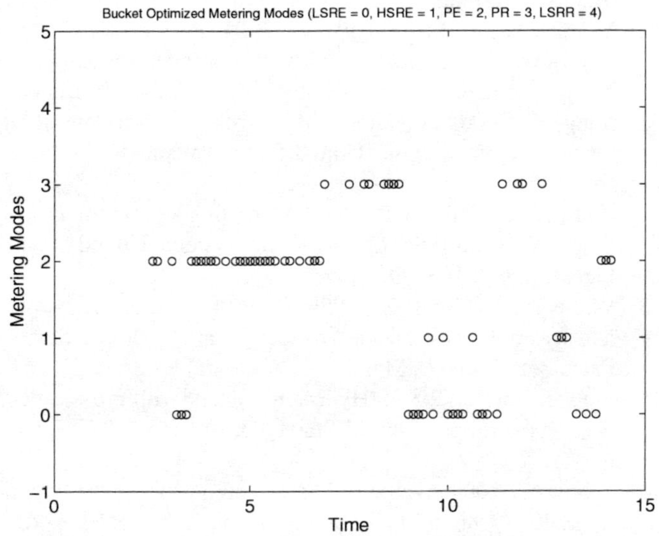

Figure 9. OPTIMIZED BUCKET MODES DURING THE DIG CYCLE

modes for boom.

Optimization is done for all the functions operating at the same time. In all three figures when the speed is almost zero, mode switching does not physically mean that valves are closing and opening because in reality the cylinder is not moving at all and thus mode switch can be ignored. Also, this frequent switching behavior noticed in this off-line optimization cannot be physically implemented, because it means that valves keep opening and closing all the time to switch modes, which is impractical

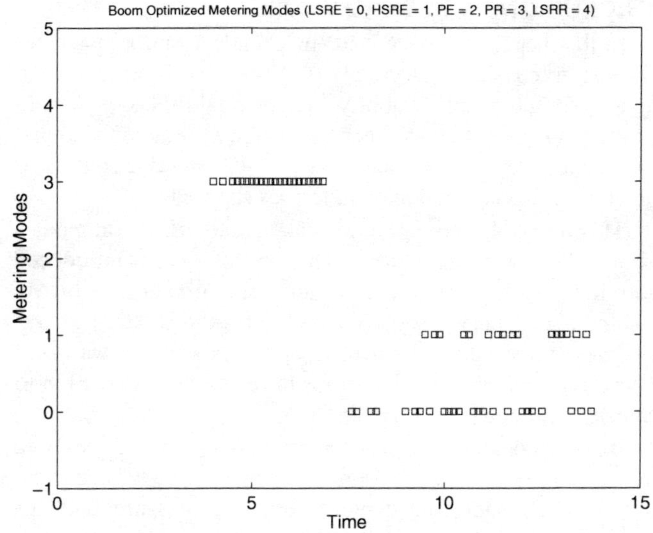

Figure 10. OPTIMIZED BOOM MODES DURING THE DIG CYCLE

and would cause jerkiness in motion of cylinders.

In the following figures modes are expressed by numbers: 0 means Low Side Regeneration Extension (LSRE), 1 means High Side Regeneration Extension (HSRE), 2 means Powered Extension (PE), 3 means Powered Retraction (PR), and 4 means Low Side Regeneration Retraction (LSRR). In fig. 8 it can be noticed that the crowd cylinder is mostly operating in powered modes probably due to high load requirements throughout the cycle and to minimize cylinder friction. Friction in the cylinders increases with increasing workport pressures, and powered modes do not require high pressures like regenerative modes. However, crowd goes into Low Side Regeneration Extension mode at some small portions in the cycle when gravity is possibly acting in the same direction.

For bucket modes also, as shown in fig. 9, powered modes are used in portions of the cycle where high load is applied. Low side regeneration extension is used when there is gravity assistance and High Side Regeneration Extension is used for a very small portion of the cycle where the bucket is in the air (not digging), and thus subjected to a small load.

Cylinder extension in the case of boom function means lowering the boom with gravity assistance and thus the force needed to lower the boom is not great. That is why in fig. 10 it can be observed that during the second half of the cycle, the cylinder is mainly operating in regenerative modes either Low Side Regeneration Extension (using gravity assistance and saving energy) or High Side Regeneration Extension for higher speed. In the first half of the cycle where the boom is raised, which is achieved by retracting its cylinder, the cylinder has to operate in Powered Retraction mode.

CONCLUSIONS & FUTURE WORK

In this paper the concept of 'independent meter in / meter out' was emphasized. Decoupling of meter in from meter out provides for more controllability and potential for energy saving in overrunning load cases when compared with a conventional spool valve controlled hydraulic system. A four-valve configuration controlling a hydraulic cylinder was stressed.

This four-valve configuration can operate in several modes, five of which were explained. The control system introduced by an industry supplier to control this configuration was briefly explained. A typical dig cycle experiment and some of the results were presented. The data from the experiment was used to perform off-line optimization for metering mode selection to achieve maximum productivity.

It was concluded that powered modes should be used through out most of the cycle since they provide high force capability and at the same time do not require high pressures and thus minimize cylinder friction. Whenever, gravity assistance is available, Low Side Regeneration modes are used since the power consumption is minimized and small flow is required from the pump at low pressure. High side regeneration mode can be advantageous when moderate force capability is required, despite high cylinder friction expected as a result of high pressure requirements.

In future work, a sensitivity analysis can be done on the data acquired from the typical dig cycle experiment to see the effect of slightly different cycles on the optimization results. The effect of the α weights can be explored as well. This optimization was done off-line and a real time algorithm is needed for implementation on the machine.

Also in future work, a dynamic model for the four-valve assembly configuration controlling a hydraulic cylinder will be developed and simulated. Switching from one mode to another mode changes the dynamics of the system. Therefore, the optimal control of switching times in switching dynamical systems will be considered.

ACKNOWLEDGMENT

This work is partially supported by HUSCO International. The cooperation of the engineering development team at HUSCO International is greatly appreciated.

REFERENCES

[1] Liu, S., and Yao, B., 2002. "Energy saving control of single-rod hydraulic cylinders with programmable valves and improved working model selection". National Fluid Power Association, **102** (2.4) May , pp. 81–91.

[2] Bu, F., and Yao, B., 1999. "Adaptive robust precision motion control of single-rod hydraulic actuators with time varying unknown inertia: a case study". *ASME International Mechanical Engineering Congress and Exposition (IMECE), FPST,* **6** May , pp. 131–138.

[3] Frankenfield, T., and Stavrou, P., 1993. "Developing trends in hydraulics tied to electronic controls". Control Engineering, **40** (5) May , pp. 62–66.

[4] Aardema, J. A., 1996. Hydraulic circuit having dual electrohydraulic control valves. United States Patent, (5,568,759).

[5] Kramer, K. D., and Fletcher, E. H., 1990. Electrohydraulic valve system. United States Patent, (Re. 33,846).

[6] Aardema, J. A., and Koehler, D. W., 1999. System and method for controlling an independent metering valve. United States Patent, (5,947,140).

[7] Garnjost, K. D., 1989. Energy-conserving regenerative-flow valves for hydraulic servo-motors. United States Patent, (4,840,111).

[8] Book, R., and Goering, C. E., 1999. "Programmable electrohydraulic valve". In SAE paper, **1** (2) , pp. 28–34.

[9] Yang, X., Stephenson, D., and Paik, M., 2001. Bidirectional pilot operated control valve. United States Patent, (6,328,275), June.

[10] Yang, X., Pfaff, J., and Paik, M., 2004. Pilot operated control valve having a poppet with integral pressure compensating mechanism. United States Patent, (6,745,992), June.

[11] Yang, X., 2000. Pilot solenoid control valve with pressure responsive diaphragm. United States Patent, (6,149,124), November.

[12] Pfaff, J., and Willard, R., 2003. Operating system for a programmable controller of a hydraulic system. United States Patent, (6,640,163), October.

[13] Wilke, R., and Yang, X., 1999. Pilot solenoid control valve and hydraulic control system using same. United States Patent, (5,878,647), March.

[14] Stephenson, D., 2003. Hydraulic system with cross function regeneration. United States Patent, (5,502,393), January.

[15] Tabor, K., 2004. Velocity based method for controlling a hydraulic system. United States Patent, (6,718,759), April.

[16] Tabor, K., 2004. Velocity based method for controlling an electro-hydraulic proportional control valve. United States Patent, (6,775,974), August.

[17] Pfaff, J., and Tabor, K., 2004. Velocity based electronic control system for operating hydraulic equipment. United States Patent, (6,732,512), May.

[18] Pfaff, J., and Tabor, K., 2004. Method of sharing flow of fluid among multiple hydraulic functions in a velocity based control system. United States Patent, (6,779,340), August.

[19] Hu, H., Zhang, Q., and Alleyne, A. "Multi-function realization of a generic programmable e/h valve using flexible control logic". *Proceedings of the Fifth International Con-*

Table 2. NOMENCLATURE

Variable Name	Variable Description
R	Cylinder area ratio for the different functions
A_a	Head area for different functions
A_b	Rod area for different functions
$Xdot$	Velocity of the different functions
$Xdot_m$	Modified function velocity based on flow sharing
$Xdot_{mm}$	Modified function velocity based on equivalent valve conductance adjustments
P_b	Rod side workport pressure for different functions
P_a	Head workport pressure for different functions
Pa_{min}	Minimum head pressure for different functions
Pa_{max}	Maximum head pressure for different functions
Pb_{min}	Minimum rod side pressure for different functions
Pb_{max}	Maximum rod side pressure for different functions
$\Delta Pmin$	Minimum valve differential pressure
$Ps - SetPoint$	Pressure set point required for each function
Keq_{Max}	Maximum equivalent valve conductance value
K_{in}	Inlet valve conductance coefficient
K_{out}	Outlet valve conductance coefficient

ference on Fluid Power Transmission and Control.

[20] Book, R., 1998. *Programmable electrohydraulic valve*. PhD Thesis, University of Illinois at Urbana-Champaign.

Appendix A

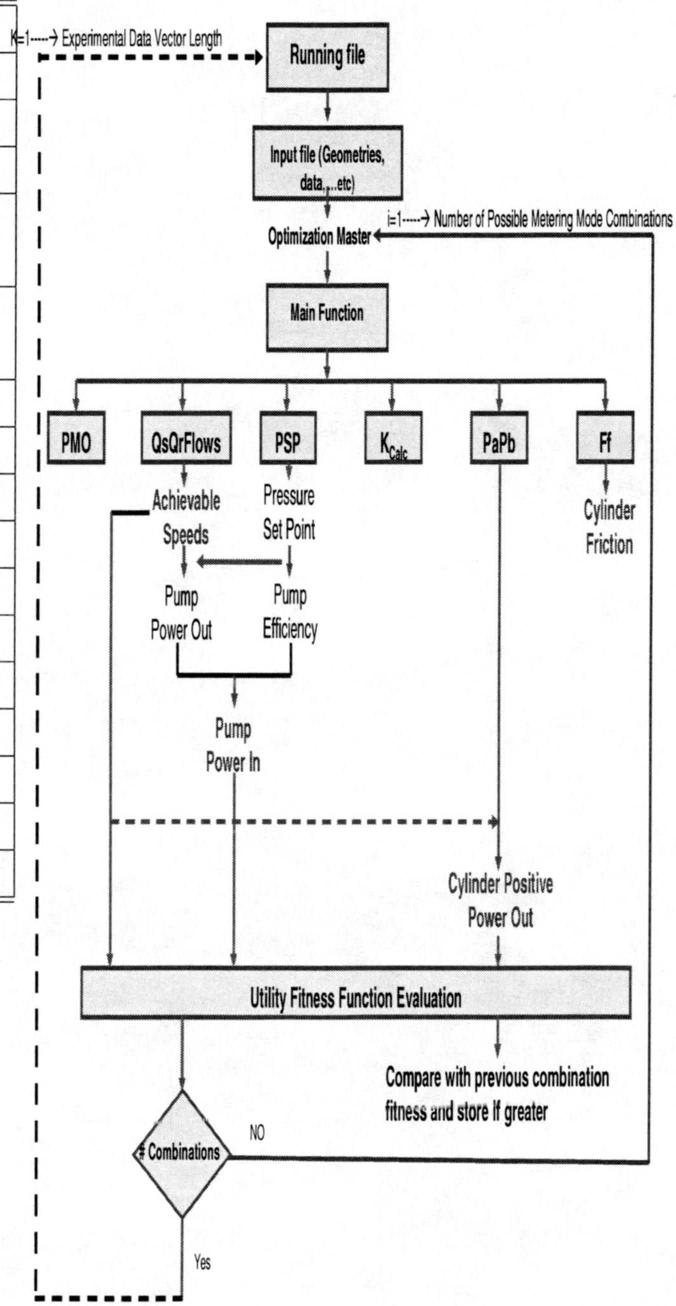

Figure 11. OPTIMIZATION CODE FLOWCHART

Proceedings of IMECE2005
2005 ASME International Mechanical Engineering Congress and Exposition
November 5-11, 2005, Orlando, Florida USA

IMECE2005-82124

ELECTRIC MOTORS COUPLED TO HYDRAULIC MOTORS AS ACTUATORS FOR HYDRAULIC HARDWARE-IN-THE-LOOP SIMULATION

Scott Driscoll†, James D. Huggins‡, Wayne J. Book*
Intelligent Machine Dynamics Laboratory
Fluid Power Motion Control Center
Department of Mechanical Engineering
Georgia Institute of Technology
Atlanta, Georgia 30332
Email: gtg137p@mail.gatech.edu†, {wayne.book*,james.huggins‡}@me.gatech.edu

ABSTRACT

Hardware-in-the-Loop (HIL) Simulation enables testing of an actual physical component of a system under a variety of conditions without the expense of full scale testing. In hydraulic systems, flows or pressures that interface with the component in question are controlled by a computer running a simulation designed to emulate a complete system under real operating conditions. Typically, servo valves are used as actuators to control the flows or pressures. This paper investigates the use of electric servo-motors coupled to hydraulic gear motors as alternative actuators, and discusses some of the advantages and disadvantages that motors have in comparison to valves. A demonstration HIL simulation involving a mobile proportional flow control valve attached to an emulated backhoe is described, and results are compared to data from a real backhoe.

1. INTRODUCTION

Despite ever increasing modeling capabilities, hydraulic systems oftentimes require full scale testing to verify performance of components and controllers. This is due to the many non-linearities contained in hydraulic systems, and the fact that their performance is dependent on parameters that may be difficult to predict or vary throughout operation, such as viscosity and the amount of entrapped air. In the case of mobile equipment valves, they will end up operating in a large variety of different machines and operating conditions, some of which may be hard to anticipate. This makes full scale testing under every possible situation infeasible, and the testing that is done can be very expensive and even present safety hazards. An alternative approach for evaluating new hydraulic components and controllers is to install the component under question within an emulated surrounding system and operating environment. This avoids the expense of full scale testing, allows for quick changes to operating conditions, and does not require modeling of the component under evaluation since the actual hardware is being used. This technique is called Hardware-in-the-Loop (HIL) Simluation, and is also referred to as load simulation or emulation.

While there are many examples of HIL systems from a variety of different industries, most are developed in an ad hoc manner, and there has been little work in the literature concerning a framework for developing HIL systems in general. Also, many of the published systems employ servo valves by default as the main actuator in the emulation system. The goal of this work is to investigate the use of electric and hydraulic motors as alternative actuators for hydraulic HIL systems through the development of a test system, and in the conclusion, offer some suggestions for choosing an appropriate actuator for a new HIL design. The test system in this study was a John Deere backhoe, with the main boom cylinder being emulated by the motors, and a proportional flow control valve as the test component, or "hardware" in the loop. A brief review of other HIL systems will be followed by motivation and design of the motor-based system in the second and third sections. Controllers and models for that system will then be discussed in sections four and five, followed by results and a comparison to data from a real backhoe in section six. Finally, an evaluation of the merits and demerits of motor-based systems is offered in the conclusion.

2. LITERATURE REVIEW

Hardware-in-the-Loop (HIL) Simulation has been used extensively in a variety of fields as a method for dealing with the inadequacies of pure simulation in a cost effective manner. Several examples from outside the hydraulics industry include a Minimally Invasive Surgical system using Haptic operations [1], hard drive control system [2], a software traffic simulator [3], test system for embedded controllers [4], and applications in space [5]. Akpolat et al created a system designed to test new electric motor controllers [6]. It consisted of two vector controlled induction motors connected via a common shaft, with one designated as the test hardware drive and the other emulating a desired load. They point out some inherent limitations of using inverse dynamics to derive load torques, and present a feed-forward based speed controller that preserves the normal causality for a load system, where torque is measured and speed displayed. It would seem that inverse dynamics, where velocity is measured and torque output, are commonly used in HIL systems, which as Akpolat points out is inherently limited in practical implementation due to a differentiation requirement. Another consequence of using inverse dynamics is that the inertia in the load actuator ends up causing disturbance torques as the test hardware attempts to change speed. This problem was addressed by Qinghe et al [7] in a hydraulic load emulator system analogous to Akpolat's electric motor system. Their hydraulic emulator consisted of a drive and load cylinder rigidly connected, each controlled by a servo valve. A leakage port was designed to reduce the disturbance torque effect, however, this also limited the emulation system's performance.

Guerrier and Edge [8] also address a cylinder-based load simulator, although theirs was designed specifically to replicate loads found in injection molding machines. A single cylinder was used in the device, where the pressure in one side was controlled via a servo valve as in a standard injection molding machine, and flow out of the other side was regulated by another valve to emulate injection pressures. A gain scheduling controller was developed that effectively replicated loads in the filling stage, although its bandwidth was too low to emulate the packing stage. It should be noted that the emulator was designed to share as many physical characteristics with real injection machines as possible (including mass, cylinder), so much of the emulation characteristics were pre-made into the test setup.

Two other servo valve based systems of note include a high speed transient dynamometer [9] and earthmoving vehicle powertrain emulator [10, 11]. The dynamometer was created to enhance engine development by emulating the torque contribution of additional cylinders when a single cylinder test engine is evaluated. Both a hydraulic motor and pump were attached to the engine's shaft, and torque was controlled by regulating flow through the hydraulic motor and pump via servo valves. The valves were high performance Moog valves with a bandwidth of 120 Hz, which enabled successful emulation of cylinder firing events. A similar setup was used in the earthmoving power train emulator, which was designed to emulate several different functions of a large earthmoving vehicle (drive, steering, implement) so that different controllers could be evaluated. The method employed by Zhang et al [10, 11] in the earthmover would not be able to emulate over running loads (a mass being lowered under gravity for example), but could be augmented as in the automotive dynamometer, although this would require an additional pressure supply.

Some of the most extensive work for designing a system that can emulate a wide variety of loads was done by Ramden and Palmberg [12–14]. They created a load simulator for hydraulic valves that allows either the flow or pressure on each work port of a test valve to be independently controlled. Each port is attached to a high performance Moog valve (80 Hz bandwidth) that can expose the port to either a high pressure source or tank with varying degrees. Instantaneous flows are ascertained from a lookup table that relates flow to spool position and pressure drop, which is effectively an orifice flow meter. The system is able to accurately recreate a wide variety of conditions, including overrunning loads and relatively stiff environments, and results were compared to an actual system to verify its accuracy. However, at low and close to steady-state flows, their system exhibits pressure offsets. This is mainly a consequence of measuring flow with an orifice, since in steady state or extremely low flow conditions there is minuscule pressure drop. It would also seem that at zero flow, it is fundamentally impossible for a servo valve of the type they used to control pressure, since there would be no pressure drop across the valve orifice. Servo valves with underlap may present a possible solution to this issue but they were not investigated in this work.

3. MOTIVATION AND TEST SYSTEM DESCRIPTION

The goal of this research was to evaluate the effectiveness of using electric and hydraulic motors as actuators for hydraulic HIL Simulation. A prototype test system was developed with the intention to both demonstrate that motors could be used as an effective HIL tool, and also to highlight any strengths and limitations of this method as compared to valve based systems that are typically used. Previous work with the electric motors used in this paper was done by Jung et al and demonstrated basic operational capabilities of the motors, simulation environment, and associated hardware [15]. Two pairs of a coupled electric motor and pump were used to emulate an internal combustion engine driven pump and excavator arm lifting under gravity, re-

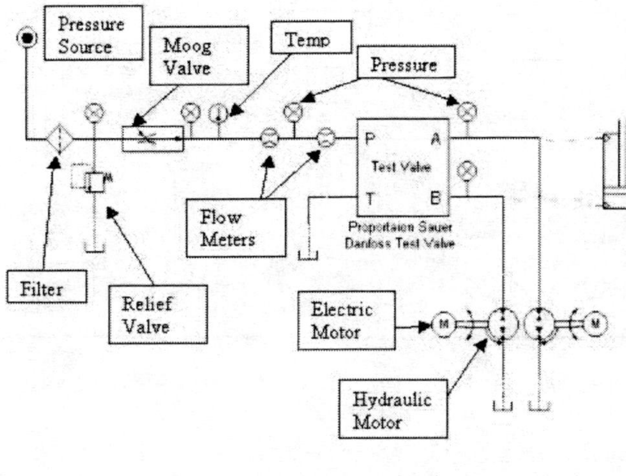

Figure 1. HYDRAULIC CIRCUIT

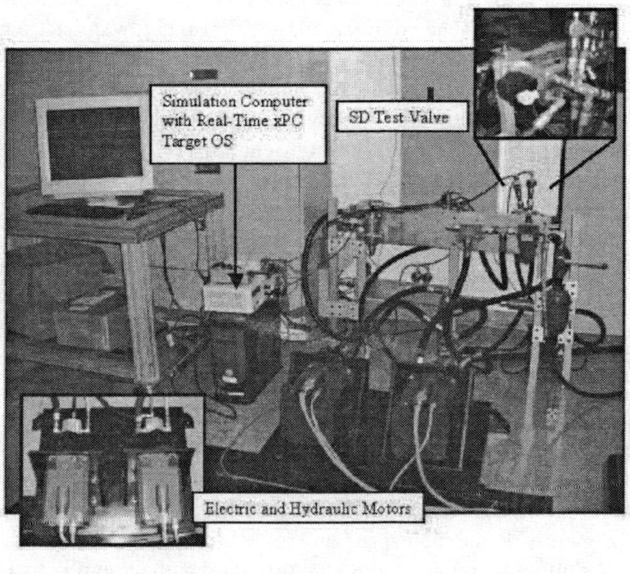

Figure 2. EXPERIMENTAL SETUP

spectively. In this manner, a standard proportional flow valve was placed in an environment where both the flow supply and load were emulated. In the current research, the goal was to achieve a much more thorough evaluation of the capabilites of the electric motors, and determine what types of HIL similuations were appropritate and possible using the motors. Towards this end, a completely new system was designed to emulate the boom cylinder on a John Deere backhoe, with a Sauer Danfoss proportional flow valve as the test valve.

There were several motivations for the the choice of the target system. For one, the lab had access to a real backhoe with identical valve type, so comparisons could be made between the HIL simulation and real system. Also, due to the load-sensing capabilities of the Sauer Danfoss valve, which enable it to hydromechanically compensate for changing loads, accurately modeling the valve or designing a closed loop controller would be very difficult, making it an ideal candidate for HIL Simulation. Finally, as was mentioned previously, low flow situations are inherently problematic for valve-based systems to emulate, so a target system that would frequently encounter zero flow conditions was desired to find out if a motor-based simulation system might be better equipped for handling those situations.

A picture of the hydraulic circuit, lab setup, and backhoe are shown in Figures 1, 2, and 3, respectively. In the hydraulic circuit, note that the dotted lines leading to the cylinder are the connections that exist on the actual backhoe, and have been replaced by lines going to hydraulic motors. One hydraulic motor is designated as the cap side of the cylinder, and the other as the rod side. In this manner, cylinders of arbitrary cap and rod side areas can be emulated since the two sides are completely independent of each other. Also, since each hydraulic motor can both act as a pump, or a motor (resisting flow), overrunning loads can

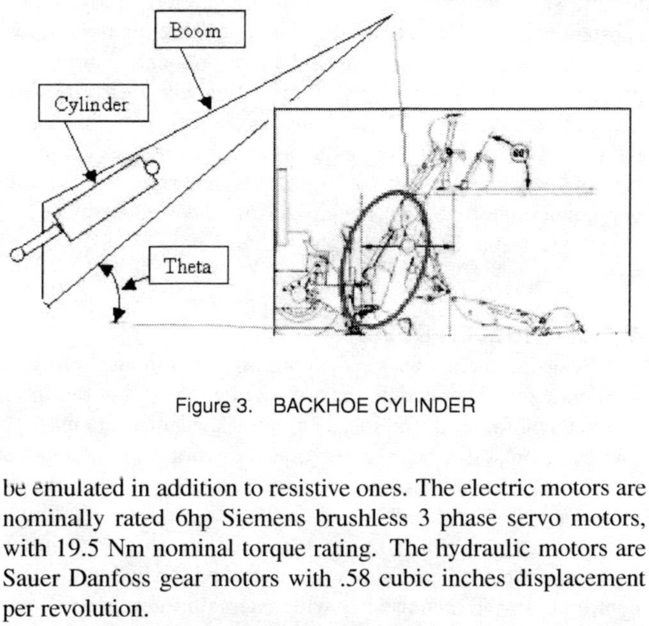

Figure 3. BACKHOE CYLINDER

be emulated in addition to resistive ones. The electric motors are nominally rated 6hp Siemens brushless 3 phase servo motors, with 19.5 Nm nominal torque rating. The hydraulic motors are Sauer Danfoss gear motors with .58 cubic inches displacement per revolution.

Since our lab did not have access to a third electric motor, the supply side (left side of Figure 1) leading to the valve's pressure (P) port was emulated using a Moog valve that regulated flow from a constant pressure source. On the actual backhoe, the supply is a constant flow, so an additional controller was developed for the Moog valve to maintain a constant flow as the pressure varied in the test valve. In the backhoe, the dynamics of the engine undoubtedly affect the valve operation, but in this work no model of the engine was developed, a controller was simply developed to maintain constant flow as best as possible.

One consequence of interacting with real hardware in a simulation is that the simulation must run in real time. Unfortunately, a fast, reliable, and constant step time is difficult to guarantee within Microsoft Windows, so we chose to use a special purpose operating system (OS) designed specifically to run realtime simulations, namely The Mathwork's xPC Target OS. Models are developed in Simulink on a "host" computer, compiled into realtime code using Realtime Workshop, and finally downloaded onto the realtime OS where they are executed. While the simulation performance was more than adequate (maximum execution time of .00049s), the PCI card that enabled the Target PC to communicate to the Siemens motors did not have drivers supported under the realtime OS. The Siemens motors received and sent information via a network called Profibus, which is one of the largest industrial communication standards. A custom sofware driver was developed to enable the realtime OS (xPC Target) to use a Profibus PCI card and communicate digitally to the motors on the high speed and reliable Profibus network. One other difficulty relating to the motors was that their amplifiers injected a significant amount of electrical noise into the room and ground. This was particularly troublesome for the many analog pressure and temperature sensors that were employed, as a substantial amount was being aliased upon sampling despite low pass filtering the signals and following recommended grounding procedures. While more expertise in dealing with electrical noise may have prevented many problems, careful consideration to this issue is advised when using large electrical motors. A combination of lowpass filtering, careful grounding, and most notably non-conductive hydraulic hosing was used to reduce electrical noise to less than 1% of full signal value.

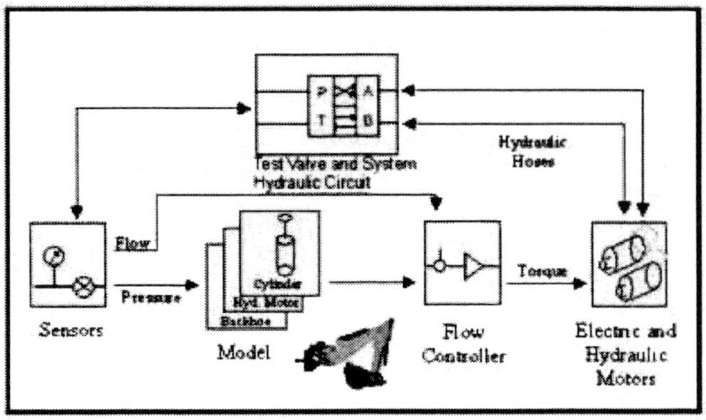

Figure 4. SIMULATION INTERACTION WITH PHYSICAL SYSTEM

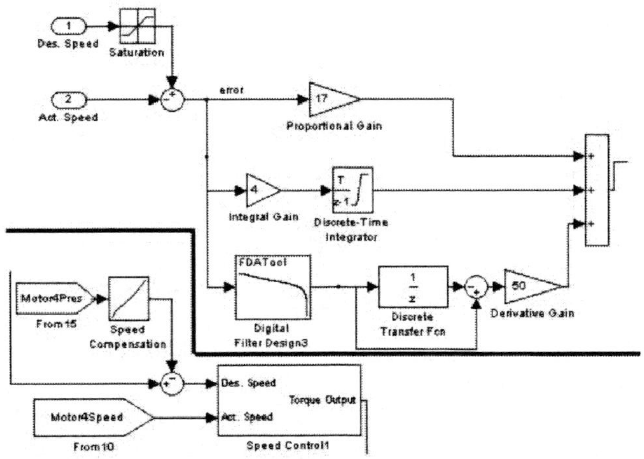

Figure 5. SPEED CONTROLLER AND LEAKAGE COMPENSATION

4. CONTROLLER DESIGN

Basic interaction between the simulation computer, physical system, and motors can be seen in 4. Figure 4 shows the interaction when forward dynamics, or causal dynamics, are used. In this case, pressures are measured at the work ports of the test valve, and then used in a model to derive appropriate flows, which are then realized by motor flow controllers. A simple cylinder model would take the measured pressures and multiply them by respective cylinder areas to find the hydraulic force contribution, and then use this with external forces, friction, and mass to calculate acceleration. Finally, integration would lead to velocity and corresponding flow rates for the cap and rod side of the cylinder. Because the model output is flow, this type of interaction requires a flow controller. Flow was experimentally found (using a standard flow meter in steady state conditions) to be highly proportional to motor speed above about 20 rpms, with the remaining leakage proportional to pressure. The final flow controller consisted of a simple PID speed controller that was tuned experimentally and compensated linearly with pressure to accommodate leakage flow (Figure 5). A bode plot showing a -90 degree bandwidth of about 11Hz for the speed controller is shown in Figure 6. Step response was between 40 and 50 ms depending on amplitude, and this seemed to be a reasonable response given the 4ms computational step time in the Siemens frequency converter (amplifier). To verify basic operation, a simple model of a cylinder with mass was created and subjected to a step change in flow from the test valve. Results are shown in Figure 7 and correspond with what intuition would predict from a 2nd order system. As flow is increased (a lower voltage given to the Sauer Danfoss valve), the pressure spikes up, and as flow is reduced, the pressure drops to zero as the mass tries to maintain velocity, or becomes an overrunning load. Although not shown, reducing the mass and friction had the expected effect of increasing natural frequency and increasing settling time, respectively.

Since inverse dynamics are commonly used in HIL Simulators, a simple cylinder model using inverse dynamics instead

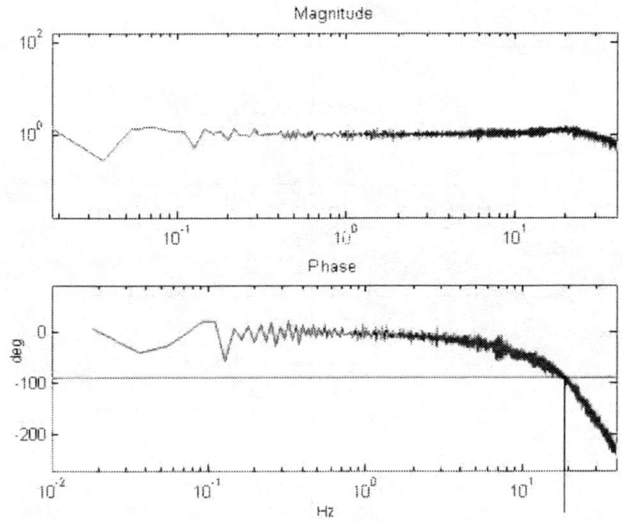

Figure 6. BODE PLOT FOR SPEED CONTROLLER

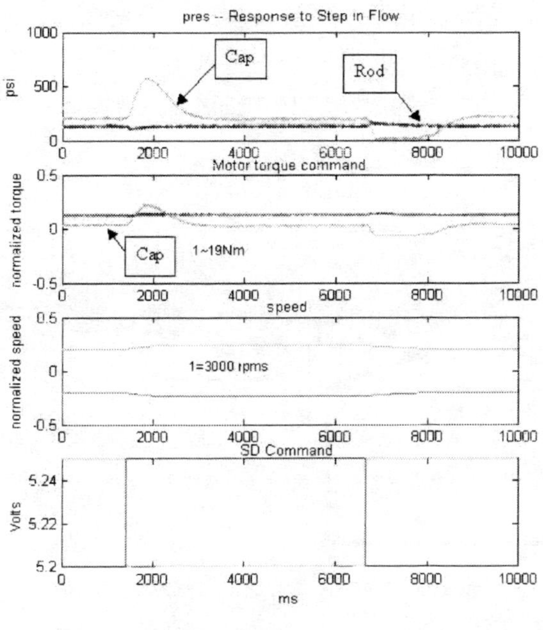

Figure 7. SIMPLE CYLINDER WITH MASS RESPONSE TO FLOW STEP

of "forward" dynamics was also investigated. In this case, flows are ascertained from motor speeds, pressure, and temperature via a lookup table determined with standard flow meters in steady state conditions. The flows are then sent to the model which calculates appropriate pressures that are then realized by pressure motor controllers. The pressure controller was much less suc-

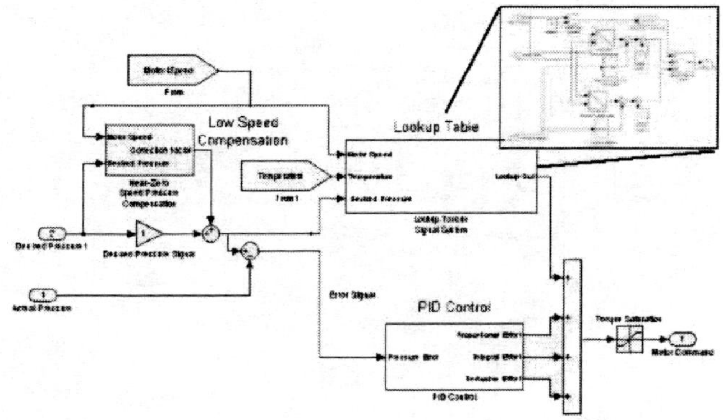

Figure 8. PRESSURE CONTROLLER BLOCK DIAGRAM

cessful than the flow controller, mainly because the control variable was an analog signal that contained substantially more noise than the digital speed signal used in the flow controller. The noise reduced the amount of gain that could effectively be applied to the derivative portion of the controller. The final controller is shown in Figure 8 and consists of a lookup-based feedforward term plus PD controller. When using inverse dynamics, a derivative is needed to go from velocity to acceleration to calculate the required pressure to output. The relatively noisy pressure signal, combined with the necessity to differentiate in the controller as well as the model resulted in an unsuccessful simple cylinder simulation. The severity of low pass filtering needed to clean the signals led to a delay that detracted from performance just as much as the noise on its own. Therefore, no results are given for the pressure controller, and forward dynamics were exclusively used in the emulation of the backhoe discussed in the following section.

A final controller was developed in order to emulate the constant supply flow entering the main pressure port of the Sauer Danfoss proportional valve. As mentioned above, flow from a constant pressure source was regulated by a Moog valve to maintain a constant flow under varying pressures. On the real backhoe, input flow is by no means constant under severe pressure spikes, and although ideally a model would be developed for the backhoe engine, it was felt that an adequate evaluation of electric motors in a HIL system could be achieved even without a completely realistic supply flow. The controller block diagram used to regulate constant flow is shown in Figure 9. One problem with the development of this controller was that flow feedback from a standard turbine based flow meter was too slow to effectively respond to fast pressure changes. A fixed orifice was placed in line with the turbine based flow meter to supplement the turbine signal with a faster flow signal derived from the pressure difference across the orifice. This is a standard flow measurement technique that was ideally suited to this application since flow was desired to remain constant. A fixed orifice can be unsatis-

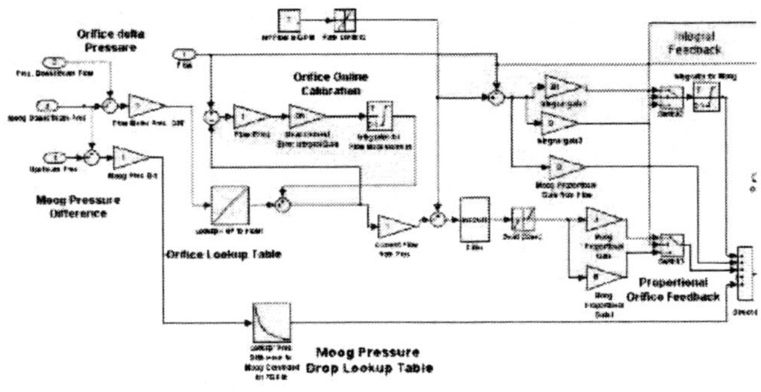

Figure 9. FLOW CONTROL BLOCK DIAGRAM

factory if flows drop too low to create a sizable pressure drop, or increase enough such that a substantial amount of energy is lost at the orifice. While the orifice provided timely flow information in our application, it also caused temperatures to rise substantially faster, which limited the amount of time experiments could be run since no heat exchanger was used. The flow response to a disturbance in pressure is shown in Figure 10, and as can be seen, the flow is restored to the desired level by the time the turbine based flow meter registers the peak in flow. The controller used the orifice based signal to provide proportional feedback, and the turbine flow meter in integral feedback to calibrate the orifice over time and temperature changes. A feedforward command was derived from a lookup table relating the delta pressure across the Moog valve to the Moog command necessary to achieve the desired flow.

BACKHOE MODEL DESCRIPTION

The backhoe model described here, primarily developed by Matt Kontz, is a somewhat simple approximation of the real backhoe boom cylinder and dynamics (Figure 11). In order to achieve an exact match in performance of the HIL emulation system and real backhoe, both an exact model of the backhoe would be needed and the motors would need to perfectly track the desired flows from that model. The creation of a perfect backhoe cylinder model is out of the scope of this work, and we believe is somewhat unnecessary to judge the capabilities of a motor based HIL system. The goal in this work is to demonstrate and investigate what behaviors are achievable using motors, and at the same time expose any shortcomings. The comparison to the real backhoe data will mainly serve as a benchmark in reality, and if the motor controllers are able to adequately track the demanded flows from the model, differences between the real system and simulation are therefore due to inadequacies in the model.

Keeping in mind the research goals, a brief description of the model is offered. Pressures on the work ports of the test valve are measured, and then multiplied by respective cylinder

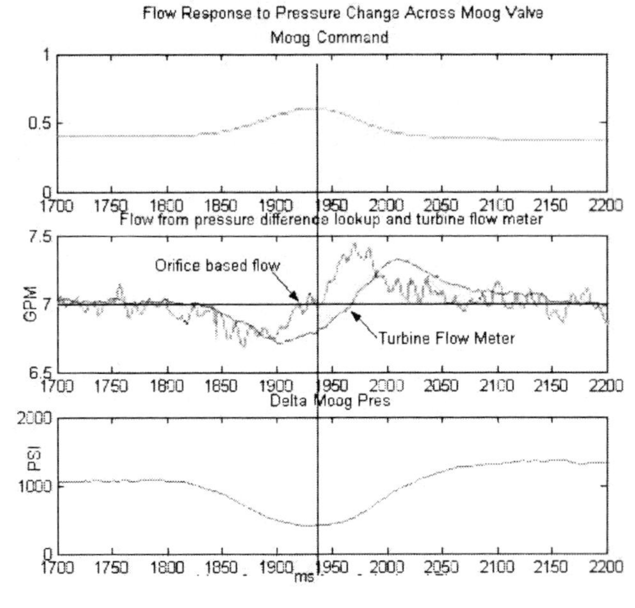

Figure 10. FLOW RESPONSE TO DISTURBANCE IN PRESSURE FOR VALVE-BASED INPUT FLOW SUPPLY CONTROLLER

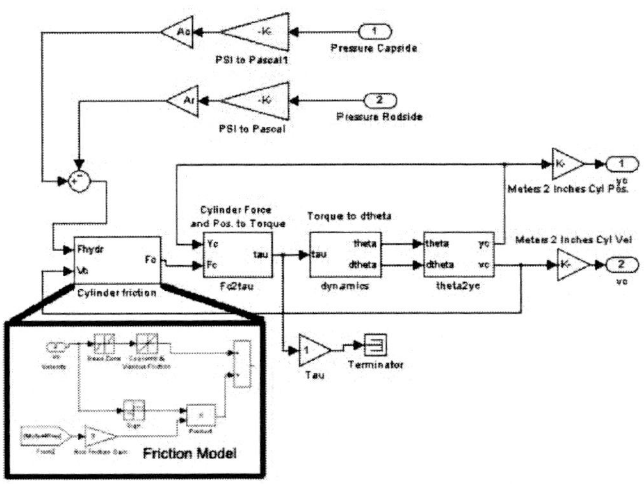

Figure 11. BACKHOE MODEL BLOCK DIAGRAM

areas to calculate hydraulic force. This is then combined with friction, gravity effects, and inertia to finally produce a boom angular acceleration and velocity. The boom velocity is used to calculate cylinder velocity and the corresponding flow rates for each chamber, which are then sent to the flow controllers. The friction model consists of a simple viscous and coulomb term, as well as a term proportional to rod chamber pressure, in an attempt to account for the increase in friction as the fluid in the chamber presses on the cylinder seal.

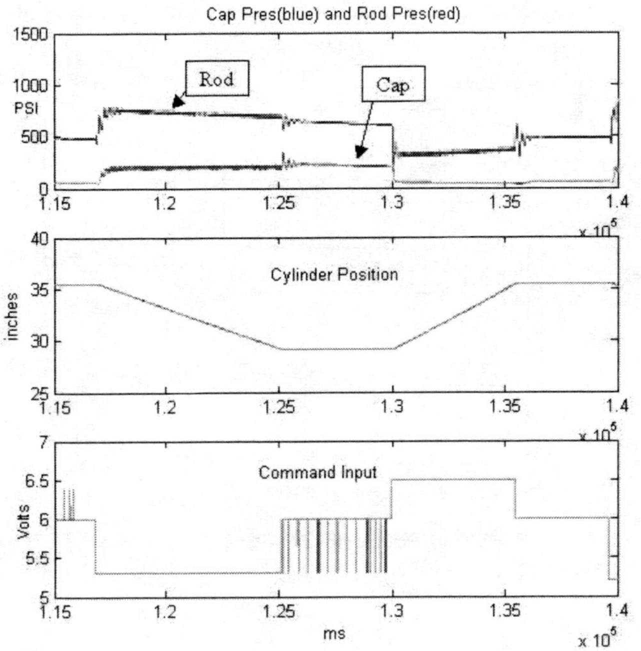

Figure 12. RAISE AND LOWER OPERATION ON REAL BACKHOE

EXPERIMENTAL RESULTS

Before examining results from the HIL Simulation, the pressure and cylinder position from a raise and lower operation on the real backhoe are given in Figure 12. Note that the cylinder lengthens as the boom is lowered and shortens as the boom raises. This is because the cylinder is installed in the hinge with the rod pointing down as shown in Figure 3. As the boom is raised, the pressure in the cap side decreases slightly as the torque from gravity is reduced. At the same time, the cylinder has increasingly less leverage on the boom, but the effect of gravity is larger, so the pressure drops slightly. A corresponding increase in pressure is seen when the boom lowers. Two other features to note are the transients at the start and stop points of motion, and the fact that the pressure maintains even when the valve is neutral (command of 6) and blocks the ports to the cylinder.

Figures 13 and 14 show the corresponding raise and lower operations on the HIL simulation. It should be noted that the flow control valve used in the HIL Simulation, while the same model, is not an exact replica of the one on the real backhoe. One notable difference is that the deadband is not exactly the same, so flow does not begin to flow at the same command on both systems. Nevertheless, the results show many positive comparisons. The overall pressures are about the same in magnitude, the rise and drop in pressure in the lower and raise operations are approximately the same, and the magnitude of the oscillations at the start and end of motions are comparable. However, these results were obtained using a rotational inertia for the backhoe boom of about twice the value expected based on geometry and mass distribution. When the inertia was set to the expected value, occasionally oscillations of about 200 psi would begin during the raise stroke and continue until the boom was lowered. Raising the inertia effectively eliminated these oscillations, but led to other problems. As the magnitude of Sauer Danfoss valve commands increased, the magnitude of the start and stop oscillations increased dramatically in the HIL simulation. Figure 15 shows the effect of a faster raise command given in the HIL simulation. When the raise motion stopped, the cap side pressure spiked to over 2900 psi, whereas the cap pressure never breaches 1500 psi in the actual backhoe data, even for faster motions. The extreme pressure is in part due to the unrealistic rotational inertia programmed into the model. But it is also a result of the motor not having enough torque to overcome its own inertia. Figure 15 shows that the actual speed of the motor was not able to track the desired speed at the overshoot moment. On the actual backhoe, gravity aids in stopping the cylinder when the valve shuts off, but in the HIL simulation, the motor must rely on its own torque to quickly stop itself after reaching a high velocity. In a positive note, when the inertia was set to the expected value, the frequency of oscillation at the start and end points of motions matched very closely to that of the real backhoe. The real backhoe exhibited end stroke oscillations of 4.7 Hz, and the simulation showed 5 Hz. The unstable oscillations that occurred when the correct inertia was used were undoubtedly partly caused by an inadequate friction model, but it is not entirely clear whether the blame can be written off to modeling errors or whether a more fundamental problem with the system is at fault, such as an excited natural frequency between the motors, valve, and connecting hosing.

One more outcome of the results to point out is that the motors were able to maintain pressures even when the flow rate dropped to zero. This is of special interest because it demonstrates a potential advantage motor based systems might have over valve based ones. However, the ability to control pressure at low or zero flows with motors is not without its own problems. In further experiments, as desired pressures increased under zero flow conditions, spikes in pressure due to the nonlinear manner that the gear teeth press on the fluid became apparent. At zero flow, the motor must spin slowly to accommodate fluid lost through leakage, and the average rate required to maintain pressure can be found by simply noting the speed necessary to maintain pressure. However, at these low speeds, the gear teeth reach configurations periodically in which the motor has almost no leverage on the fluid, and the motor jerks forward causing a pressure spike. These spikes become more problematic as pressures increase. The addition of a negative pulse of torque upon detection of a spike onset was helpful in reducing this effect, although the detection of a genuine spike, and not a sharp command change is somewhat difficult. Manring et al describe a model of gear teeth in [16] that may be helpful in countering this

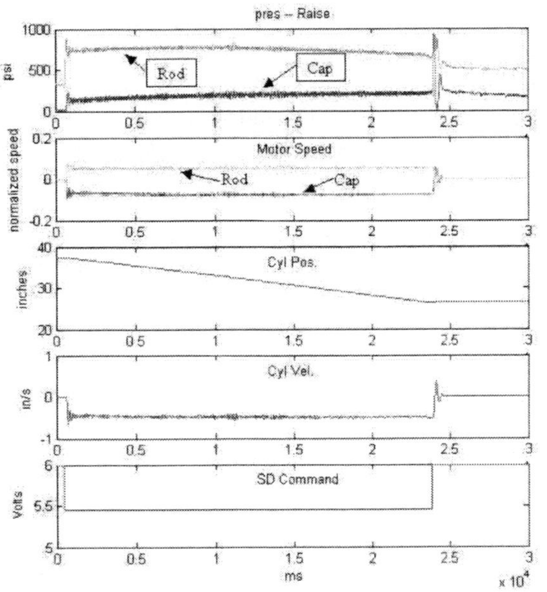

Figure 13. RAISE OPERATION ON HIL SYSTEM

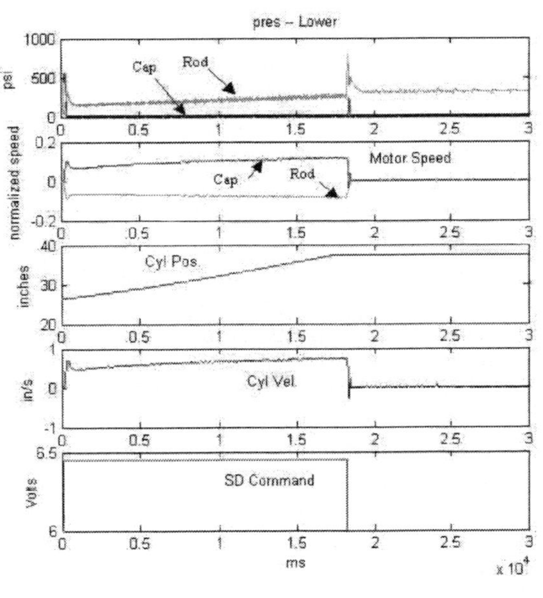

Figure 14. LOWER OPERATION ON HIL SYSTEM

effect, but this was not addressed in the current research.

CONCLUSION AND FUTURE DIRECTIONS

A Hardware-in-the-Loop simulation system using electric and hydraulic gear motors as actuators has been implemented.

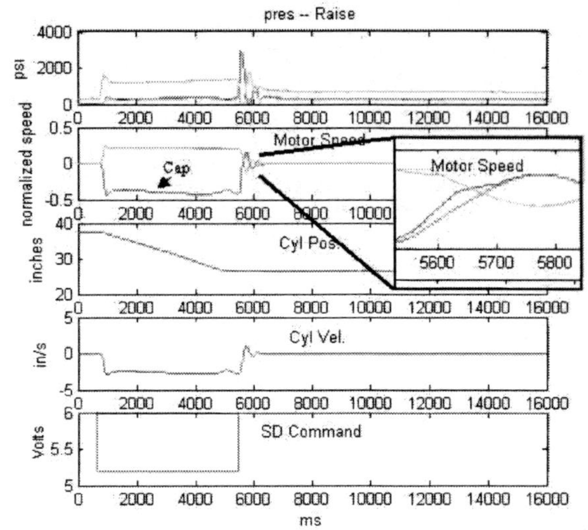

Figure 15. CAPSIDE PRESSURE OVERSHOOT AT END OF STROKE IN HIL SYSTEM

The target system was a John Deere backhoe, with a Sauer Danfoss proportional flow valve as the test test "hardware" in the loop. The HIL simulation, although it did not create an exact replica of data from a real backhoe, was able to emulate many of the same behaviors found in the real system, including the maintenance of pressure at zero flow conditions. This capability is inherently problematic for systems based on valves, as a typical servo valve has no control over the pressure drop across its orifice if no flow is passing through it. On the other hand, high performance servo valves typically have bandwidths over 60 Hz, so extremely stiff systems can be emulated with them. The motors and controllers developed in this study had a speed control bandwidth of 11Hz, mainly due to the large inertia contained in the electric motor, coupling, and hydraulic motor. This bandwidth may not be as limiting as it first seems, though, as many hydraulic systems exhibit a natural frequency of less than 10 Hz. The main oscillations in the backhoe used for this study were around 5 Hz.

Another less obvious benefit that a motor-based system has is that reasonably accurate flow measurement can be achieved simply by recording the motor speed and accounting for a small amount of leakage. In this study, the leakage was typically less than a few percent of the main flow, and mostly linearly dependent on pressure. In systems without a fixed displacement device, some sort of flow meter is required, and it is not a simple matter to find a flow meter that can measure a large range of steady state flows, fast transients, and at same time avoid introducing a substantial energy sink into the system. A motor based system can also simulate over running loads without an additional pressure supply.

A few remaining comments will be made on the design of a hydraulic HIL simulation system. If motors are chosen as the appropriate actuator, careful consideration should be given to the maximum rate the flows will need to change at, which is in essence a measure of the stiffness of the target environment. For a given available torque, and combined inertia of electric motor, coupling, and hydraulic motor, there is a trade off between the maximum pressure that can be displayed and maximum rate of change in flow. If a lower displacement per revolution is chosen for the hydraulic gear motor, a higher maximum pressure will be achievable, but faster motor speeds will be required for a given flow, requiring more acceleration. Finally, a HIL System is only as beneficial as its ability to correctly emulate a given environment, and this is dependent on the designer's capability to model that environment and the system's ability to follow that model.

Future work would include a theoretical analysis and modeling of motor-based systems, as well work toward building a framework for selecting actuators and designing HIL systems from the ground up.

ACKNOWLEDGMENT

This research has been supported by companies of the FPMC Center and the HUSCO/RAMIREZ Chair, and in particular by Siemens, which donated the high performance electric servo motors.

REFERENCES

[1] J. Kim, M. A. S., 2002. "Computationally efficient technique for real time surgical simulation with force feedback". In Proc. 10th Sym. On Haptic Interfaces For Virtual Environment & Teleoperator Systems.

[2] Y. Senta, E. O., 2002. "Hil simulation system for hdd servo firmware". IEEE Transactions on Magnetics, **38** (5), pp. 2204–2207.

[3] Zhen Li, M. Kyte, B. J., 2004. "Hardware-in-the-loop real-time simulation interface software design". In Proc of The 7th International IEEE Conference on Intelligent Transportation Systems, pp. 1012–1017.

[4] W. K. N. Anakwa, H. P. Roca, J. L., 2002. "Environments for rapid implementation of control algorithms and hardware-in-the-loop simulation". In IEEE 28th Annual Conference of the Industrial Electronics Society, vol. 3, pp. 2288–2293.

[5] Leitner, J., 1996. "Space technology using hardware in the loop simulation". In Proc of IEEE Aerospace Applications Conference, vol. 2, pp. 303–311.

[6] Z. H. Akpolat, G. M. Asher, J. C. C., 1999. "Dynamic emulation of mechanical loads using a vector-controlled induction motor-generator set". IEEE Transactions on Industrial Electronics, **46** (2), pp. 370–379.

[7] L. Qinghe, P. Zongcai, W. S., 1999. "A study on producing mechanism and eliminating methods of the disturbance torque in electrohydraulic load simulator". In Fourth JHPS International Symposium on Fluid Power.

[8] P. K. Guerrier, K. A. E., 2001. "Hydraulic emulation of injection moulding". In Proc ASME International Mechanical Engineering Congress & Exposition.

[9] J. L. Lahti, S. J. Andrasko, J. J. M., 2003. "A transient hydrostatic dynamometer for single cylinder engine research with real time multi-cylinder dynamic simulation". In Proc ASME International Mechanical Engineering Congress & Exposition, pp. 223–230.

[10] R. Zhang, D. E. Carter, A. G. A., 2003. "Multivariable control of an earthmoving vehicle powertrain experimentally validated in an emulated working cycle". In Proc ASME International Mechanical Engineering Congress & Exposition.

[11] E. A. Prasetiawan, R. Zhang, A. G. A., 1999. "Modeling and control design of a powertrain simulation testbed for earthmoving vehicles". *Journal of Fluid Power Systems and Technology, Transactions of the ASME,* **6**, pp. 99–105.

[12] C. Ramden, A. Jansson, J. O. P., 1998. *Design and analysis of a load simulator for testing hydraulic valves*. Research Studies Press Ltd., Baldock, Hertfordshire, England. In C. R. Burrows and K. A. Edge, editors, Fluid Power Engineering: Challenges and Solutions, Tenth Bath International Fluid Power Workshop.

[13] C. Ramden, J. O. P., 1998. "Load simulation-an effective tool for testing hydraulic valves". In Proceedings of The Sixth Scandinavian International Conference on Fluid Power.

[14] C. Ramden, J. O. P., 1999. Experimental investigation of the accuracy of a load simulator for testing fluid power valves. Tech. Rep. LiTH-IKP-R-1040, Division of Fluid and Mechanical Engineering Systems, Department of Mechanical Engineering, Linkoping University, Linkopoing. SE-581 83.

[15] S. Jung, Y. J. Lee, W. J. B., 2003. "Hardware-in-the-loop simulation for control development in ehpv applications". In Proc ASME International Mechanical Engineering Congress & Exposition.

[16] N. D. Manring, S. B. K., 2002. "The theoretical flow ripple of an external gear pump". In Proc ASME International Mechanical Engineering Congress & Exposition, pp. 19–27.

Appendix A

Electric Motors:

(2) Siemens 3 phase brushless servomotor:
1FT6102-8AF71-3AK1

Nominal torque: 19.5 Nm

Nominal speed rating: 3000 rpms
Inertia: 9.9e-3 kgm2

Hydraulic Motors:

(2) Sauer Danfoss gear motors: SNM2/95 C106 MER1/1D
Displacement: .58 cubic inches / rev
Inertia: 35e-6 kgm2
Coupling Inertia: .0181 in-lbs-s2

Proportional Valve:

Sauer Danfoss Proportional Valve: PVG 32 with
PVES electrical actuator

Backhoe:

John Deere 4410 series tractor and model 47 backhoe

Proceedings of IMECE2005
2005 ASME International Mechanical Engineering Congress and Exposition
November 5-11, 2005, Orlando, Florida USA

IMECE2005-82237

MICRO-PRECISION HYDROSTATIC ACTUATION SYSTEM

S.R. Habibi, R. Burton and E. Sampson
University of Saskatchewan,
57 Campus Drive, Saskatoon, SK, Canada

ABSTRACT

This paper reports on an important discovery that is: hydrostatic actuation systems are able to manipulate heavy loads with sub-micron precision and large stroke. In this relation, the design of a high-precision hydrostatic actuation system referred to as the ElectroHydraulic Actuator (EHA) is presented. A laboratory prototype of this system has achieved an unprecedented level of performance by being able to move a large load of 20Kg with a precision of 50 nanometers and a stroke of 12cm. This level of performance places the hydrostatic actuation concept in competition with piezoelectric platforms in terms positional accuracy. Experimental results from this prototype are reported and analyzed.

***Index terms*—hydrostatic, electro-hydraulic, actuator**

1. INTRODUCTION

This paper summarizes the findings that were previously reported in publications [1-4] and augments them with new experimental results that collectively lead to a very significant discovery pertaining to hydrostatic actuation systems. The hypothesis that initiated this research was that with inner-loop flow control, hydrostatic actuation can achieve the same level of positional accuracy as servo-valve controlled hydraulic systems. The element of discovery is that they can go well beyond this target and achieve sub-micron and possibly nano precision. This manuscript reports on recent experimental results that demonstrate for the first time the sub-micron precision capability of hydrostatic actuation systems and supports these results with theoretical analysis.

In section 2 of this paper, an overview of the EHA design is provided. The mathematical model of the EHA is reviewed in Section 3. The EHA prototype and the latest experimental results are presented and analyzed in section 4. Section 5 contains the concluding remarks. The nomenclature is listed in Table 1.

1.1 Nomenclature

Symbol	Definitions
I_c, V_c	Control input current and voltage.
G_1, G_2, G_{CS}, G, T, G_{pm}, G_h	Transfer functions.
K, K_1, K_2	Constants.
$DEN_T(s)$	Function.
β_e	Effective bulk modulus of hydraulic oil.
$K_{fric}, K_{P_{visc}}$	Coefficient of friction and viscosity.
K_c	Motor gain.
K_{a_1}	Conversion factor linking maximum actuator displacement to maximum motor speed.
K_{a_2}	Conversion factor linking input to the motor to motor speed.
K_ω	Back emf.
R_c, L_c	Motor line to line resistance and inductance.
J_{pm}	Motor/pump inertia.
K_{PCS}, K_{ICS}	Inner-loop proportional and integral gains.
K_{pos}	Outer-loop proportional gain.
ξ	Pump cross-port leakage coefficient.
Q_a, Q_b, Q_1, Q_2	Pump/actuator flows.
Q_L	Load flow.
V_a, V_b	Pipe plus actuator chamber volumes.

V_o	Pipe plus mean actuator chamber volume.
$V_{acc}, V_{acc_{initial}}$	Accumulator volume.
x_o, x	The mean position and displacement from the mean position.
A	Pressure area in symmetrical actuators.
P_1, P_2	Actuator chamber pressures.
P_a, P_b	Pump port pressures.
P_{case}, Q_{case}	Pressure and flow at the case drain.
$Q_{acc_{in}}, Q_{acc_1}, Q_{acc_2}$	Flow terms to and from the accumulator.
$P_{acc}, P_{acc_{initial}}$	Accumulator pressure.
$\Delta P_{line}, \Delta P_{ele}$	Pressure drops due to line, bend, and elevation difference.
D_p	Pump volumetric displacement.
s	Laplace variable.
T_{DB}	Torque loss due to dead-band.
F	Force.
T_m, T_{load}	Motor/load torque.
T_c	Friction torque including dead-band $T_c = K_{fric}\omega_p + T_{DB}$.
θ_m	Motor angular position.
B	Damping.
ω_{n_h}, ζ_h	Hydraulic subsystem natural frequency and damping ratio.
ω_p	Pump angular velocity.
$\omega_{P_{op}}$	Operating angular velocity for linearization.
ω_d	Demanded pump angular velocity.
L, L_e	External leakage coefficients
K_{pipe}	Coefficient relating pressure drop to flow.
κ	Hydraulic sub-system gain.
ω_{max}	Maximum motor angular velocity.
P_{pipe}	Minimum system pressure in the outer-circuit.
$P_{valve_{break}}$	Valve break pressure.
P_r	Pressure drop across pipe.
Acc	EHA positional accuracy.
Acc_{DB}	Maximum accuracy limited by dead-band.
$X_{sensor_{acc}}$	Accuracy of the outer-loop position sensor.
Ω_p	Accuracy of the inner-loop pump-speed.
τ_{pm}	Time constant.
γ	$K_{P_{CS}} / K_{I_{CS}}$ ratio.

Table 1: Nomenclature

2. THE ELECTROHYDRAULIC ACTUATOR (EHA)

In hydrostatic systems, the pumping action controls the motion of the actuator by moving oil from one actuator chamber to the other (Figure 1a), [5-8]. Traditionally, pump controlled hydraulic circuits were used as an alternative to valve controlled systems in applications that did not have stringent requirements for positional accuracy, or where energy efficiency was paramount in the application. [9,10]. Pump controlled systems are generally configured in a hydrostatic form in which actuation is achieved by direct coupling of hydraulic pump and hydraulic motor. As a result, the actuator displacement varies with the pump flow and there are no requirements for oil reservoirs or electrohydraulic servo-valves, [9-13]. Most hydrostatic systems use a variable displacement piston pump, [9, 14]. The piston pump rotates at a constant speed and its pumping action is regulated by moving a swash plate that changes the magnitude and direction of flow. Use of piston pumps reduces the energy efficiency of these systems to a small degree, as the pump continuously runs irrespective of the control action. Furthermore, piston pumps result in ripple effects that can degrade high precision motion, [9, 15].

An alternative strategy is to vary the flow by changing the speed and direction of a fixed displacement gear pump, [11, 13]. Gear pumps produce a smaller amount of ripple because only one gear is creating the pumping action at any one time. This alternative pumping strategy has been used in the ElectroHydraulic Actuator (EHA) as shown in Figure 1. The EHA consists of: (i) an electric motor, (ii) a gear pump, (iii) a symmetrical actuator, (iv) pressure and position sensors, (v) accumulator, (vi) pressure relief sub-circuit, and (vii) optional filtering sub-circuits.

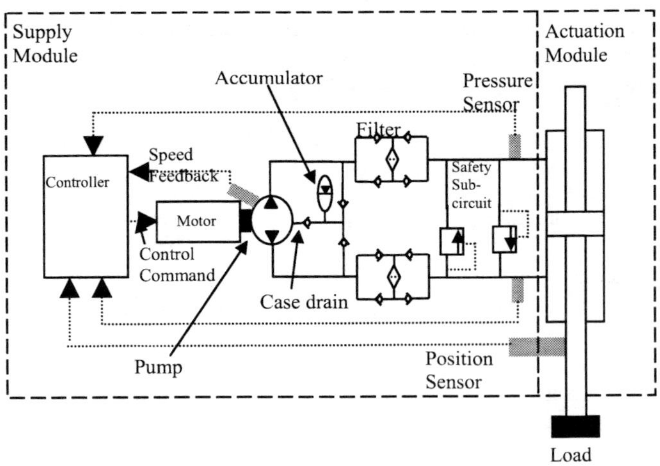

Figure 1: The EHA Circuit Diagram, [1]

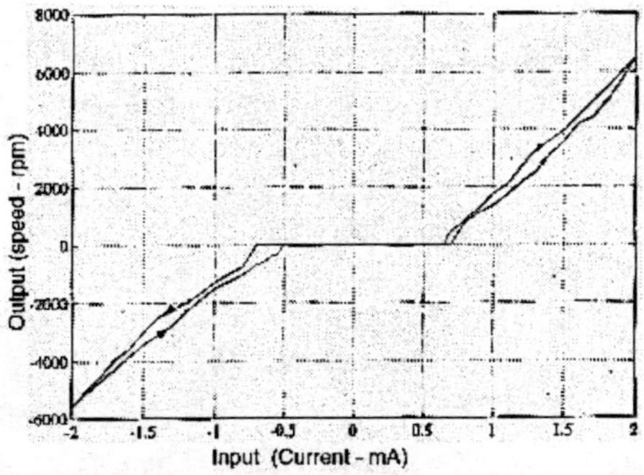

Figure 2: Deadband in Hydrostatic Actuation, [13]

In the EHA, the pump operates only when control action is needed therefore resulting in an overall better energy efficiency in comparison to swash plate hydrostatic and valve controlled hydraulic systems. The disadvantage of using a variable-speed fixed-displacement pump in hydrostatic actuation is that the volumetric efficiency can be compromised at very low pump speeds; in addition, the presence of a dead-band with an effect similar to backlash as shown in Figure 2, can severely limit the positional accuracy and stability of these system types, [13]. The effect of this dead-band was alleviated in the EHA through the implementation of a high-gain inner-loop control strategy involving motor speed feedback as discussed in the next section and as initially reported in [1].

To achieve the circuit simplicity while using linear actuation, a special three chambered linear actuator has been designed and prototyped as shown in Figure 3, [3]. Here the two working chambers are C1 and C2. The solid rod of the conventional piston is replaced by a hollow cylinder that is closed at one end and has a circular disc around the opening at the other end. The surface area of the closed end of the cylinder, A1, is made equal to the area A2 of the disc. A1 and A2 are the active areas of the two pressure chambers and are made equal for symmetry. Chamber C3 can be opened to the atmosphere or could be filled with pressurized gas or fluid in applications where a bias is required to counter balance a dead weight acting under gravity.

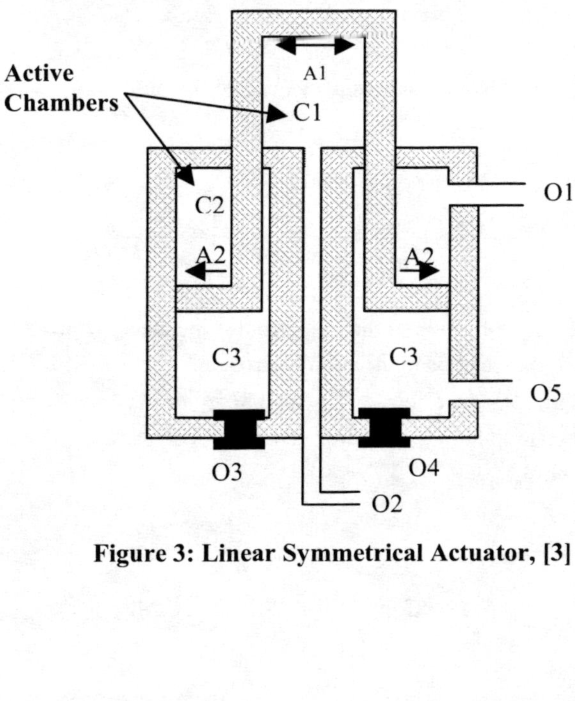

Figure 3: Linear Symmetrical Actuator, [3]

3. A Physical Model of the EHA

The EHA system in Figure 1 can be divided into two hydraulic sub-systems:

- a high pressure outer circuit connected to the Symmetrical Linear Actuator, and
- an inner circuit that employs the low-pressure accumulator.

The inner-circuit (or accumulator sub-circuit) prevents cavitation and replaces leakage losses. During normal operation, the inner-circuit will have negligible effect on the dynamic performance of the EHA and can be neglected in modeling. The mathematical model of the EHA can be constructed by considering its components as follows.

3.1 Linear Symmetrical Actuator

The new Linear Symmetrical Actuator simplifies the hydrostatic circuit by having equal volumetric flows to and from its chambers. Referring to Table 1 for nomenclature, and noting that there is no internal cross-chamber leakage in this actuator, its mathematical model is described by the following flow and force equations:

$$Q_a = A\dot{x} + \frac{A(x_o + x)}{\beta_e}\frac{dP_1}{dt} + LP_1 \qquad (1.)$$

$$Q_b = A\dot{x} - \frac{A(x_o - x)}{\beta_e}\frac{dP_2}{dt} - LP_2 \qquad (2.)$$

Assuming that the EHA is connected to a horizontal sliding mass M with damping of B. The equation of motion for the actuator is obtained as:

$$F = (P_1 - P_2)A = M\ddot{x} + B\dot{x} \qquad (3.)$$

3.2 Hydraulic pump

The hydraulic pump in the EHA is the main control element and converts mechanical energy received from the electric motor to fluid flow. It is a bi-directional fixed displacement gear pump with three ports. Two large diameter input-output ports are directly connected to the Symmetrical Linear Actuator by steel tubing. The third port is a case drain provided for pump leakage. The case drain line is connected to the inner circuit as shown in Figure 1. EHA's positioning accuracy is dependent on the pump volumetric displacement. Intuitively if the pump has a very small displacement per revolution, it will discharge a small amount of fluid per revolution that will result in a very fine movement of the cylinder rod. Therefore, a pump with the smaller volumetric displacement will provide higher control resolution. On the other hand rod speed is proportional to the pump displacement. Consequently, there is a trade-off between maximum speed and positioning accuracy - a higher pump displacement would provide a higher maximum speed but a lower positioning accuracy. The main factors for pump selection are: pump size, efficiency (small internal leakage), working pressure and volumetric displacement.

A small bi-directional gear pump with a volumetric displacement of $D_p = 1.7 \times 10^{-7} m^3 / rad$ is used in the EHA prototype as described later in this paper. The maximum speed of the pump is 418 rev/s (4000 rpm), producing a nominal output pressure of up to 207Bar and a maximum intermittent output pressure of 276Bar. The internal seals of the pump are modified to withstand a maximum case pressure of 6.9 Bar (100 psi). This allows the attachment of the pump's case drain to EHA's accumulator which is maintained at 3 to 5 Bar. This connection is made via a check valve of $P_{valve_{break}} = 6.9$ kPa (1psi) break pressure.

The model of this pump is constructed by calculating the in/out flows (Q_a / Q_b), case drain flow (Q_{case}), and load torque (T_{load}) by using the following equations:

$$Q_a = D_p\omega_p - \xi(P_a - P_b) - L_e(P_a - P_{case}) + \frac{V_a}{\beta_e}\frac{dP_a}{dt} \quad (4.)$$

$$Q_b = D_p\omega_p + \xi(P_a - P_b) - L_e(P_b - P_{case}) + \frac{V_b}{\beta_e}\frac{dP_b}{dt} \quad (5.)$$

$$Q_{case} = L_e(P_a + P_b - 2P_{case}) \quad (6.)$$

$$T_{load} = D_p(P_a - P_b) + K_{visc}\omega_p + T_c \quad (7.)$$

where, pipe elasticity is considered negligible due to the use of solid tubing and, $V_a = V_o + Ax$ and $V_b = V_o - Ax$.

3.3 Accumulator

A small bladder-type accumulator is used to prevent:

- cavitation, and
- excessive case pressure build up in the pump.

The model used for the accumulator sub-circuit is as follows:

If $P_{acc} - P_a > P_{valve_{break}}$, $Q_{acc_1} = K_{valve}(P_{acc} - P_a)$

If $P_{acc} - P_a < P_{valve_{break}}$, $Q_{acc_1} = 0$

If $P_{acc} - P_b > P_{valve_{break}}$, $Q_{acc_2} = K_{valve}(P_{acc} - P_b)$

If $P_{acc} - P_b < P_{valve_{break}}$, $Q_{acc_2} = 0$

If $P_{case} - P_{acc} > P_{valve_{break}}$, $Q_{acc_{in}} = Q_{case}$

If $P_{case} - P_{acc} < P_{valve_{break}}$, $Q_{acc_{in}} = 0$

The total in/out flows and the pressure of the accumulator are obtained as:

$$Q_{acc} = Q_{acc_{in}} - Q_{acc_1} - Q_{acc_2} \quad (8.)$$

Assuming isothermic expansion and compression of gas:

$$P_{acc} = \frac{P_{acc_{initial}} V_{acc_{initial}}}{V_{acc_{initial}} + \int Q_{acc}\,dt} \quad (9.)$$

3.4 Control Strategy

The control strategy is critical to the performance of the EHA. Here, two control loops are used as follows.

- A high-gain inner-loop controller (loop 1) is used as shown in Figure 4 to regulate the speed of pumping and hence provide an accurate control over pump flow. This flow control strategy desensitize the system to dead-band caused by friction at the pump motor interface, [1].
- An outer-loop controller (loop 2) allows precision control of the output variable which could be speed and/or position.

In the EHA prototype, inner-loop control of the pump velocity is accurate to within 1 revolution per minute (or $\Omega_p = 0.1 rad/s$). Furthermore, the load position is measured by a linear optical encoder with a resolution of 50 nanometers. The Controller sampling rate is 1 kHz.

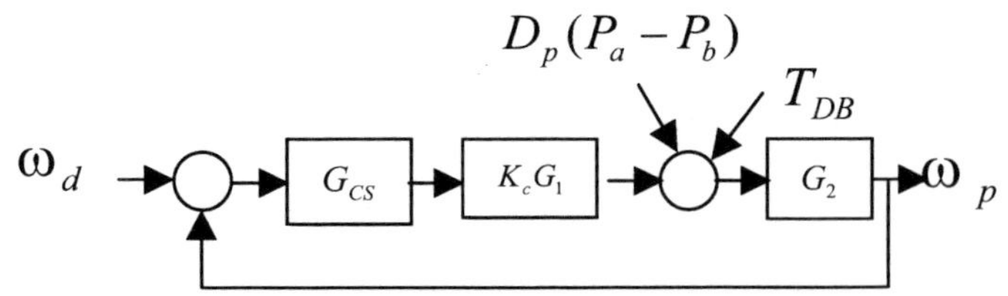

Figure 4: Control Block Diagram

3.5 Removal of Pump/Motor Dead-band by Using High-gain Inner-loop Speed Control

The hydraulic pump is driven by a high performance three phase AC star-connected electrical motor. Careful selection of the motor in conjunction with a high-gain inner-loop proportional-integral motor-speed control strategy can substantially reduce the dead-band of Figure 2. An analysis of the inner-loop controller is presented in this section.

Referring to Table 1 for nomenclature, the relationship between the field current and the input voltage of this three phase motor can be modeled by a first order transfer function such that: $I_c = G_1(V_c - K_\omega\dot{\theta})$ where: $G_1 = \frac{1/R_c}{L_c/R_c s + 1}$. The torque generated by the motor is characterized by $T_m = K_c I_c$. Given that the motor is connected directly to the pump, the torque exerted on the pump is given by, [9]:

$$T_m = J_{pm}\dot{\omega}_p + (K_{P\,visc} + K_{fric})\omega_p + T_{DB} + D_p(P_a - P_b) \quad (10.)$$

Rearranging the above equations, the speed of the pump is obtained as:

$$\omega_p = \frac{K_c G_1 V_c}{J_{pm}s + (K_{P_{visc}} + K_{fric} + K_\omega K_c G_1)} - \frac{D_p(P_a - P_b)}{J_{pm}s + (K_{P_{visc}} + K_{fric} + K_\omega K_c G_1)} - \frac{T_{DB}}{J_{pm}s + (K_{P_{visc}} + K_{fric} + K_\omega K_c G_1)} \quad (11.)$$

T_{DB} denotes static friction at the pump/motor interface, which contributes to the dead band effect of hydrostatic circuits as shown in Figure 2. The effect of T_{DB} can be minimized by using an inner-loop high-gain motor velocity-feedback as shown in Figure 4. In this figure,

$$G_2 = \frac{1/(K_{P_{visc}} + K_{fric} + K_\omega K_c G_1)}{\frac{J_{pm}}{(K_{P_{visc}} + K_{fric} + K_\omega K_c G_1)} s + 1} \quad (12.)$$

A proportional-integral strategy is used for speed control such that $G_{CS} = \left(K_P + \frac{K_I}{s}\right)$. The overall closed loop function is then obtained as:

$$\omega_p = \frac{G_{CS} K_C G_1 G_2}{1 + G_{CS} K_C G_1 G_2} \omega_d - \frac{G_2}{1 + G_{CS} K_C G_1 G_2} T_{DB} - \frac{G_2}{1 + G_{CS} K_C G_1 G_2} D_P (P_a - P_b) \quad (13.)$$

This equation reveals that the effect of T_{DB}, which is highly undesirable, can be removed by using a high gain inner loop velocity controller (such that $G_2 << G_{CS} K_C G_1 G_2$) and careful selection of the motor such that $|K_c G_1| >> |G_2|$. The above analysis treats T_{DB} as a disturbance input.

3.6 Filters and Pressure Drop in the Connecting Lines

Filtering is desirable in the industrial version of the EHA to extend the service life of hydraulic elements by preventing oil contamination. As the pump is bi-directional, none of the pressure lines can be designated specifically as forward or return. Filtering can be achieved by using rectifier circuits that ensure one-way flow through their corresponding filter elements. The filtering sub-circuits are not used in the EHA prototype.

3.7 Hydraulic Subsystem Model

During normal operation, the inner-circuit will have negligible effect on the dynamic performance of the EHA and the pump model may be simplified to the following:

$$Q_a = D_p \omega_p - \xi (P_a - P_b) - \frac{V_a}{\beta_e} \frac{dP_a}{dt} \quad (14.)$$

$$Q_b = D_p \omega_p - \xi (P_a - P_b) + \frac{V_b}{\beta_e} \frac{dP_b}{dt} \quad (15.)$$

where, pipe elasticity is considered negligible and, $V_a = V_o + Ax$ and $V_b = V_o - Ax$.

The dynamic significance of the accumulator flow can be assumed negligible, resulting in a simplification of the model such that $Q_a = Q_1$ and $Q_b = Q_2$. Given that the in/out flows from the pump are equal to the out/in flows of the actuator, the load flow, Q_L, is obtained as:

$$Q_L = \frac{Q_1 + Q_2}{2} = \frac{Q_a + Q_b}{2} \quad (16.)$$

In the prototype, the filtering sub-circuit is not implemented. Due to the symmetry of EHA, the pump/actuator pipe connection is modeled as a pressure drop P_{pipe} which is a function of flow. P_{pipe} is approximately modeled by considering the elevation of the lines (connecting the pump and the actuator) and, by using Darcy's pipe flow equation as $P_{pipe} \approx 2K_{pipe} D_p^2 (\omega_p^2) + P_{ele}$. In a linearized form it can be expressed at the operating flow ω_{Pop} as:

$$\Delta P_{pipe} \approx 2K_{pipe} D_p^2 \omega_{Pop} (\Delta \omega_p) \quad (17.)$$

The relationship between the pump port pressures and actuator chamber pressures can be approximated to: $P_a = P_1 + P_{pipe}$, $P_b = P_2 - P_{pipe}$. Substituting and equating from the pump and actuator flow equations (1), (2), (14) to (16), the following is obtained:

$$A\dot{x} + \frac{V_o}{2\beta_e}\left(\frac{dP_1}{dt} - \frac{dP_2}{dt}\right) + \frac{Ax}{2\beta_e}\left(\frac{dP_1}{dt} + \frac{dP_2}{dt}\right) + \frac{L}{2}(P_1 - P_2) = D_p \omega_p - \frac{V_o}{2\beta_e}\left(\frac{dP_a}{dt} - \frac{dP_b}{dt}\right) - \frac{Ax}{2\beta_e}\left(\frac{dP_a}{dt} + \frac{dP_b}{dt}\right) - \xi(P_a - P_b) \quad (18.)$$

Since $\frac{dP_a}{dt} \approx \frac{dP_1}{dt}$ and $\frac{dP_b}{dt} \approx \frac{dP_2}{dt}$, and due to the symmetry of the actuator $\frac{dP_1}{dt} \approx -\frac{dP_2}{dt}$, a simplified pump/actuator model is obtained as:

$$D_p \omega_p = A\dot{x} + \frac{V_o}{\beta_e}\left(\frac{dP_1}{dt} - \frac{dP_2}{dt}\right) + \xi(P_1 - P_2) + 2\xi P_{pipe} + \frac{L}{2}(P_1 - P_2) \quad (19.)$$

The prototype of the EHA is connected to a horizontal sliding mass M with damping of B as specified by equation (3). As such, substituting $(P_1 - P_2)$ by $(M\ddot{x} + B\dot{x})/A$, and rearranging, then:

$$x(s) = \frac{Dp\omega_p(s)}{s^3\left(\frac{MV_o}{\beta_e A}\right) + s^2\left(\frac{LM/2 + \xi M + BV_o/\beta_e}{A}\right) + s\left(\frac{A^2 + LB/2 + \xi B}{A}\right)} - \frac{2\xi P_{pipe}(s)}{s^3\left(\frac{MV_o}{\beta_e A}\right) + s^2\left(\frac{LM/2 + \xi M + BV_o/\beta_e}{A}\right) + s\left(\frac{A^2 + LB/2 + \xi B}{A}\right)} \quad (20.)$$

Substituting from equation (17), the transfer function of the hydraulic part in a linearized form is obtained as:

$$G_h(s) = \frac{\Delta x(s)}{\Delta \omega_p(s)} = \frac{Dp(1 - 2K_{pipe}\xi D_p \omega_{P_{op}})}{s^3\left(\frac{MV_o}{\beta_e A}\right) + s^2\left(\frac{LM/2 + \xi M + BV_o/\beta_e}{A}\right) + s\left(\frac{A^2 + LB/2 + \xi B}{A}\right)} \quad (21.)$$

This transfer function describes the hydraulic system and can be presented in a general form as:

$$\frac{\Delta x(s)}{\Delta \omega_p(s)} = \frac{\kappa_h \omega_{n_h}^2}{s\left(s^2 + 2\zeta_h \omega_{n_h} s + \omega_{n_h}^2\right)} \quad (22.)$$

where:

- $\omega_{nh} = \sqrt{\left(\frac{A^2 + LB/2 + \xi B}{MV_o}\right)\beta_e}$ is the hydraulic undamped natural frequency,

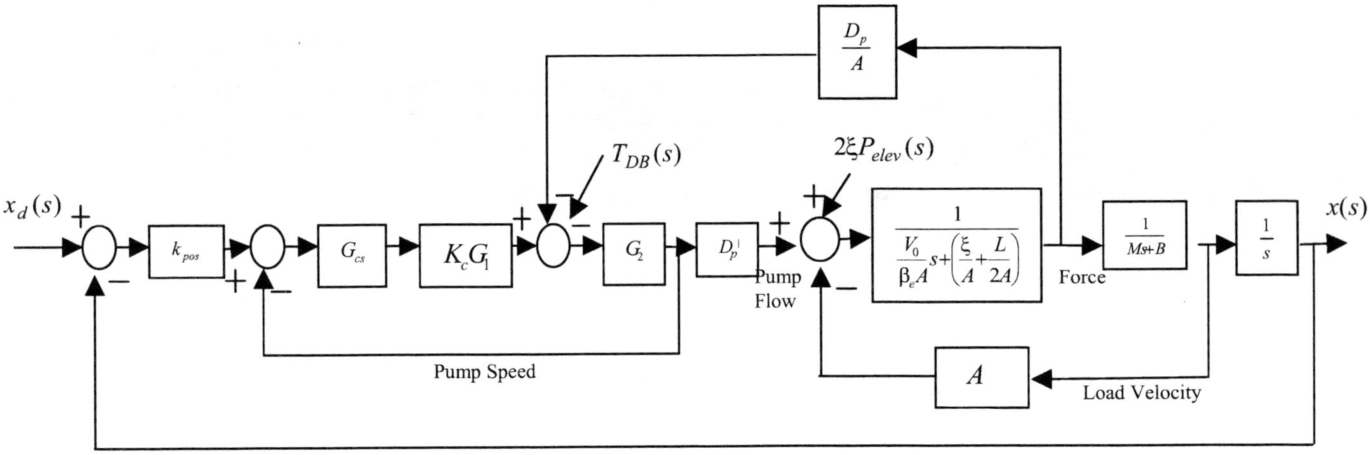

Figure 5: Linearized EHA Model Block Diagram

- $\zeta_h = \left(\dfrac{LM/2+\xi M+BV_o/\beta_e}{2\omega_{n_h} MV_o}\right)\beta_e$ is the hydraulic damping ratio, and
- $\kappa = \dfrac{D_p\left(1-2K_{pipe}\xi D_p\omega_{p_{op}}\right)A}{A^2+LB/2+\xi B}$ is the hydraulic system gain.

The hydraulic system gain is therefore proportional to the ratio of the pump displacement over piston area. Furthermore, since $2K_{pipe}\xi D_p\omega_{p_{op}} << 1$, the hydraulic system gain κ can be approximated to $\kappa = \dfrac{D_p}{A}$. This confirms the earlier intuitive analysis indicating that a faster system response can be obtained by a higher pump displacement.

3.8 Overall Linearized Model

Using an outer loop position controller such that input, $\omega_d = K_{pos}(x_d - x)$ and substituting:

- ω_p from equation (13) in (19);
- $(P_1 - P_2)$ by $(M\ddot{x}+B\dot{x})/A$; and
- rearranging, and defining:

$$DEN(s)=s^3\left((1+G_{CS}K_CG_1G_2)\frac{MV_o}{\beta_e A}\right)+s^2\left((1+G_{CS}K_CG_1G_2)\left(\frac{LM/2+\xi M+BV_o/\beta_e}{A}\right)+MG_2D_p^2/A\right)$$
$$+s\left((1+G_{CS}K_CG_1G_2)\left(\frac{A^2+LB/2+\xi B}{A}\right)+G_2BD_p^2/A\right)+D_pG_{CS}K_CG_1G_2K_{pos} \quad (23.)$$

the following model is obtained:

$$x=\frac{D_pG_{CS}K_CG_1G_2K_{pos}x_d}{DEN(s)}-\frac{G_2D_pT_{DB}+(1+G_{CS}K_CG_1G_2)2\xi P_{pipe}+2P_{pipe}G_2D_p^2}{DEN(s)} \quad (24.)$$

Assuming that the term $2K_{pipe}\xi D_p\omega_{p_{op}} << 1$ is negligible (resulting in $P_{pipe} \approx P_{e;ev}$), a linearized mathematical model of the EHA may be summarized by the block diagram in Figure 5.

4. THE EHA PROTOTYPE AND ANALYSIS OF EXPERIMENTAL RESULTS

A prototype of the EHA with the new Symmetrical Linear actuator has been produced. This prototype is shown in Figure 6 and has been able to move a load of 20Kg with a stroke of 12 cm at a precision of 50 nm. Figure 7 illustrates the step response of the EHA prototype. This unprecedented combination of precision, large load and large stroke places this concept in competition with piezoelectric platforms. Other performance specifications are as follows:

Maximum working Pressure	207 Bar (3000 psi)
Maximum Output Force	10.5 kN
Maximum Piston Stroke	12 cm
Maximum speed	14 cm/s
Closed loop position accuracy	50 nanometers
System Bandwidth	22 Hz (load of 20kg)
Force to weight ratio	1.1 kN/kg (actuator)
Deadband	Negligible
Rise time	0.3 sec (load of 20 kg)

The physical parameters associated with the EHA prototype are listed in Table 2 and were obtained through measurements and estimation as described in [16].

Figure 6: The EHA Prototype

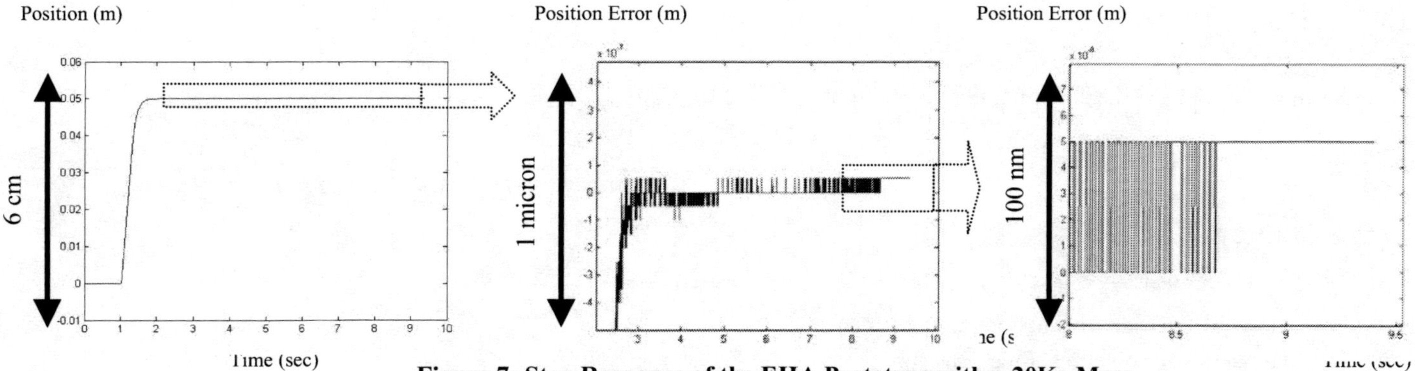

Figure 7: Step Response of the EHA Prototype with a 20Kg Mass

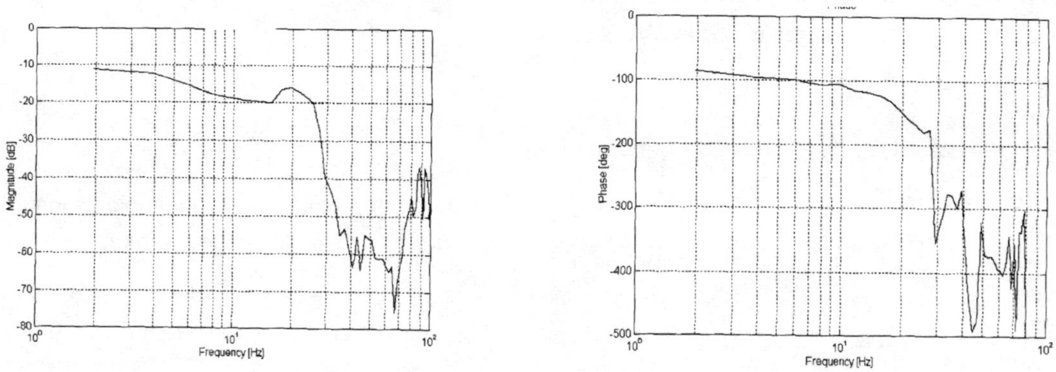

8a) Estimated Gain Plot

8b) Estimated Phase Plot

Figure 8: ETFE Gain and Phase plots for the ibnput amplitude of 1 Volts, [4]

The following sections provide a report on important findings observed during experimentation on the EHA prototype. These observations pertain to the EHA's precision, dynamic order of significance, and linearity. They are then explained by using the theoretical model of the EHA as presented in the previous section.

4.1 Dynamic Characteristics

Observation and Method: Experimental analysis of the EHA by using methods used in system identification have revealed that the EHA is piecewise linear and is dominantly characterized by a third order type 1 transfer function. The model structure was determined experimentally by considering the separation of the singular values from a Hankel matrix of Markov parameters generated by taking the derivative of the step response from the EHA. Piecewise linearity was investigating by generating a succession of Empirical Transfer Function Estimates (ETFE) for different ranges of input. Essentially an ETFE consists of estimated gain and phase characteristics obtained by taking the ratio of the discrete Fourier transforms of the output and input signals. To obtain this transfer function, a series of open loop tests were performed on the EHA. A typical ETFE estimate of the amplitude and phase plots for the input amplitude range of –1 to 1 Volts is presented in Figure 8 and confirms a dominant third order type 1 characteristic.

Explanation: The model structure of the EHA can be confirmed by considering its design strategy that uses a high gain inner-loop controller. The gain of the inner-loop controller, its dynamic complexity and its speed of response have a large impact on the dynamic performance of the EHA.

The transfer function relating the demanded pump speed to the actual pump speed can be obtained from equation (13) as:

$$G_{pm} = \frac{G_{CS}K_C G_1 G_2}{1+G_{CS}K_C G_1 G_2} = \frac{G_{CS}K_C / R_c}{J_{pm}s + (K_{P_{visc}} + K_{fric} + K_{\omega}K_C / R_c + G_{CS}K_C / R_c)} \qquad (25.)$$

In the prototype, proportional-integral action is used in the inner-loop control strategy in order to improve the overall accuracy of the system such that $G_{CS} = K_{P_{CS}} + \frac{K_{I_{CS}}}{s}$. As such, the transfer function G_{pm} becomes second order as follows:

$$G_{pm} = \frac{(K_{P_{CS}}s + K_{I_{CS}})K_C / R_c}{J_{pm}s^2 + (K_{P_{visc}} + K_{fric} + K_{\omega}K_C / R_c + K_{P_{CS}})s + K_{I_{CS}}K_C / R_c)} \qquad (26.)$$

The time constant of the overall pump speed mechanism is considerably faster in the prototype than the hydraulic/ mechanical time constant at the load. This results in reduction of the order of our model and the simplification of G_1 from $G_1 = \frac{1/R_c}{L_c/R_c s + 1}$ to $G_1 = \frac{1}{R_c}$, and G_2 from $G_2 = \frac{1/(K_{p_{visc}} + K_{fric} + K_\omega K_c G_1)}{\frac{J_{pm}}{(K_{p_{visc}} + K_{fric} + K_\omega K_c G_1)} s + 1}$ to $G_2 \approx 1/(K_{p_{visc}} + K_{fric} + K_\omega K_c / R_{c1})$.

In the EHA prototype, the transfer function G_1 is obtained as $G_1 = \frac{2}{0.008s + 1}$. Given that the electrical time constant $L_c / R_c = 0.008$ is approximately 100 times faster than the time constant of the hydraulic circuit in this system, G_1 may thus be simplified to $G_1 \approx \frac{1}{R_c} = 2$ and, G_2 correspondingly to $G_2 \approx \frac{0.83}{0.002s + 1} \approx 0.83$. Model reductions of G_1 and G_2 result in a simplified motor transfer function:

$$G_{pm} = \frac{G_{CS} K_C G_1 G_2}{1 + G_{CS} K_C G_1 G_2} \approx \frac{G_{CS} K_C / R_c}{(K_{p_{visc}} + K_{fric} + K_\omega K_C / R_c + G_{CS} K_C / R_c)} \quad (27.)$$

Symbol	Values
K_c	$1Nm/A$
K_ω	$0.6V/rad/s$
R_c	0.5Ω
L_c	$0.004H$
J_{pm}	$1.6 \times 10^{-3} Kgm^2$
M	$20Kg$
K_p	$1.2 \times K_{a_2}$
K_I	$2.4 \times K_{a_2}$
K_{pos}	$1 \times K_{a_1}$
V_o	$6.85 \times 10^{-5} m^3$
A	$5.05 \times 10^{-4} m^2$
$P_{sensor_{acc}}$	$1 \times 10^{-6} m$
$V_{sensor_{acc}}$	$1rad/s$
K_{a_1}	$6980 rad/s/m$
K_{a_2}	$41.87 rad/s/V$
D_p	$1.7 \times 10^{-7} m^3/rad$
M_a	$8Kg$
$x_{\max}$	$12cm$
β_e	$2.1 \times 10^8 Pa$
K_{fric}	$2 \times 10^{-4} Nm/rad/s$
B	760 N/m/s
$\frac{L}{2} + \xi$	$5 \times 10^{-13} m^3/s/Nm^{-2}$
K_{pipe}	$2.5 \times 10^{12} Pa/m^6/s^2$

Table 2: Physical Parameters of the Prototype

Substituting for $G_{CS} = K_{P_{CS}} + \frac{K_{I_{CS}}}{s}$, G_{pm} is obtained as:

$$G_{pm} = \frac{G_{CS} K_C G_1 G_2}{1 + G_{CS} K_C G_1 G_2} \approx \frac{(\gamma s + 1)}{((K_{p_{visc}} + K_{fric}) R_c / (K_{I_{CS}} K_c) + K_\omega / K_{I_{CS}} + \gamma) s + 1)} \quad (28.)$$

where $\gamma = \frac{K_{P_{CS}}}{K_{I_{CS}}}$.

For high inner loop gains and $\frac{K_{p_{visc}} R_c / K_C + K_{fric} R_c / K_C + K_\omega}{K_{I_{CS}}} << \gamma$, these result in:

$$G_{pm} = \frac{(\gamma s + 1)}{(\frac{K_{p_{visc}} R_c / K_C + K_{fric} R_c / K_C + K_\omega}{K_{I_{CS}}} + \gamma) s + 1)} \approx \frac{(\gamma s + 1)}{(\gamma s + 1)} = 1 \quad (29.)$$

In the EHA prototype, the motor and its inner-loop controllers are fast enough to allow the approximation of G_{pm} to $G_{pm} \approx 1$. Furthermore, for large inner-loop gains, a dynamic coupling effect in the system termed as $\frac{D_p(Ms + B)}{AG_{CS} K_c G_1}$ becomes insignificant and, a simplified transfer function for our system can then be expressed as:

$$G_h(s) = \frac{x(s)}{\omega_p(s)} = \frac{Dp}{s^3\left(\frac{MV_o}{\beta_e A}\right) + s^2\left(\frac{LM/2 + \xi M + BV_o/\beta_e}{A}\right) + s\left(\frac{A^2 + LB/2 + \xi B}{A}\right)} \quad (30.)$$

$$\equiv \frac{\kappa_h \omega_{n_h}^2}{s(s^2 + 2\zeta_h \omega_{n_h} s + \omega_{n_h}^2)} \approx \frac{13.2}{s(s^2 + 40s + 39150)}$$

The corresponding natural frequency, damping ratio, and DC gain are obtained from equation (30):

$$\omega_{nh} = \sqrt{\left(\frac{A^2 + LB/2 + \xi B}{MV_o}\right)\beta_e} = 198 rad/s \quad (31.)$$

$$\zeta_h = \left(\frac{LM/2 + \xi M + BV_o/\beta_e}{2\omega_{n_h} MV_o}\right)\beta_e = 0.1 \quad (32.)$$

$$\kappa = \frac{D_p A}{A^2 + LB/2 + \xi B} = 0.34 \times 10^{-3} \quad (33.)$$

This theoretical transfer function of the EHA prototype confirms the experimental result of Figure 8 that indicates a third order type 1 system characteristic, a DC gain of approximately –70db, and a relatively large resonant peak (indicating a low damping ratio) at around 30Hz corresponding approximately to the natural frequency of 198 rad/s. The bandwidth of the EHA is comparable to that of a servo-valve driven hydraulic system, but significantly lower than piezoelectric actuators.

4.2 Nonlinearity

<u>*Observation:*</u> Further to its mathematical model, the EHA is a nonlinear system. Nonlinearity in most systems is manifested

according to their operating envelope. In this case, one of the main determinants of the operating envelope is the magnitude of the input signal to the motor that in the case of the EHA prototype is in the range of +/-10V. The nonlinear characteristic can be observed by comparing the amplitude and phase plots generated experimentally by using ETFEs for different input amplitudes. ETFE plots were obtained for coarse movement of the EHA starting from an input amplitude to the electric motor of 1 Volts to 10 Volts in 1 Volt steps. The ETFE characteristics changed at 1 Volts and 7 Volts. This points to three piece-wise linear regions corresponding to input voltage levels of less than 1 Volts, intermediary level of 1 to 7 Volts and high range exceeding 7 Volts, [4].

Explanation: ETFE experimentation revealed that the nonlinearity occuring at 7 Volts is due to the electrical motor of the EHA. The nonlinearity occurring at 1 Volts is due to the inner circuit of Figure 1 that limits the minimum pressure in the return line to that of the accumulator. This can be confirmed by considering this effect on the mathematical model of the EHA as follows. Assuming a constant accumulator pressure P_{acc} and a minimum system pressure P_r that is approximately the accumulator pressure minus the pressure drop across the check valve ($P_{valve_{break}}$ = 6.9 KPa), the governing equations of the actuator change to:

$$\left(Q_a = D_p\omega_p - \xi(P_a - P_r + P_{pipe}) - \frac{V_a}{\beta_e}\frac{dP_a}{dt}\right) = \left(Q_1 = A\dot{x} + \frac{(V_o + Ax)}{\beta_e}\frac{dP_1}{dt} + LP_1\right) \quad (34.)$$

$$\left(Q_b = D_p\omega_p - \xi(P_a - P_r)\right) = \left(Q_2 = A\dot{x} - LP_r\right) \quad (35.)$$

$$D_p\omega_p = A\dot{x} + \frac{V_1}{2\beta_e}\left(\frac{dP_1}{dt}\right) + \xi(P_1 - P_r) + \xi P_{pipe} + \frac{L}{2}(P_1 - P_r) \quad (36.)$$

Further to this saturation effect, the resulting hydraulic transfer function is obtained as:

$$G_h(s) = \frac{x(s)}{\omega_p(s)} = \frac{Dp}{s^3\left(\frac{MV_1}{\beta_e A}\right) + s^2\left(\frac{LM/2 + \xi M + BV_1/\beta_e}{A}\right) + s\left(\frac{A^2 + LB/2 + \xi B}{A}\right)} = \frac{\kappa_h\omega_{n_h}^2}{s\left(s^2 + 2\zeta_h\omega_{n_h}s + \omega_{n_h}^2\right)} \quad (37.)$$

The corresponding natural frequency, damping ratio and DC gain differ from those of equations (29) to (31) and are:

$$\omega_{nh} = \sqrt{\left(\frac{A^2 + LB/2 + \xi B}{MV_1}\right)\beta_e} \quad (38.)$$

$$\zeta_h = \left(\frac{LM/2 + \xi M + BV_1/\beta_e}{2\omega_{n_h}MV_1}\right)\beta_e \quad (39.)$$

$$\kappa = \frac{D_p A}{A^2 + LB/2 + \xi B} \quad (40.)$$

The three coarse level piece-wise linear regions of the EHA are thus explained.

It should be noted that the ETFEs generated for small signals associated with micro-precision reveal that piecewise linearity applies also to the sub-micron region, with a general trend that points to decreasing bandwidth. This effect can be associated with un-modeled friction effects. Gain scheduling was used in the outer loop control scheme in the interval of +/- 5 micron error range to amplify the outer-loop proportional-integral gains.

4.2 Precision

Observation: Further to Figure 7, a precision of 50 nm is achieved and reached within 2 seconds of the step demand. It should be noted that the increase in the step response time in comparison to the actuator's rise time is due to saturation of the input voltage to the motor, thus limiting the speed of response.

Explanation: The outer-loop positional accuracy of our system is affected by the performance of the pump-speed inner-loop controller at low angular velocities and its effectiveness in dealing with the dead-band T_{DB}. The highest pump velocity control (inner-loop) accuracy possible in the EHA prototype was 1 rev/minute or $\Omega_p = 0.1 rad/s$.

Given that a proportional-integral controller was used in the inner-loop controller, the effect of dead-band on positional accuracy (Acc_{DB}) can be calculated for *s=j0* from equation (24) as: $Acc_{DB} = 2\xi P_{pipe}/D_p K_{pos}$. From equation (17), P_{pipe} can be modeled as $K_{pipe}Q_L^2 + \Delta P_{elev} \approx K_{pipe}D_p^2\omega_p^2 + \Delta P_{elev}$ and the expected positional accuracy is obtained as:

$$Acc_{DB} = 2\xi K_{pipe}D_p\Omega_p^2/K_{pos} + 2\xi\Delta P_{elev}/K_{pos} \quad (41.)$$

Further to equation (41), the positional accuracy of the EHA is critically dependent on the pump (its volumetric displacement D_p and its cross-port leakage ξ) and its control precision, Ω_p. This is true provided that the theoretical effect of dead-band on the positional accuracy of EHA is greater than the resolution of the position sensor $X_{sensor_{acc}}$ that is 50 nanometers. The outer-loop maximum position control accuracy is then stated as:

$$Acc = max(X_{sensor_{acc}}, Acc_{DB}) \quad (42.)$$

Substituting from Table 2, Acc_{DB} predicts nano-precision in the absence of sensor noise. The EHA prototype has achieved a precision 50nm that is its predicted maximum precision with its present position sensor resolution.

5. CONCLUSIONS

Until recently hydrostatic actuation systems were dismissed as being inaccurate. This paper presents a hydrostatic actuation concept that is capable of moving large inertial loads with large stroke at extreme precision (sub-micron scale). In this relation, a novel actuation system referred to as the ElectroHydraulic Actuator (EHA) has been prototyped and has achieved an unprecedented level of performance by being able to move a large load of 20Kg with a precision of 50 nanometers and a stroke of 12cm. This level of performance places the hydrostatic actuation concept in competition with piezoelectric platforms that are capable of similar positional accuracies and load capability, but are much restricted in their stroke (typically less than 100 microns, with some stacked systems reaching as high as 1mm).

The theoretical foundation for the design of the EHA is provided in this paper and used for an analysis of experimental results obtained from its prototype. Further to this analysis, it is concluded (from equation (41)) that the control accuracy in the pumping action is central to the high performance of the EHA. It should be noted that the EHA prototype used an off-the-shelf conventional gear pump. It is therefore not unreasonable to further conclude that with a customized pump, the EHA concept has the potential to go beyond sub-micron precision actuation. Our results project that with specialized components and control strategies, the EHA can reach nano-precision. As such, the EHA opens a new research domain that is the design of fluid power components and systems for micro and nano manipulation.

References

[1] Habibi S.R. and Goldenberg A.A.,(2000), *Design of a New High Performance ElectroHydraulic Actuator,* IEEE/ASME Transactions on Mechatronics,. Vol. 5 , No 2, pp 158-164.

[2] Habibi S.R. and Singh G.,, (2000) *Derivation of Design Requirements for Optimization of a High Performance Hydrostatic Actuation Systems,* Int. Journal of Fluid Power, Vol. 1, Number 2, pp 11-29.

[3] Habibi S.R. and Goldenberg A.A., (1999) , *Design and analysis of a new symmetrical linear actuator for hydraulic and pneumatic systems.,* Transaction of the CSME, Vol. 23, No. 3 & 4, pp 377-397.

[4] Habibi S.R., Pastrakuljic V. and Goldenberg A.A. (2001), *Experimental Analysis of a High Performance Hydrostatic Actuation System,* SAE Transactions, Journal of Commercial Vehicles, 109, pp 367-377.

[5] Blackburn J. F., Reethof G. and Shearer J. L. (1960), *Fluid Power Control*, The M.I.T. Press.

[6] Merritt H.E. (1967), *Hydraulic Control Systems*, John Wiley & Sons Inc., 1967.

[7] Viersma T. J. (1980), *Analysis, Synthesis and Design of Hydraulic Servo-systems and Pipelines*, Elsevier Scientific Publishing Company.

[8] Habibi S. and Richards R., 1991,*Computed-torque and variable-structure multivariable control of a hydraulic industrial robot:* Pr. Inst Mech Eng., Part I, Vol 205.

[9] Watton J., (1989), *Fluid Power Systems*, Prentice Hall.

[10] Korn J., (1972), *Hydrostatic Transmission Systems*, Int. Textbook Ltd.

[11] Manring N.D. and Luecke G.R., (1998), *Modeling & Designing a Hydrostatic Transmission with a Fixed-Dis. Motor*, J. Dyn. Sys. Meas. & Cont. Vol 120, pp 45-50.

[12] Desai J. and Bobrow J.E., (1989), *Modeling and Analysis of a High Torque Hydrostatic Actuator for Robotic Applications*, Exp. Robotics.

[13] Arnautovic S., (1993), *Electrohydraulic Actuator,* Report,Univ. Toronto.

[14] Lennevi J., (1995), *Hydrostatic Transmission Control*, Linkoping Studies in Sc.&Tech., No.395.

[15] Manring N., (2000), *The discharge flow ripple of an axial-piston swash-plate type hydrostatic pump*, J. Dyn. Sys. Meas.&Cont., V122.

[16] Chinniah Y., (2004), *Fault Detection in the EHA using the EKF,* Ph.D. Thesis, University of Saskatchewan.

Proceedings of IMECE2005
2005 ASME International Mechanical Engineering Congress and Exposition
November 5-11, 2005, Orlando, Florida USA

IMECE2005-82350

EFFECT OF PARAMETER UNCERTAINTY ON RELIABILITY OF HYDRAULIC TRANSMISSION SYSTEM SIMULATION

A. ELLMAN
Institute of Machine Design
Tampere University of Technology
P.O. Box 589
FIN-33101, Tampere, Finland
asko.ellman@tut.fi

H. KAURANNE
Department of Machine Design
Helsinki University of Technology
P.O. Box 4400
FIN-02015 HUT, Espoo, Finland
heikki.kauranne@tkk.fi

J. KAJASTE
Department of Machine Design
Helsinki University of Technology
P.O. Box 4400
FIN-02015 HUT, Espoo, Finland
jyrki.kajaste@tkk.fi

M. PIETOLA
Department of Machine Design
Helsinki University of Technology
P.O. Box 4400
FIN-02015 HUT, Espoo, Finland
matti.pietola@tkk.fi

ABSTRACT

Simulation of fluid power systems has become a tool widely used for designing and testing purposes. The usefulness of this approach is however highly dependent on the results that have uncertainty to some extent. In the theoretical approach uncertainty depends mainly on the simplifications made in the modeling, algorithms used and parameters. In the experimental approach uncertainty depends on the transducers and the construction of the measurement system and data transfer.

This paper discusses the effect of parameter uncertainty on the reliability of virtual testing of hydrostatic transmission systems in general. The definition of the total uncertainty by using a sum of variances is presented and applied to basic equations of fluid power. As an application a simplified hydrostatic transmission circuit is investigated. Importance of grade of models in reducing the computational inaccuracy is discussed.

INTRODUCTION

A virtual prototype based on simulation models together with parameter information forms the core of the product model that can be utilized in the three stages of the product life cycle: design, manufacture and use.

At the design stage a virtual prototype can be used for optimal product design in two types of simulation. First type, the man-in-the-loop-simulation can be used for improving the man-machine interface and also for training new machine operators in a safe environment. The second type, the hardware-in-the-loop-simulation can be used for software and hardware testing of digital control systems. At the manufacturing stage hardware-in-the-loop-simulation can be used for rationalizing the product inspection. During the normal operation simulation models combined to condition monitoring methods can be used for monitoring the service need.

The usefulness of the theoretical simulation models is however highly dependent on their output that have uncertainty to a specified extent. Reliable decisions can only be made when results are above the uncertainty limits. The uncertainty of simulation results depends mainly on:

- Simplifications made in the modeling
- Algorithms used
- Component and system parameters

The first two issues have been dealt with in many papers, but there are fewer studies on the influence of parameter inaccuracy. This can be divided into three classes:

- Accuracy of the parameter information given by component manufacturers
- Statistical dispersion between similar components
- Change of parameter values during the course of time (wear).

When the parameter information is determined with measurements also this information embodies uncertainty which depends on the characteristics of transducers and the construction of the measurement system and data transfer.

The basic theory of positive displacement pumps and motors was published by Wilson [1, 2]. These flow and torque models have since been developed at least by Schlösser [3, 4], Thoma [5], McCandlish and Dorey [6], Dorey [7] and Puormovahed [8] to be suitable for various types of pumps and various operating conditions.

Zarotti and Nervegna [9], Bavendiek [10], Huhtala and Vilenius [11], Shi and Manring [12], Ivantysynova [13] and Ortwig [14] used modeling techniques that were not based on the Wilson models but on more complicated functions or on fitting functions on the measurement results. These models need typically a lot of measurement data to give accurate results. The basic idea in using function fits is to avoid the determination of uncertain parameters, such as derived capacity and loss coefficients used in Wilson-type models, from the measurement data and thus achieve more reliable calculation results.

The literature dealing with parameter sensitivity of pumps and the models of the components is very few. Pourmovahed [8] studied the effect of source data on the reliability of calculation results with modified Wilson-type flow and torque models. The values of loss coefficients determined on basis of data provided by the component manufacturer and the measurement data provided by a third party was found to differ significantly from each other. The modeling accuracy was found to be good at operating points of high efficiencies, but at points of low efficiencies the calculation results deviated considerably from the measured values. The conclusion was that the efficiency information available for a pump/motor can contain considerably uncertainty. Kim et al [15], [16] studied the parameter sensitivity of a complex non-linear fourth-order dynamic model of an axial piston pump in order to analyze the effect of system parameters on the output of the model.

Kauranne [17] has previously studied the uncertainty associated with the measurement results of the characteristics of an axial piston pump. These and the efficiencies of an axial piston pump have been dealt in [18]. Suitability of models for axial piston pump for varying operational conditions has been dealt in [19].

NOMENCLATURE

a	maximum deviation
c, d	constant
C_f	Coulomb friction coefficient
C_h	hydrodynamic friction coefficient
C_s	slip coefficient
C_{st}	turbulent slip coefficient
C_v	viscous friction coefficient
k	coverage factor
m	mean value
n	rotational speed
p_1	pressure at pump inlet
p_2	pressure at pump outlet
Δp	pressure difference (p2 - p1)
q_{V2}	effective pump outlet flow
q_{Vs}	flow loss
s	value of standard deviation
T	torque need of pump
T_c	torque loss (constant)
T_s	torque loss
TR	transmission ratio
u	uncertainty
u_c	standard deviation
u_c^2	variance
U	expanded uncertainty
V_i	displacement of pump
x	input estimate
y	output estimate
ε	setting of displacement
σ	standard deviation
η_{hm}	hydromechanical efficiency
η_t	total efficiency
η_v	volumetric efficiency
ν	kinematic viscosity of hydraulic fluid
ρ	density of hydraulic fluid

UNCERTAINTY ANALYSIS

The value of a quantity, which is dependent on several random effects, can be determined using normal probability distribution (Gaussian). Evaluation of uncertainty can be made by the statistical analysis of series of observations x_i. This is represented by a statistically estimated standard deviation σ. The standard deviation may be estimated by

$$\sigma = \sqrt{\frac{\sum_{i=1}^{N}(x_i - \mu)^2}{N-1}} \quad (1)$$

$$\mu = \frac{1}{N}\sum_{i=1}^{N} x_i \quad (2)$$

where μ is the average of the random variable and N is number of samples.

The uncertainty of a measurement is often defined as $m \pm$ ks interval, inside of which an individual measurement is at certain level of confidence [12]. The value of the coverage factor k is chosen to give a desired level of confidence. Typically, k is in the range 2 to 3. If k is chosen to be 2 the level of confidence is approximately 95 % with normally distributed signals and circa 99.7 % if k has the value 3. Figure 1 illustrates the probability density function for a continuous distribution of a variable.

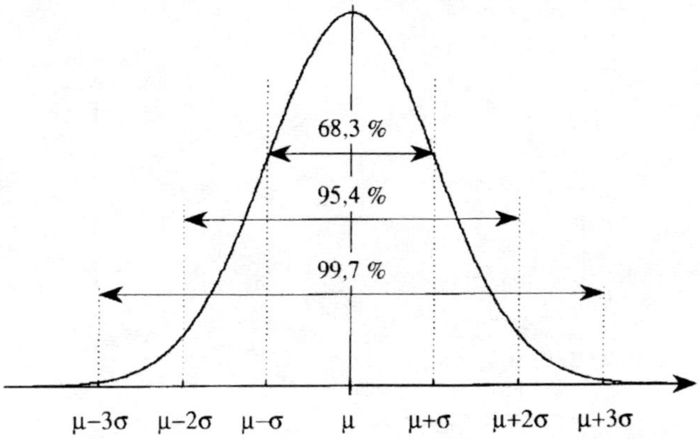

Figure 1: The density function and confidence limits for a normally distributed variable.

If a quantity consists of several variables, which are normally distributed, the uncertainty of the measured results can be expressed by combined uncertainty. Its numerical value is obtained by using a method for the combination of estimated variances.

If two independent quantities x_1 and x_2 are summed, then mean values and variances will be summed:

$$y = x_1 + x_2 \tag{3}$$

$$\mu_y = \mu_1 + \mu_2 \tag{4}$$

$$\sigma_y^2 = \sigma_1^2 + \sigma_2^2 \tag{5}$$

If two independent quantities x_1 and x_2 are multiplied, divided or are raised to powers c and d the values for the mean and the relative standard deviation can be determined by

$$y = x_1^c x_2^d \tag{6}$$

$$\mu_y = \mu_1^c \mu_2^d \tag{7}$$

$$\left(\frac{\sigma_y}{\mu_y}\right)^2 = \left(\frac{c\sigma_1}{\mu_1}\right)^2 + \left(\frac{d\sigma_2}{\mu_2}\right)^2 \tag{8}$$

Simulation model of a hydrostatic transmission system consist of theoretical equations such as pump and motor models. If these parameter values are assumed to be normally distributed similar approach can be used for evaluating uncertainty of the simulation results.

Standard deviation of parameter values is however, rarely available. Usually range of the parameter value is known. According to [20] standard deviation of the parameter can be evaluated

$$\sigma = \sqrt{\frac{1}{3}a^2} \tag{9}$$

where a is maximum deviation from parameter's mean value.

MODELS OF HYDRAULIC PUMP AND MOTOR

Tab. 1 presents the Fixed estimate (10.1), Wilson (10.2) and Schlösser (10.3) flow rate models for pumps and motors used in this study and Tab. 2 the corresponding torque models.

$q_{V2} = \varepsilon V_i n - q_{Vs}$	(10.1)
$q_{V2} = \varepsilon V_i n - C_s \dfrac{V_i \Delta p}{2\pi\nu\rho}$	(10.2)
$q_{V2} = \varepsilon V_i n - C_s \dfrac{V_i \Delta p}{2\pi\nu\rho} - C_{st} V_i^{2/3} \sqrt{\dfrac{2\Delta p}{\rho}}$	(10.3)

Table 1. Flow rate models.

$T = \varepsilon \dfrac{V_i \Delta p}{2\pi} + T_s$	(11.1)
$T = \varepsilon \dfrac{V_i \Delta p}{2\pi} + C_f \dfrac{V_i \Delta p}{2\pi} + C_v V_i n \nu\rho + T_c$	(11.2)
$T = \varepsilon \dfrac{V_i \Delta p}{2\pi} + C_f \dfrac{V_i \Delta p}{2\pi} + C_v V_i n \nu\rho + \dfrac{C_h V_i^{5/3} n^2 \rho}{4\pi}$	(11.3)

Table 2. Torque models.

The Fixed estimate models feature the simplest form of losses that do not depend on any operating parameter. Since this assumption does not hold this kind of models are supposed to be accurate at one operating point only. The basic form of the Wilson and Schlösser models is between the same, but due to the additional pressure dependent term in flow model and to the replacement of constant friction term T_c with rotational speed dependent term in torque model the Schlösser models are capable to somewhat more precise modeling of the flow and torque losses.

Figs. 2 and 3 present measured and modeled volumetric and hydromechanical efficiencies as a function of outlet pressure and rotational speed. The measurement system and measurements are presented in [17] and [19].

As supposed the Fixed estimate model gives accurate results only at one operating point, but it does however give quite accurate results at wider operating range with speed dependent hydromechanical efficiency since this it is not very rotational speed dependent. The values of fixed estimate efficiencies shown in Figs. 2 and 3 are chosen based on a long series of measurements.

The Wilson and Schlösser models are capable to model the pressure and speed dependency of efficiencies quite accurately and the deviations between model and measurements are minimal. In modeling the speed dependency of volumetric efficiency both models are prone to exaggeration while in modeling the speed dependency of hydromechanical efficiency only the Schlösser model tends to exaggerate.

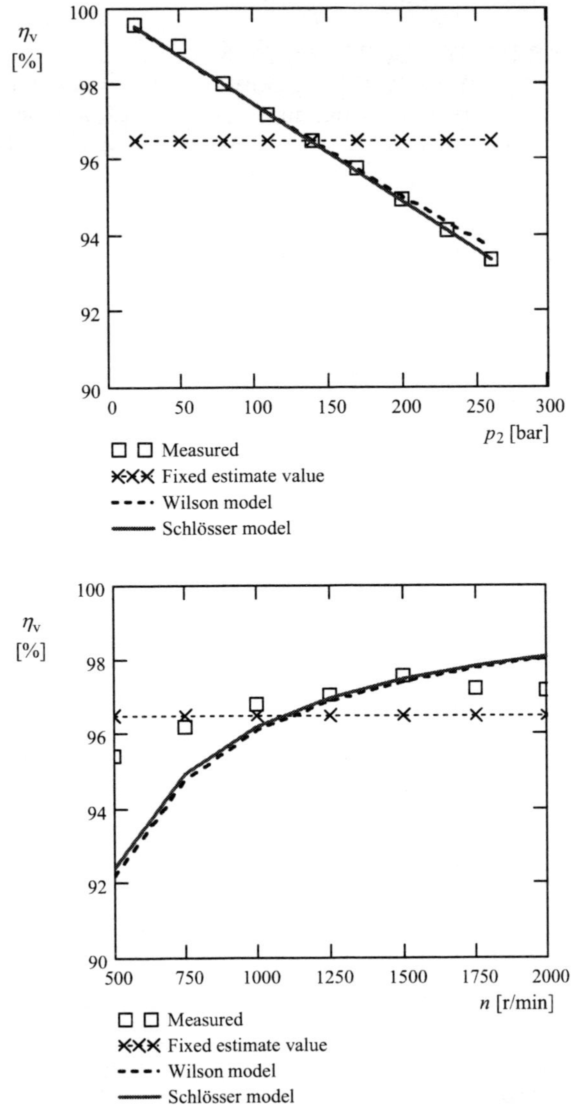

Figure 2: Volumetric efficiency, measured and modeled values $f(p_2)$ and $f(n)$.

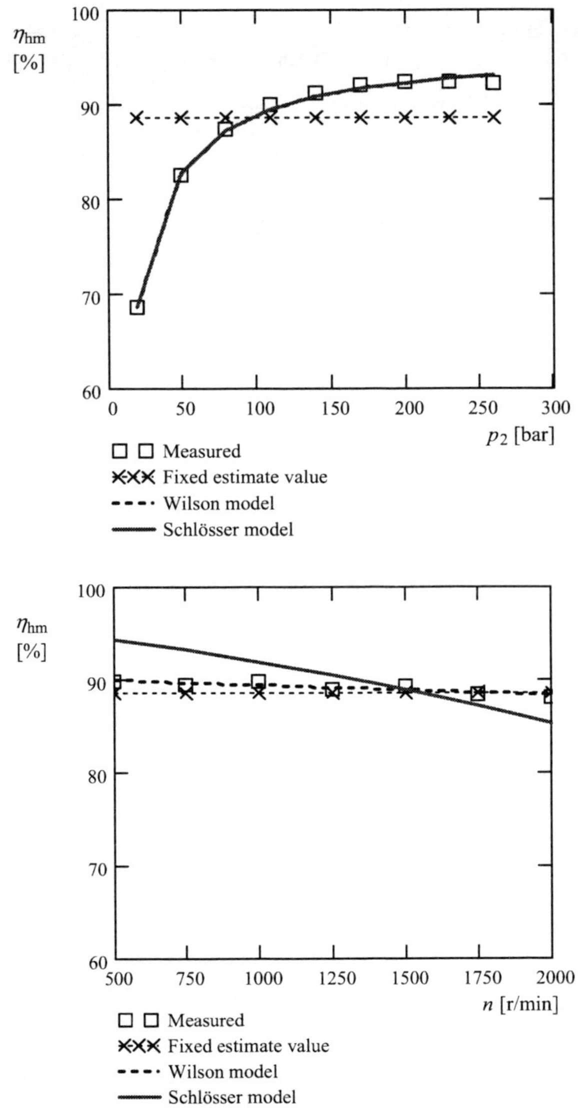

Figure 3: Hydromechanical efficiency, measured and modeled values $f(p_2)$ and $f(n)$.

Figs. 4 and 5 present calculated deviations between measured and modeled efficiency values. From these operating point dependent values the maximum values are selected and used to calculate the standard deviation i.e. uncertainty of each efficiency with (9), Tabs. 3 and 4.

Model	$f(p_2)$ [%]	$f(n)$ [%]
Fixed estimate	1.9	0.64
Wilson	0.17	1.7
Schlösser	0.13	1.7

Table 3. Maximum uncertainties of volumetric efficiencies $u(\eta_v)$ as function of p_2 and n.

Model	$f(p_2)$ [%]	$f(n)$ [%]
Fixed estimate	12	0,69
Wilson	0.55	0.20
Schlösser	0.55	2.6

Table 4. Maximum uncertainties of hydromechanical efficiencies $u(\eta_{hm})$ as function p_2 and n.

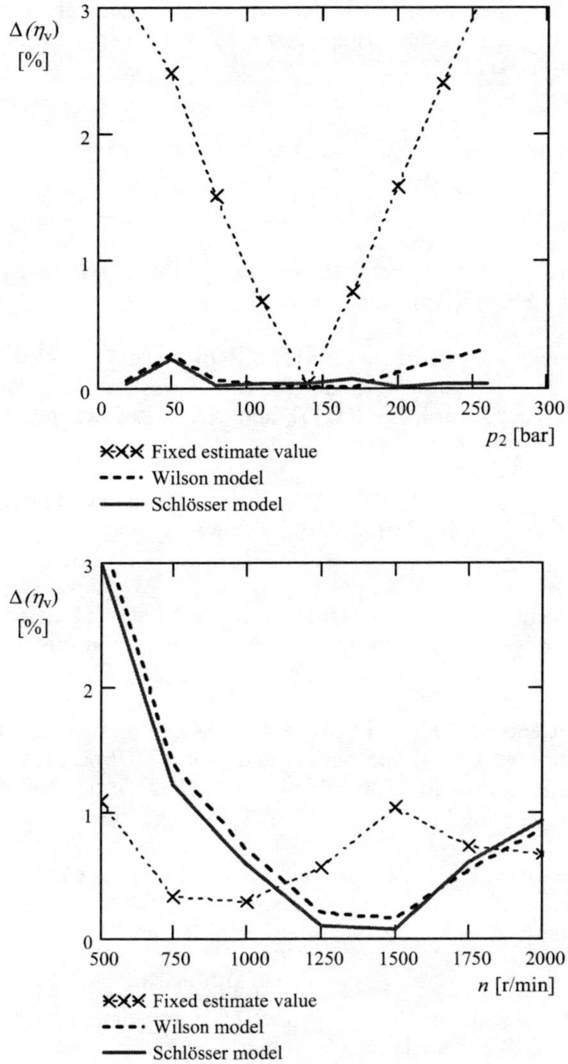

Figure 4: Deviations of volumetric efficiency $f(p_2)$ and $f(n)$.

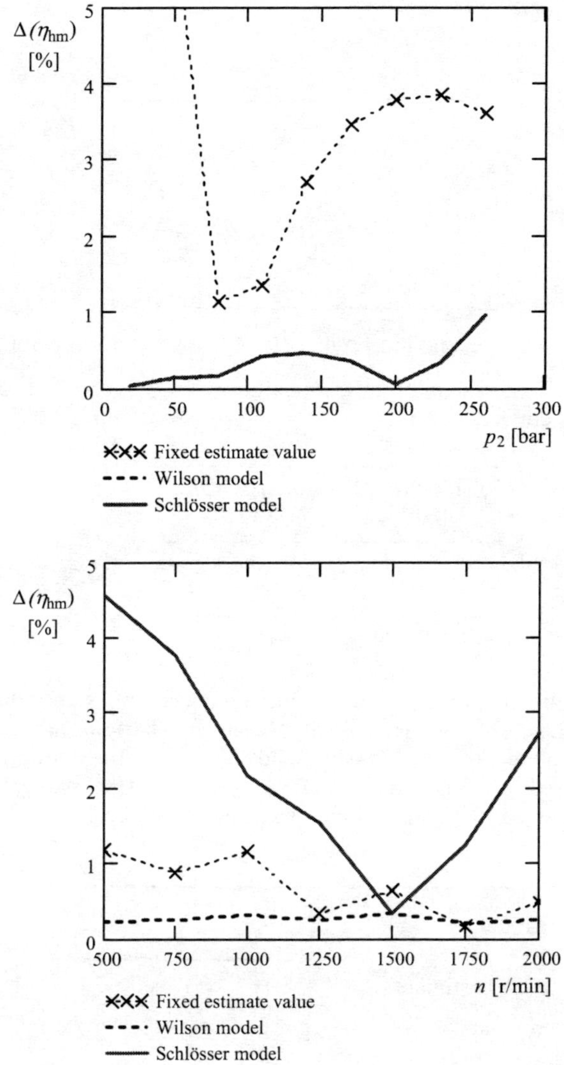

Figure 5: Deviations of hydromechanical efficiency $f(p_2)$ and $f(n)$.

Due to the good capability of Wilson and Schlösser models to represent the actual strong pressure dependent characteristics of efficiencies the uncertainties of these are minor while the assumption of fixed efficiency values leads to higher uncertainties. In the case of speed dependency of efficiencies the situation is reversed because of the speed dependency exaggeration of the Wilson and Schlösser models.

Generally Schlösser model is more advanced than Wilson model. Therefore modeling uncertainty is expected to be smaller using Schösser model. However, in this case with limited measurement data the results were slightly opposite.

RELIABILITY OF EFFICIENCY SIMULATION IN HYDRAULIC TRANSMISSION SYSTEM

A simplified hydrostatic transmission system shown in Fig. 6 is studied. Computation inaccuracy of major design quantities such as total efficiency and transmission ratio of angular velocity and torque are considered as an example.

The total efficiency η_t, transmission ratios between angular speed of motor and pump TR_w and torque of motor and pump TR_T of transmission system is defined with volumetric and hydromechanical efficiencies of the components

$$\eta_{t12} = \eta_{v1}\,\eta_{hm1}\,\eta_{v2}\,\eta_{hm2} \tag{12}$$

$$TR_{\omega} = \frac{\omega_2}{\omega_1} = \frac{V_{i1}}{V_{i2}}\eta_{v1}\,\eta_{v2} \tag{13}$$

$$TR_T = \frac{T_2}{T_1} = \frac{V_{i2}}{V_{i1}}\eta_{hm1}\,\eta_{hm2} \tag{14}$$

where the subscripts 1 and 2 refer to pump and motor respectively.

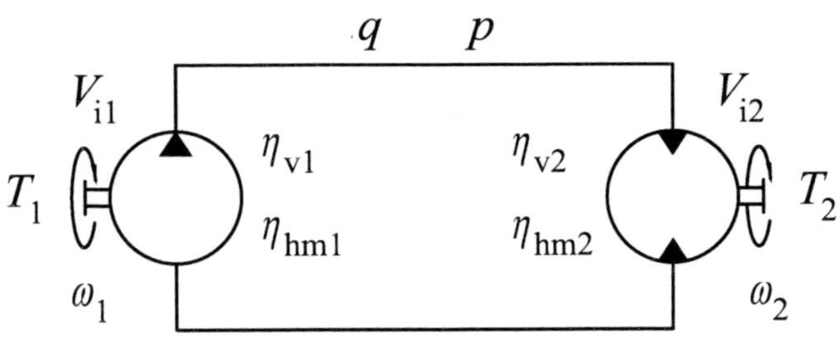

Figure 6: Simplified hydrostatic transmission system.

The uncertainty of these quantities with 95 % confidence level is conducted from (8)

$$u(\eta_t) = 2\sqrt{u(\eta_{v1})^2 + u(\eta_{hm1})^2 + u(\eta_{v2})^2 + u(\eta_{hm2})^2}$$

$$u(TR_\omega) = 2\sqrt{u(\eta_{v1})^2 + u(\eta_{v2})^2 + u(V_{i1})^2 + u(V_{i2})^2} \quad (15)$$

$$u(TR_T) = 2\sqrt{u(\eta_{hm1})^2 + u(\eta_{hm2})^2 + u(V_{i1})^2 + u(V_{i2})^2}$$

Tab. 5 presents the resulting uncertainties of these quantities with the measured efficiency values for an axial piston pump presented in tabs 3 and 4. Values defined for pump are used as an estimate for a motor. Uncertainty of displacement of pump and motor is estimated to be 1 %.

Model	$u(\eta_t)$ [%]	$u(TR_{tw})$ [%]	$u(TR_{tT})$ [%]
Fixed estimate	35	6.1	34
Wilson	5.8	5.6	3.2
Schlösser	9.2	5.6	7.9

Table 5. Uncertainties for total efficience and transmission ratio for speed and torque.

CONCLUSIONS

Uncertainty analysis is a useful tool for evaluating reliability of virtual testing. This can be done using the same method as in experimental research. This requires that uncertainty level of each parameter is known.

Generally it is a problem to know the amount of the uncertainty in the values of parameters. In addition environmental changes such as temperature also have an effect on parameter values. Furthermore disturbances such as oil type or wear also affect the parameter values.

In virtual testing of hydrostatic transmission circuits it is hard to achieve simulation results which uncertainty is less than 5% expanded uncertainty of 95 % confidence level, which is generally used, in experimental research. The computational uncertainty can be reduced by improving model grade and identifying the models using data from whole operation area that is to be studied.

REFERENCES

[1] Wilson, W.E., 1946, "Rotary Pump Theory," Transactions of the ASME, May, 1946, pp. 371-383.

[2] Wilson, W.E., 1950, "*Positive Displacement Pumps and Motors*", Pitman Publishing Corporation.

[3] Schlösser, W.M.J., 1961, "Mathematical model for displacement pumps and motors," Hydraulic Power Transmission, April, pp. 252-57 and 269, and May, pp. 324-328.

[4] Schlösser, W.M.J., 1968, "The overall efficiency of positive displacement pumps," BHRA Fluid Power Symposium, pp. 34-48.

[5] Thoma, J., 1969, "Mathematical models and effective performance of hydrostatic machines and transmissions", Hydraulic Pneumatic Power. November, pp. 642-651.

[6] McCandlish, D. and Dorey, R.E., 1984, "The mathematical modelling of hydrostatic pumps and motors," Proceedings of the Institution of Mechanical Engineers, Part B: Management and Engineering Manufacture, **198**, No 10, pp. 165-174.

[7] Dorey, R.E., 1988, "Modelling of losses in pumps and motors," First Bath International Fluid Power Workshop, University of Bath, England, 8th September, pp. 71-97.

[8] Pourmovahed, A., 1992, "Uncertainty in the efficiencies of a hydrostatic pump/motor," ASME Winter Annual Meeting, November 8-13, Anaheim, California, USA, 11 p.

[9] Zarotti, G.L., and Nervegna, N., 1981, "Pump efficiencies approximation and modelling," 6th International Fluid Power Symposium, Cambridge, England, April 8-10, pp. 145-164.

[10] Bavendiek, R., 1987, "Neue Methode zur Bestimmung der Verluste (Wirkungsgrade) an hydrostatischen Maschinen," O+P "Ölhydraulik und Pneumatik" 31 Nr. **11**, pp. 861-866. In German.

[11] Huhtala, K. and Vilenius, M., 1997, "Comparison of Steady-State Models of Hydraulic pump," The Fifth Scandinavian International Conference on Fluid Power, Linköping, Sweden, May 28-30, pp. 139-156.

[12] Shi, X. and Manring, N.D., 2001, "A torque efficiency model for an axial-piston, swash-plate type, hydrostatic pump," Bath Workshop on Power Transmission and Motion Control, PTMC 2001, Bath, England, pp. 3-20.

[13] Ivantysynova, M., 2001, "Energy losses of modern displacement machines – a new approach of modelling," The

Seventh Scandinavian International Conference on Fluid Power, Linköping, Sweden May 30 – June 1, pp. 377-395.

[14] Ortwig, H., 2002, "Wirkungsgradsimulation von Verdrängereinheiten," O+P "Ölhydraulik und Pneumatik" 46 Nr. **4**, pp. 240-246. In German.

[15] Kim, S.D., Cho, H.S. and Lee, C.O., 1987, "A parameter sensivity analysis for the dynamic model of a variable displacement axial piston pump," Proceedings of the Institution of Mechanical Engineers, Part C: Mechanical Engineering Science, **201**, pp. 235-243.

[16] Kim, S.D., Cho, H.S. and Lee, C.O., 1987, "A parameter sensivity analysis for the dynamic model of a variable displacement axial piston pump," ASME Winter Annual Meeting, December 13-18, Boston, Massachusetts, USA, 9 p.

[17] Kauranne, H.O.J., 2003, *Determination of static and dynamic characteristics of pressure controlled pump*, Licentiate's Thesis. Helsinki University of Technology, Finland, 2003, 119 p. (In Finnish)

[18] Kauranne, H.O.J., Kajaste, J.T., Ellman, A.U. and Pietola, M.T., 2003, "Dependency of an axial piston pump characteristics on variation of operating conditions," Proceedings of The Eighth Scandinavian International Conference on Fluid Power SICFP'03, May 7-9, Tampere, Finland, pp. 191-205.

[19] Kauranne, H.O.J., Kajaste, J.T., Ellman, A.U. and Pietola, M.T., 2003, "Applicability of pump models for varying operational conditions," ASME International Mechanical Engineering Congress and Exposition IMECE 2003, November 16-21, Washington, D.C., USA, 10 p.

[20] *Guide to the expression of uncertainty in measurement*, International Organization for Standardization, Switzerland, 1993, 101 p. ISBN 92-67-10188-9.

Proceedings of IMECE2005
2005 ASME International Mechanical Engineering Congress and Exposition
November 5-11, 2005, Orlando, Florida USA

IMECE2005-79154

EXPERIMENTAL AND NUMERICAL INVESTIGATION OF CENTRIFUGAL PUMP PERFORMANCE WHEN HANDLING VISCOUS FLUIDS

M. H. Shojaee Fard
Professor
Department of Mechanical Engineering,
Iran University of Science and Technology
Narmak, Tehran, Iran
Email: mhshf@iust.ac.ir

M. B. Ehghaghi
assistance Professor
Department of Mechanical Engineering,
Tabriz University
Tabriz, Iran
Email: Ehghghi@navidsahand.com

F.A. Boyaghchi
Ph.D. Student[1]
Department of Mechanical Engineering
Iran University of Science and Technology
Narmak, Tehran, Iran
Email:Fateme_Ahmadi@mecheng.iust.ac.ir

ABSTRACT

On the test bed of centrifugal pump, the centrifugal pump performance has been investigated using water and viscous oil as Newtonian fluids, whose kinematic viscosities are 1×10^{-6}, 43×10^{-6} and 62×10^{-6} m^2/s, respectively. Also, the finite volume method is used to model the three dimensional viscous fluids for different operating conditions. For these numerical simulations the SIMPLEC algorithm is used for solving governing equations of incompressible viscous/turbulent flows through the pump. The k-ε turbulence model is adopted to describe the turbulent flow process. These simulations have been made with a steady calculation and using the multiple reference frame (MRF) technique to take into account the impeller- volute interaction. Numerical results are compared with the experimental characteristic curve for each viscous fluid. The data obtained allow the analysis of the main phenomena existent in this pump, such as: head, efficiency, power and pressure field changes for different operating conditions. Also, the correction factors for oils are obtained from the experimental for part loading (PL), best efficiency point (BEP) and over loading (OL) and the results are compared with proposed factors by American Hydraulic Institute (HIS) and Soviet Union (USSR). The comparisons between the numerical and experimental results show a good agreement.

INTRODUCTION

When a fluid of high viscosity –such as heavy oil- is pumped by a centrifugal pump the performance is impaired in comparison to service with water due to increased losses. Viscosity is defines as resistor to pouring with higher viscosity fluids affecting centrifugal pump performance by increasing the power, reducing the flow rate , head and efficiency, making trouble for mechanical seals and causing more load on bearings.

Many researches, such as Daugherty (1926) [1], Stepanoff(1940) [2], Telow (1942) [3], Ippen (1946) [4], and Itaya and Nishikawa (1960) [5], have tested the performance of the centrifugal oil pumps as a function of the viscosity of the viscous oil without knowing the phenomena of viscous fluid flow within the pump. They have proposed some correction factors when pump handles the viscous fluid for determining the performance. These typical results have provided great insights into the effects of the oil viscosity on the

performance of centrifugal oil pumps, which have been applied in the important guidelines used today for the design of these pumps. However, these experimental results are based on the oil pump products available in the 1920s and 1950s. On the other hand, some researches, such as Stoffel (1980) [6], Li (1996) [7] and Li and Hu (1997) [8], conducted experiments using commercial centrifugal oil pumps.

Theoretical analysis of viscous fluid flow through turbo-machines is impossible or very difficult, and it needs more time for solving the equations by use of super computers. Within the last few years, several computer codes for analyzing the effects of viscosity and boundary layers on turbo machinery performance have been published. The codes use different numerical schemes, boundary conditions and turbulence models for predicting the viscosity effects, impeller tip leakage and secondary flow through blade row. By using this code, Denton (1986) [9] simulated the viscous effects in the passage of blades by use of a distributed body force. Miner (1992) [10,11] has done several researches on two dimensional flow analysis and turbulence measurements of a centrifugal pump.

Croba and Kueny (1996) [12], simulated the 2D, unsteady water flow in centrifugal pump by using the CFD code. Denus and Gode (1999) [13], studied a mixed

– flow pump impeller when handling low viscous fluid flow by using the CFD analysis. However, due to the difficulties of the task, most of studies have done heavy simplifications of the problem either in the geometry or in the flow characteristics. Research slowly trends towards more complete simulations. Blanco (2000) [14,15] simulated the fluid flow in the centrifugal pump for investigating the pressure fluctuates in the volute by using the CFD code. Shojaee Fard and Ehghaghi (2002) [16, 17] studied the performance of the centrifugal pump experimentally for different geometries of impeller in order to optimizing the centrifugal pump when handling the viscous oil and simulated the fluid flow in a passage of impeller by use of a CFD code. Gugau (2003) [18], simulated the flow in the centrifugal pump by using the sliding meshes for analyzing the main phenomena existent in these pumps.

The purpose of the present study is to show a numerical study of a centrifugal pump taking into account the whole 3D geometry of pump which is used in experiments and solving the 3D Navier-Stockes equations using the finite volume method for occurrence of numerical simulation.

NOMENCLATURE

BEP	best efficiency point
D_H	hydraulic diameter, m
f	correction factor
H	head, m
I	turbulence intensity
K	turbulent kinetic energy, j
ℓ	mixing length scale, m
P	power, W
P_o	total pressure, Pa
Q	flow rate, m^3/h
Re	Reynolds number
S	centrifugal force source, N
u	flow velocity, m/s

Subscripts

in	inlet
i	component
out	outlet
H	head
Q	flow rate
w	water
υ	viscous fluid
avg	average
T	turbulent

Greek letters

ε	dissipation, j
η	efficiency
μ_T	eddy viscosity, $N.s/m^2$
ν	kinematics viscosity, m^2/s
ρ	density, kg/m^3
τ	shear stress, Pa
φ	potential function
Δ	difference

KEYWORDS

Centrifugal pump, Newtonian fluids, viscous fluid flow, correction factors, 3D numerical simulation

EXPERIMENTAL FACILITY

Test rig

A special centrifugal pump test rig, shown in Fig. 1, was used to test the pump performance and to measure the flow within the pump impeller when the pump was pumping viscous oil and water. The pump tested is driven by a three – phase AC electric motor, whose rated power is 5.5 KW and speed is 1450 rev/min.

Performance test method

The pump suction pressure and discharge pressure is measured by a pressure gage. The capacities of the pump are measured by a turbine or nuzzle flow meter. Torque meter measures the shaft torque and speed of the pump. By means of a rotary viscosity meter the viscosity of oil can be given. The density of oil is checked using a floating glass tube density meter.

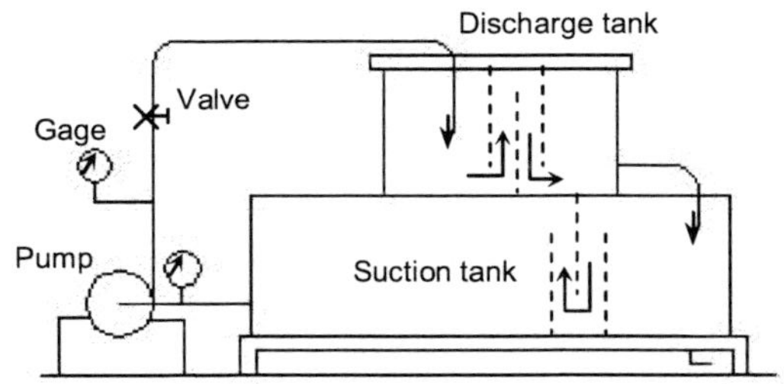

Fig. 1. Centrifugal pump test rig

Impeller geometry

The centrifugal pump simulated is a 65-200, single axial suction and vane less volute casing; equipped with an impeller of 209 mm in outside diameter and 6 backwards curved blades.

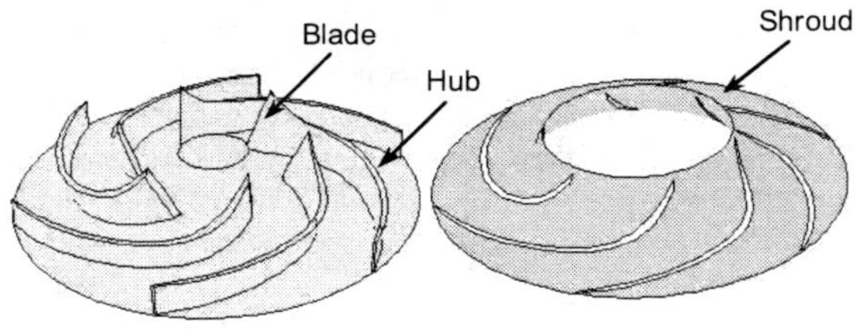

Fig.2. Impeller configuration

The blade outlet and wrapping angles of the impeller are 27° and 20°. When the pump is run at 1450 rev/min the best efficiency point corresponds to 50 m^3/h flow rate and 14 m height.

The shroud of the impeller made of metal is machined. The roughness of the impeller and volute is 100 μm. The Fig. 2 and 3 show the impeller configuration and components of pump respectively.

Working fluid

Working fluids are the special transparent viscous oil refined from crude oils and tap water, respectively. They are Newtonian fluids verified by using the rotary viscosity meter. The density and kinematic viscosity of the oils are 875 kg/m^3 and 43×10^{-6} m^2/s and 878 kg/m^3 and 62×10^{-6} m^2/s, respectively, at 20°C.The density and the kinematic viscosity of the tap water are 998.2 kg/m^3 and 1×10^{-6} m^2/s, respectively, at 20°C. First, the centrifugal pump performance pump performance and flow fields in the impeller were measured using the tap water as a working fluid. Then the performance and flow fields were measured again using the viscous oils as a working fluid.

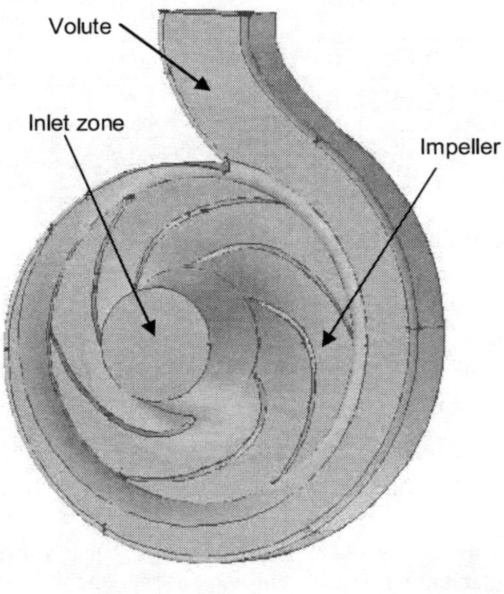

Fig. 3. The components of pump

Uncertainty

The uncertainties in head, capacity, torque, speed and efficiency are 0.4%, 0.5%, 0.3%, 1% and 1.15%, respectively.

MODEL DESCRIPTION AND COMPUTATIONAL METHOD

Pump, geometry and grid

In turbo-machines the geometry is highly three-dimensional. Due to the transfer of energy is brought about through the centrifugal forces that the impeller passes on to fluid, and this forces have radial and tangential directions, it is possible to consider that the model captures the fundamental phenomena that occur in pump.

With the three-dimensional model there is a useful approach for investigation of flow behavior in different parts of pump.

The blade number and shape has great importance. A pump with few backward blades gives less distortion in the flow and it is possible to simulate it with a relatively rough mesh. A large number of blades require a great number of cells to correctly simulate the flow passages. Forward blades, because of the severe stall that is usual in the suction side, needs a fine mesh in this area if we want to capture this phenomenon.

Figure 4 shows the unstructured grid generated. There are 45499 cells in the inlet zone, 72591 cells in the impeller and 144206 cells in the volute. This is enough for precise boundary layer simulation and it gives correct values for the pump performance and allows to analysis the details of the main phenomena involved.

Fig. 4. Sketch of the pump grid

Surface between inlet-impeller and impeller-volute correspond to grid interfaces. The MRF technique allows the relative motion of the impeller grid with respect to the inlet and the volute during steady simulation. Grid faces do not need aligned on both sides.

Mathematical model

Continuity and 3-D incompressible Navier-Stockes equations, including the centrifugal force source in the impeller and steady terms are used in centrifugal pump to analyzing the turbulent viscous fluid flow. Turbulence is simulated with the standard *k-ε* model. Although grid size is not adequate to correctly take account of boundary layers on the blades, wall functions, based on the logarithmic law, has been used. The pressure – velocity coupling are calculated through the SIMPLEC algorithm. Second order, upwind discretizations have been used for convection terms and central difference schemes for diffusion terms.

Continuity equation:

$$\frac{\partial}{\partial x_j}(\rho u_j)=0 \tag{1}$$

Momentum equation:

$$\begin{cases} \dfrac{\partial}{\partial x_j}(\rho u_j \varphi)=\dfrac{\partial}{\partial x_j}\left(\mu\dfrac{\partial \varphi}{\partial x_j}\right)+S \\ \rho u_j\dfrac{\partial \varphi}{\partial x_j}=\dfrac{\partial}{\partial x_j}\left(\mu\dfrac{\partial \varphi}{\partial x_j}\right)+S \end{cases} \tag{2}$$

Turbulence Modeling: For modeling phenomena the $k-\varepsilon$ modeling is applied, therefore two transport equations for k and ε are used as following:

$$\frac{\partial}{\partial x_i}(\rho u_i k)=\frac{\partial}{\partial x_i}\frac{\mu_T}{\sigma_k}\frac{\partial k}{\partial x_i}+G_k+G_b-\rho\varepsilon \quad (3)$$

$$\frac{\partial}{\partial x_i}(\rho u_i \varepsilon)=\frac{\partial}{\partial x_i}\frac{\mu_T}{\sigma_k}\frac{\partial \varepsilon}{\partial x_i} \quad (4)$$

$$+C_{1\varepsilon}\frac{\varepsilon}{k}\left(G_k+(1-C_{3\varepsilon})G_b\right)-C_{2\varepsilon}\rho\frac{\varepsilon^2}{k}$$

The parameters in Eq.(3) and (4) are defined as following:

$$G_k=\mu_T\left(\frac{\partial u_j}{\partial x_i}+\frac{\partial u_i}{\partial x_j}\right)\frac{\partial u_j}{\partial x_i}$$

$$G_b=-g_i\frac{\mu_T}{\rho\sigma_h}\frac{\partial \rho}{\partial x_i}$$

$$\sigma_h=\frac{\mu_T C_P}{k_t}$$

$$\mu_T=\rho C_\mu\frac{k^2}{\varepsilon}$$

$$\sigma_\varepsilon=1.3,\ \sigma_k=1.0,\ C_\mu=0.09,\ C_{2\varepsilon}=1.92,\ C_\varepsilon=1.44$$

BOUNDARY CONDITIONS

Inlet boundary conditions

At the inlet, a steady, uniform, axial velocity distribution is imposed. Table 1 shows the amounts of inlet velocities for different operating conditions. These velocities specify the flow rates.

	Inlet Velocity , (m/s)		
	PL	BEP	OL
Water ($\nu=1\times10^{-6}$ m²/s)	2.65	3.31	3.98
Oil ($\nu=43\times10^{-6}$ m²/s)	2.57	3.21	3.8
Oil ($\nu=62\times10^{-6}$ m²/s)	2.46	3.08	3.69

Table 1. Amounts of inlet velocities for different operating conditions.

Outlet boundary conditions

A t the outlet, constant static pressure is maintained.

$$P_{gage}=0$$

Wall boundary conditions

Non slip boundary conditions have been imposed over the impeller blades and walls, the volute casing and the inlet pipe wall and the roughness of all walls is considered 100 µm.

Turbulence boundary conditions

a) Turbulence Intensity

The turbulence intensity, I, is defined as ratio of the root-mean-square of the velocity fluctuations, u' to the mean flow velocity, u_{avg}.
The turbulence intensity at the inlet is totally depends on the upstream history of flow. Since the fluid in the suction tank is undisturbed, the turbulence intensity for all conditions is considered 1%. For outlet, the flow is assumed fully developed then the turbulence intensities are estimated by empirical relation blow:

$$I\equiv\frac{u'}{u_{avg}}=0.16(\mathrm{Re}_{DH})^{-1/8} \quad (5)$$

$D_{H,out}=0.065m$	$Q\ (m^3/hr)$	$u_{avg}\ (m/s)$	$I(\%)$
Water (ν=1 Cst)	60	5.025	3.2
Oil (ν=43 Cst)	58.2	4.874	5.2
Oil (ν=62 Cst)	55.8	4.673	5.5

Table 2. Outlet turbulence intensities

b) Turbulence Length Scale

The turbulence length scale , ℓ, is a physical related to the size of the large eddies that contain the energy in turbulent flows. In fully developed duct flows, ℓ is restricted by the size of the duct. An approximate relationship between ℓ and the physical size of the duct is.

$$\ell=0.07D_H \quad (6)$$

Using the above relation, the turbulence length scales at the inlet and outlet are estimated 5.6 mm and 4.55 mm, respectively.

Numerical solution control

The iterations number has been adjusted to reduce the residual below an acceptable value in each step.
The numerical results dependence on the grid size has been tested with a lower number of cells. The overall performance of the pump is the same even with less than a half the cells. Flow main characteristics do not change.

RESULTS

Performance

When increasing the viscosity the hydraulic losses in inlet, impeller and volute are made up by friction losses and by losses created by exchange of momentum whenever the velocity distribution is non uniform. Such non uniformities or mixing losses are causes by the work transfer from the blades, deceleration of the fluid, incidence or even local flow separation. Also, disk friction losses on impeller side walls, in labyrinths of the impeller and a possible device for axial thrust balancing grow with increasing viscosity.
From the above it follows that the disk friction losses and friction losses in the hydraulic channels of the pump are

the main cause for the performance impairment caused by viscous fluids compared with water.

Table 3 indicates the experimental values of head, efficiency and power for viscous fluids at 20°C.

Figure 5 shows the effects of viscosity on the performance of centrifugal pump.

Water ($\nu=1\times10^{-6}$ m²/s)						
Flow rate (m^3/h)	0	36	48	60	72	75
Head (m)	15.5	14.3	13.8	12.1	11	10
Power (Kw)	---	2	2.45	2.65	2.8	3.06
Efficiency (%)	0	70	75	76	73	68

Oil ($\nu=43\times10^{-6}$ m²/s)						
Flow rate (m^3/h)	0	34.9	46.5	58.2	68.8	72.2
Head (m)	14	13.8	13.1	11.6	10.2	9.1
Power (Kw)	---	2.16	2.7	2.9	3.1	3.22
Efficiency (%)	0	48	52	53	50	47

Oil ($\nu=62\times10^{-6}$ m²/s)						
Flow rate (m^3/h)	0	33.5	44.6	55.8	66.9	69.7
Head (m)	13.8	13.5	12.8	11.1	9.9	8.8
Power (Kw)	---	2.34	2.8	3	3.3	3.6
Efficiency (%)	0	46	49	50	48.2	40.8

Table 3. The experimental values of performance

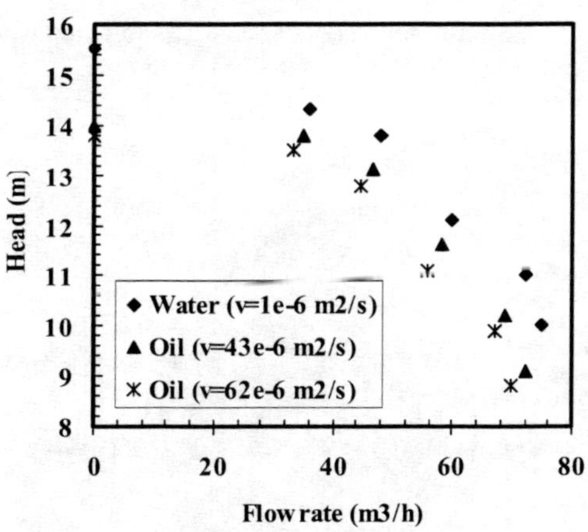

(a)

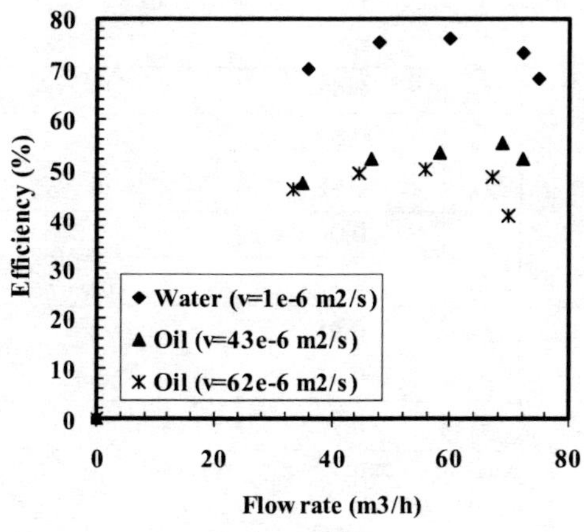

(b)

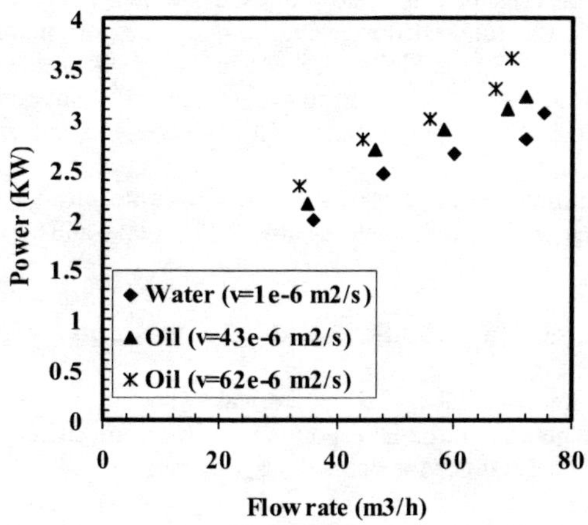

(c)

Fig. 5. Oil viscosities effect on the performance of centrifugal pump in comparison with water: (a) head, (b) efficiency and (c) power

Correction factor

In order to determine pump performance with viscous oil from known or measured performance with water, it is necessary to establish performance correction factors for the flow rate, head and efficiency as a function of viscosity. The flow rate, head and efficiency correction factors f_Q, f_H and f_η are defined as:

$$f_Q = \frac{Q_\upsilon}{Q_w}; \quad f_H = \frac{H_\upsilon}{H_w}; \quad f_\eta = \frac{\eta_\upsilon}{\eta_w} \qquad (5)$$

Where Q_w, H_w and η_w represent the water flow rate, head and efficiency at the best efficiency point, Q_υ is the viscous oil flow rate at that point. H_υ and η_υ are the viscous oil head and efficiency under any operating condition.

Oil , m^2/s		43×10^{-6}	62×10^{-6}
Experimental	f_Q	0.97	0.93
	f_H	0.95	0.91
	f_η	0.69	0.65
HIS	f_Q	0.98	0.95
	f_H	0.96	0.93
	f_η	0.75	0.71
USS	f_Q	0.95	0.92
	f_H	0.97	0.93
	f_η	0.72	0.65

Table 4. Comparison of experimental results with existing correction factors

At present there are two series of performance correction factors in use; one is that proposed by the American Hydraulic Institute (HIS) in 1969; the other was formulated by the former Soviet Union (USSR) in the 1950s. Table 4 shows how the correction factors determined in the study described in this paper compare with the two existing sets of these factors.

The maximum relative errors of the flow rate, head and efficiency correction factors are 2.1%, 2% and 8%, respectively.

3D numerical results and Comparison with experimental

For comparison the 3D numerical simulations with experimental head, the BEP points are chosen for different viscosities. Total head is defined as:

$$H = \frac{\Delta P_o}{\rho g} \qquad (6)$$

In Eq.(6) ΔP_o is the total pressure of outlet and inlet of the pump.

Table 5 shows the values of head obtained using experimental and 3D numerical model.

Fluid (viscosity)	Experimental head (m)	Numerical head (m)	Error (%)
Water ($\nu= 1\times10^{-6}$ m^2/s)	12.1	13.35	10.3
Oil ($\nu=43\times10^{-6}$ m^2/s)	11.6	12.71	9.4
Oil ($\nu=62\times10^{-6}$ m^2/s)	11.1	12.50	12.6

Table 5. Comparison between experimental and numerical.

Figures 6, 7 and 8 show the static pressure contours in BEP (Best Efficiency Point) for various viscous fluids. Due to these figures the suction pressure increases by increasing the viscosity which it causes decreasing head.

For finding the accuracy of 3D model, the numerical simulations were done for different flow rates for each fluid. Figures 9, 10 and 11 show the comparison of experimental head and 3D numerical model for PL, BEP and OL for water and oils, respectively. According to these figures as viscosity increases the suction pressure decreases.

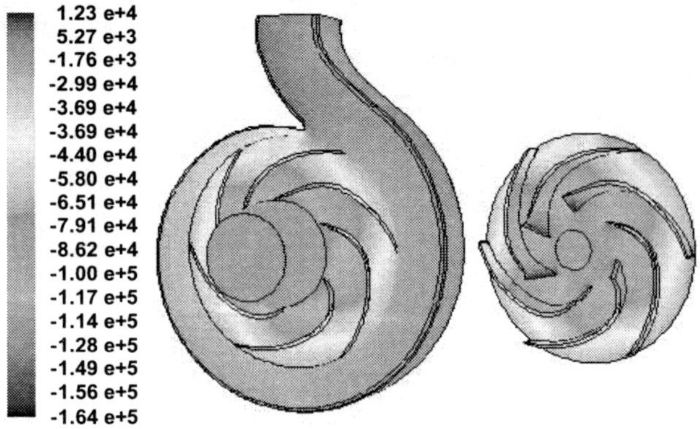

Fig. 6. The static pressure (Pa) contours in pump when handling the water.

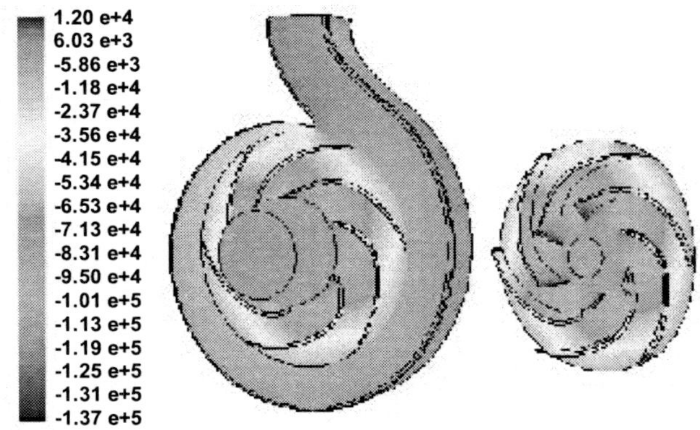

Fig. 7. The static pressure (Pa) contours in pump when handling the oil ($\nu=43\times10^{-6}$ m^2/s).

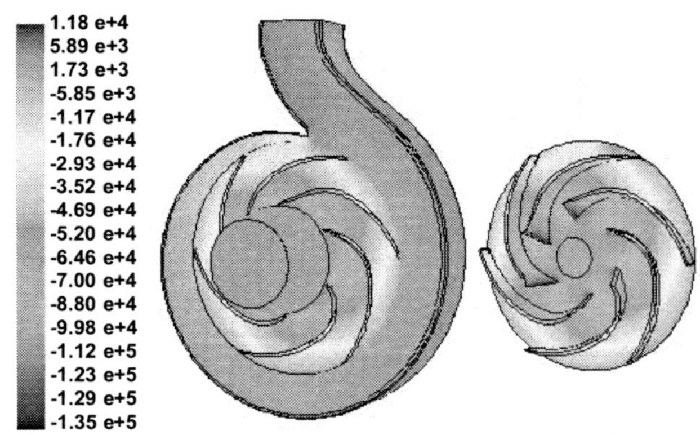

Fig.8. the static pressure (Pa) contours in pump when handling the oil ($\nu=62\times10^{-6}$ m^2/s).

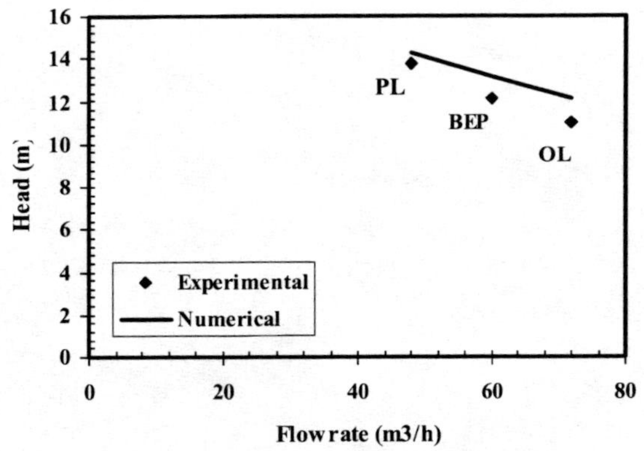

Fig. 9. Comparison between experimental and numerical head for PL, BEP and OL for water.

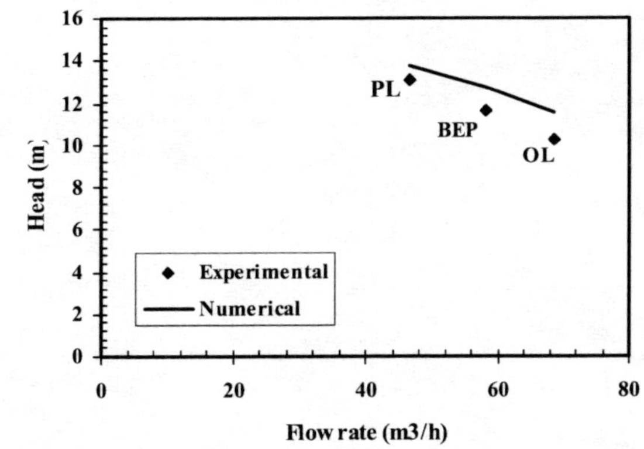

Fig. 10. Comparison between experimental and numerical head for PL, BEP and OL for oil (ν=43×10^{-6} m^2/s).

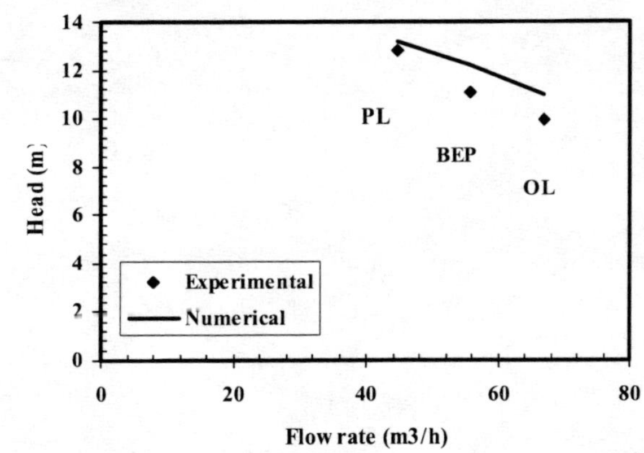

Fig.11. Comparison between experimental and numerical head for PL, BEP and OL for oil (ν=62×10^{-6} m^2/s).

CONCLUSION

With regards to the test pump and numerical simulation in the present study the following conclusions can be drawn.

(1) The reason why a centrifugal pump performance goes down when the pump handles high viscosity working fluids is that high viscosity results in a rapid increase in the disc friction losses over outsides of the impeller shroud and hub as well as in hydraulic losses in flow channels of pump.

(2) The head correction factors are related to the pump operating conditions; when determining the performance of liquid with high viscosity from known performance with water, the correction factors must be chosen based on the pump operating conditions.

(3) If the existing correction factors are used to determine the viscous oil performance from known performance in water for current centrifugal oil pumps, the relative errors of up to 8% will occur.

(4) The inlet velocity and pressure outlet conditions can be a good choice for estimating the head of pump after convergence.

(5) The three dimensional view of the pressure over the shroud, blades and part of the volute for the nominal flow point shows that pressure in the hub side is higher than in shroud, due to the transition of flow from the axial to the radial direction. Also, the flow impact over the tongue varies from the center –where the flow comes directly from the impeller- to the sides.

(6) Pressure distribution over the suction and pressure side of the blades when flow rate is the nominal one is clearly appreciated.

REFERENCES

[1] Dauherty, R. L.,1926 "A Further Investigation of Performance of Centrifugal Pumps When Pumping Oils", *Bulletin 130, Gould Pumps, Inc.*

[2] Stepanoff, A. J.,1940 "Pumping Viscous Oils with Centrifugal Pumps", *Oil and Gas Journal*, No. 4 .

[3] Telow, N., 1942 "A Survey of Modern Centrifugal Pump Practice for Oilfield and Oil Refinery Services", *The Institution of Mechanical Engineering*, No. 121.

[4] Ippen, A. T., 1946 "The Influence of Viscosity on Centrifugal Pump Performance", *Trans ASME*, pp. 823.

[5] Itaya, S. and Nishikawa, T., 1960 "Studies on the Volute Pumps Handling Viscous Fluids", *Bulletin of JSME*, Vol. 26, No. 162, pp. 202.

[6] Stoffel, B., 1980 "Tests on Centrifugal Pumps for Handling Viscous Liquids", *Institute of Chemical Engineers*, paper No. 3.

[7] Li, W. G., 1996 "LDV Measurements and Calculations of Internal Flows in the Volute and Impellers of a Centrifugal Oil Pump", *Fluids Machinery*, Vol. 22, No. 2.

[8] Li, W. G. and Z. M. Hu, 1997 "An Experimental Study on Performance of Centrifugal Oil Pump", Fluids Machinery, Vol. 25, No. 2, pp. 3-7.

[9] Denton, D.,1986 "The use of a Distribution Body Force to Simulate Viscous Effects in 3D flow

Calculation", *ASME Journal of Turbo machinery*, Vol. 86, GT-144.

[10] Miner, S. M., 1992 "Two Dimensional Flow Analysis of a Laboratory Centrifugal Pump", *ASME Journal of Turbo machinery*, Vol. 114.

[11] Miner, S. M., 1992 "Turbulence Measurement in a Centrifugal Pump with a Synchronously Orbiting Impeller", *ASME Journal of Turbo machinery*, Vol. 114.

[12] Croba, D. and Kueny, J. L., 1996 "Numerical Calculation of 2D, Unsteady Flow in Centrifugal Pumps: Impeller and Volute Interaction", *International Journal for Numerical Methods in Fluids,* Vol. 22, pp.467-481.

[13] Denus, C. K. and Gode, E., 1999 "A Study in Design and CFD Analysis of a Mixed-Flow Pump Impeller", *ASME-FEDSM-99-6858.*

[14] Blanco, E., 2000 "Numerical Flow Simulation in a Centrifugal Pump with Impeller-Volute Interaction", *ASME-FEDSM200-11297.*

[15] Blanco, E., 2000 "Numerical Simulation of Centrifugal Pumps", *ASME-FEDSM200-11162.*

[16] Shojaee Fard, M. H. and Ehghaghi, M. B., 2002 "Experimental and Numerical Study of Centrifugal Pump in the Performance of Viscous Flow", *International Journal of Engineering Science*, Vol. 13, No. 3, pp. 35-52.

[17] Shojaee Fard, M. H. and Ehghaghi, M. B., 2002 "Experimental Study of the Increase of Efficiency in Centrifugal Oil Pump", *Ph. D. thesis, Tabriz University of Iran.*

[18] Gugau, M.,2003 " Transient Impeller-Volute Intraction in a Centrifugal Pump", *Technische Universitat Darmstadt, FG Turbomaschinen und Fluidantriebstechnik.*

Proceedings of IMECE2005
2005 ASME International Mechanical Engineering Congress and Exposition
November 5-11, 2005, Orlando, Florida USA

IMECE2005-79627

Torque Ripple Attenuation of an Axial Piston Pump by Continuous Swash Plate Adjustment

Viral S. Mehta and Noah D. Manring
Mechanical and Aerospace Engineering Department
University of Missouri – Columbia
Columbia, MO 65211

Abstract

One of the major drawbacks of a hydraulic system is the disturbing noise generated by the hydraulic pump. Based on the accepted theory about noise generation in an axial piston pump, various studies suggesting modification of the port-plate relief groove geometry and addition of hydraulic attenuators have shown a limited success in reducing the noise. The noise level is still high and may not be acceptable for future applications. A recent industrial study shows that the noise apparently has relation with the torque acting on the input shaft of the axial piston pump. The primary objective of this paper is to describe a new method to reduce the noise level by varying the swash plate of the pump continuously to eliminate the torque ripple. The paper begins by deriving the equation of torque acting on the input shaft as a function of the average torque and the total number of pistons that are used within the rotating group. A control law is derived according to which if the swash plate is dithered, should give a constant shaft torque. By attenuating the torque ripple characteristics of the pump, other vibrational aspects of the pump are also expected to be reduced. The secondary objective of this paper is to present a Simulink® model of a nine piston pump to describe a functional pump. The swash plate control law earlier derived will be applied to this model to see the behavior of the output parameters like the shaft torque, output flow and pressure. The results will be investigated to see if any relationship exist between the pump torque ripple and pressure ripple, and the theory of pump noise attenuation by reduction in torque ripple will be corroborated.

Introduction

Hydraulic power transmission has been in use for more than a century. It has many advantages over other modes of power transmission like electrical motors, such as fast response and high torque to inertia ratio. Also, hydraulic systems are very compact and offer higher efficiency across a wide range of operating conditions. In hydraulic systems, power is transmitted and controlled through liquid under pressure. Positive displacement pumps convert shaft rotation into hydraulic power. During this process pumps generate a loud noise which is undesirable and unacceptable to many applications.

Noise reduction has been a key issue to most hydraulic pump manufacturers and considerable effort has been put to alleviate this problem. It is very important to understand what causes this noise in order to reduce it. Literature contains various theories [1-10] associated with the noise generation in axial piston pumps. A widely accepted theory describes the pressure variation caused by the flow variation as the main source of noise. Flow variation is basically caused by (a) the periodic variation in geometric displacement and (b) the oil compression and expansion processes at transitions between high and low pressure. Also, the reason for geometric displacement variation is that the total flow is a summation of the flows from individual pumping elements. Having said so, it is very intuitive to suggest that "the smoother the flow, lower the noise".

As stated above, various theories are associated with the noise generation of axial piston pumps. Over a period of time, some of these techniques have become industry standards while the others are still in investigation stage. Following is a list of some of the traditional techniques or sort of guiding principles used by the pump industry to reduce the noise level of axial piston pumps. (1) Use multiple pumps in parallel. (2) Use a large oil volume at the discharge port. (3) Use hydraulic acoustic filter. (4) Use the lowest practical pump speed. (5) Do system tuning- selection of size, type, quality of pump etc. (6) Use flexible hose and supports.

Edge and Darling [1] suggest a few techniques to noise reduction. One of the most common techniques is to optimize the port plate geometry. Axial piston pumps now come with relief grooves at the leading edges of the inlet and outlet ports. They eliminate cylinder pressure undershoot at top dead center (TDC) and overshoot at bottom dead center (BDC) which help reduce the noise level. The use of the capped pistons to attenuate noise is an established industry practice. A couple of recommendations regarding the port geometry are that the slop of the groove should not be more than 30 degree while groove length should be kept at minimum.

An extensive review of the progress that has been made over the last 25 years in the understanding of noise is presented by Harrison and Edge [2]. Following is a brief overview of techniques mentioned in the article and possible problems associated with each. (1) Use of an anti noise devices is cost prohibitive, more ever it introduces many other technical problems. (2) It is generally recognized that the compression and decompression phenomenon and the resulting swash plate vibration can be controlled by varying valve plate timing, i.e., the initiation of communication between the kidney port and the fluid inlet or fluid outlet in the valve plate. However, varying valve plate timing during operation of an axial piston pump requires rotating the valve plate while it is under heavy axial loading from the cylinder barrel, which is not very practical. Also it reduces volumetric efficiency of the pump and introduces the cavitations problem. (3) Use of independent suction and delivery timing is also unaffordable by high cost. (4) Use of a check valve is not possible because of very rapid switching action required in operation. Also it is ineffective at high delivery pressure and pump speed. (5) Use of a heavily damped check valve (HDCV) prevents the rapid switching action, tunes itself according to the changing operating conditions and reduces the flow ripple. This technique seems promising and is under more investigation.

In order to reduce the flow ripple, it is extremely important to know the mechanism of source flow ripple. Kojima et al. [3] introduced the '2 Pressures/ 2 System' method to predict the flow ripple and experimentally verified the same to a great accuracy. This model includes the compressibility of the fluid, the fluid momentum in the chamber and relief grooves, and the pressure dependent effective bulk modulus of the fluid containing air. The results supports the development of quieter pumps with better fluid borne noise characteristics. Ivantysynova et al. [4] presents a new computer based design method for valve plate design using the simulation program CASPAR and the extension tool AVAS. The valve plate design strongly affects the instantaneous cylinder pressure, the profile of oscillating forces applied on the machine parts and the effective flow and pressure pulsation. Having known that these oscillating forces exerted on machine parts cause structure borne noise and the flow/pressure pulsation causes the fluid borne noise, these software tools could be used to evaluate the effectiveness of design measures for noise reduction during the design phase, i.e. before prototype production.

Petterson [5] made a theoretical mapping of the noise and vibration associated and presented many schemes to attenuate noise. Three schemes namely pressure relief groove, pre-compression and ideal compression suggested by Petterson are already discussed by Harrison. The fourth scheme requires a special mention. Variable cylinder pre-compression can be achieved by using a pre-compression filter volume (PCFV). At present results from the PCFV mechanism have only been published for fixed-displacement pumps. Schutten [6] suggests putting kidney ports at different circumferences and thereby varying the timing of the communication between the leading edges of the kidney ports and the fluid inlet and fluid outlet. This design called 'random porting' shifts the sound energy from high to low frequency. Since high frequency noise is more annoying than low frequency; even though sound energy remains same, noise quality is improved. Another way of achieving the same is keeping cylinders at different circumferential dimensions but it encounters the balancing problem.

In a very different concept, Achten [7] proposes a 'floating cup principle' where in two pumps connected by a through shaft make a tandem pump. The pistons are arranged in two rings of twelve pistons each with both rings connected to the central rotor in a mirrored back-to-back configuration. It is shown that the flow and pressure pulsations are greatly reduced resulting in low noise. Along with a few general recommendations for noise reduction, Wilkes [8] proposes the use of an 'in-line nitrogen charged noise suppressor' right at the outlet of the pump. Strunk [9] developed a new type of reactive silencer called a cross-loop attenuator which is essentially a side branch resonator but differs in design.

In summary, the hydraulic community has been emphasizing on reworking port-plate geometry, adding hydraulic silencers/attenuators etc. There is a limited success in lowering down the noise level. Also, pumps with relief groove can generate a significant increase in fluid borne noise (FBN) when operated away from the design point.

Correlation Between Torque on the Input Shaft and Noise

A recent industrial study [10] shows that there is an apparent relationship between the torque acting on the input shaft of a hydrostatic pump and resulting pump noise. Like flow variation, this torque ripple results from the discrete piston locations within the pump which create oscillating moments on the shaft.

In this study, two 250cc pumps were tested where the pump speed was swept from low to high. Among other parameters, case vibration and sound were recorded. The sound level for the pump showed a prominent high operating region from 1000 to 1250 rpm. A first-look observation of the waveforms of the sound pressure and fluid pressure supported the popular theory of fluid pulsation as the main source of noise but a look at the excitation levels for the entire sweep told a different story. At excitation level, waveform of the case acceleration confirmed to the waveform of the sound pressure. It was also proved that case vibration is driven by the input shaft velocity fluctuations. The shaft velocity fluctuations also show a peak that correlate very well with sound pressure. The fluid pressure fluctuation, however, do not. A follow on test showed an in-phase correlation between velocity fluctuations and torque fluctuations. This hence showed a path of torque fluctuations to velocity fluctuations to case vibration to sound emission. The fluid pressure fluctuations did not have any physical path like that and are not a significant cause of the case noise. The fluid fluctuations are normally a significant cause of noise in the structure where a hose or valve is hard mounted to the frame.

By attenuating the amplitude of the torque ripple the vibrational characteristics of the pump will also be reduced thereby providing a machine that operates more smoothly with less noise generation.

Nomenclature

A_p cross sectional area of a single piston

C_d coefficient of discharge

d_n normal distance of the nth piston away from the centerline of the pump shaft

F_n side load produced on the cylinder block by the nth piston

K	Leakage coefficient
N	total number of pistons
N_o	odd number of pistons
N_e	even number of pistons
n	dummy variable for identifying the nth piston
n'	number of pistons pressurized by the discharge port
P_d	pump discharge pressure
P_i	pump intake pressure
P_n	fluid pressure within the nth piston chamber
r	piston pitch radius
T	net instantaneous shaft torque
T_{o1}	torque component generated due to pistons on high pressure side
T_{o2}	torque component generated due to pistons on low pressure side
$\overline{T}$	average shaft torque produced the pump
$\overline{T}_s$	average starting shaft torque produced by the pump
T_n	torque produced by the nth piston
V_o	piston chamber volume when $\alpha = 0$
α	swash plate angle
β	bulk modulus
ρ	density of fluid
ς	ratio $\frac{P_i}{P_d}$
ξ	number defined in Equation (18)
φ_n	angular position of the nth piston
φ_1	angular position of piston number one
ω	angular velocity of the pump shaft

Torque Ripple of an Input Shaft of an Axial Piston Pump

Swash plate type axial piston pump is one of the most widely used pumps due to high operating efficiencies (85 to 95%), high operating pressure (up to 42 MPa) and easily variable discharge flow. Its design and construction has been discussed by Manring [11] at great length. A typical industrial pump consists of an odd number of pistons.

The analysis for the instantaneous torque that is exerted on the shaft of an axial-piston pump by a single rotating group has been presented in previous literature [11]. Ivantysyn and Ivantysynova [12] also present some of the discussions and analysis. In this study the torque on the shaft of a swash plate type axial piston pump is derived from the component level of the machine. The net input torque is calculated by summing all the idealized forces exerted on the components. Here the frictional losses resulting due to the sliding conditions within the pump have been neglected. Also piston inertia is ignored since the pistons are evenly spaced in a pump and inertia effectively cancels out. The other assumptions are: fluid flow is considered incompressible and leakage is negligible. A constant pressure profile is assumed for this analysis as pressure is constant in the cylinder except at the TDC and BDC where compression and expansion occurs. While the dynamic component is concerned with pressure variation in the system, the aim here is to minimize the kinematic component of torque ripple. Similarly the effects of true port plate geometry are not considered.

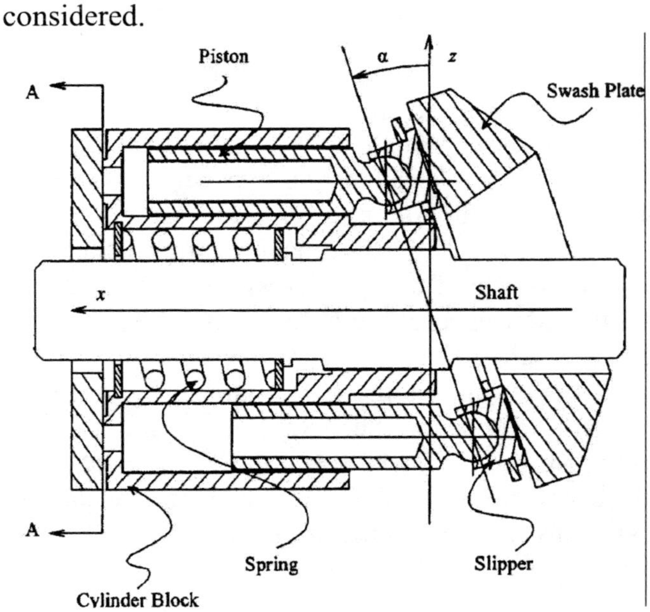

Figure 1 General Pump Configuration

Figure (1) shows a general pump configuration with a sectioned view AA showing the horse-shoe porting and reaction forces in Figure 2. From the section view it can be shown that the torque exerted on the input shaft by a single piston is

$$T_n = R_n \sin(\alpha)\, d_n \tag{1}$$

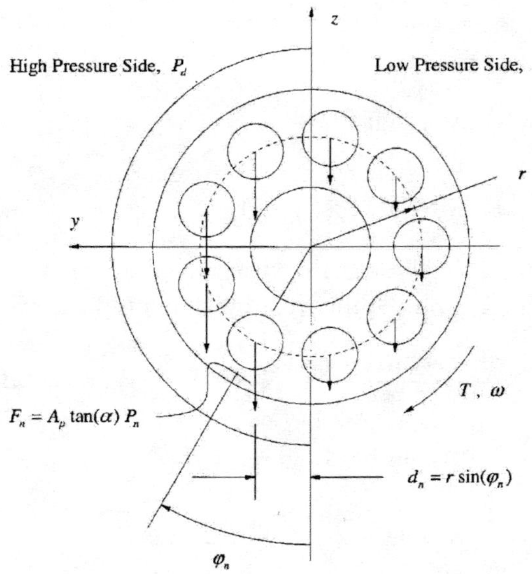

Figure 2 Reaction forces acting on the shaft

where, $R_n \sin(\alpha)$ is the vertical component of the swash plate reaction R_n and d_n is the horizontal distance of the nth piston away from the z-axis. R_n and d_n are given by the Equations (2) and (3),

$$R_n = \frac{P_n A_p}{\cos(\alpha)} \quad \textbf{(2)}$$

where, P_n is the pressure acting on the nth piston while A_p is the area of a single piston. Also it can be shown that

$$d_n = r\,\sin(\varphi_n) \quad \textbf{(3)}$$

where r is the piston pitch radius and φ_n is the angular position of the nth piston as shown in figure (2).

Hence, the net instantaneous torque is,

$$T = A_p \tan(\alpha)\, r \sum_{n=1}^{N} P_n \sin(\varphi_n) \quad \textbf{(4)}$$

where, N is the total number of pistons of the pump.

Closed-Form Expression for Average Torque

Assuming that n' is the number of pistons over the discharge port, equation for the torque at the discharge side T_d and equation for the torque at the intake side T_i can be written as,

$$T_d = A_p \tan(\alpha)\, r\, P_d \sum_{n=1}^{n'} \sin(\varphi_n) \quad \textbf{(5)}$$

and,

$$T_i = A_p \tan(\alpha)\, r\, P_i \sum_{n=1}^{n'} \sin(\varphi_n) \quad \textbf{(6)}$$

Also if the pistons are evenly spaced, then Equation (7) describes the angular position of the nth piston.

$$\varphi_n = \varphi_1 + \frac{2\pi}{N}(n-1) \quad \textbf{(7)}$$

where φ_1 is the position of the first piston. It can be easily shown that the closed-form expression of $\sin(\varphi_n)$,

$$\sum_{n=a}^{b} \sin(\varphi_n) = \csc\left(\frac{\pi}{N}\right) \sin\left(\frac{\pi}{N}(1-a+b)\right) \sin\left(\varphi_1 + \frac{\pi}{N}(a+b-2\right) \quad \textbf{(8)}$$

For the high pressure side, $a = 1$ and $b = n'$ while for the low pressure side $a = n' + 1$ and $b = n$. Summing T_d and T_i to give the total torque,

$$T = A_p \tan(\alpha)\, r \csc\left(\frac{\pi}{N}\right) \sin\left(\frac{\pi n'}{N}\right) P_d \begin{pmatrix} \sin\left(\varphi_1 + \frac{\pi}{N}(n'-1)\right) \\ + \varsigma \sin\left(1 + \frac{\pi}{N}(N+n'-1)\right) \end{pmatrix} \quad \textbf{(9)}$$

where, $\varsigma = \frac{P_i}{P_d}$. Also based on the Equation (9), it can be shown that the average torque acting on the input shaft is given by

$$\bar{T} = \frac{N}{\pi} A_p \tan(\alpha)\, r\, P_d (1-\varsigma) \quad \textbf{(10)}$$

Torque for Pumps with Odd and Even Number of Pistons

For odd number of pistons, the number piston passing over the intake and discharge port vary by Equations (11) and (13).

$$n' = \frac{N_o + 1}{2} \quad \textbf{(11)}$$

$$T_{o1} = A_p \tan(\alpha)\, r\, P_d \frac{(1-\varsigma)}{2} \csc\left(\frac{\pi}{2N_o}\right) \cos\left(\frac{\pi}{2N_o} - \varphi_1\right) \quad \textbf{(12)}$$

which is valid for $0 < \varphi_1 < \frac{\pi}{N_o}$.

Similarly,

$$n' = \frac{N_o - 1}{2} \quad \textbf{(13)}$$

$$T_{o2} = A_p \tan(\alpha)\, r\, P_d \frac{(1-\varsigma)}{2} \csc\left(\frac{\pi}{2N_o}\right) \cos\left(\frac{\pi}{2N_o} - (\varphi_1 - \frac{\pi}{N_o})\right) \quad \textbf{(14)}$$

which is valid for $\frac{\pi}{N_o} < \varphi_1 < \frac{\pi}{2N_o}$. These two solutions match at $\varphi_1 = \frac{\pi}{N_o}$.

Similarly for an even number of pistons,

$$n' = \frac{N_e}{2} \quad \textbf{(15)}$$

$$T_e = A_p \tan(\alpha)\, r\, P_d (1-\varsigma) \csc\left(\frac{\pi}{N_e}\right) \cos\left(\frac{\pi}{N_e} - \varphi_1\right) \quad \textbf{(16)}$$

which is valid for $0 < \varphi_1 < \frac{2\pi}{N_e}$

Torque Ripple Attenuation by Continuous Swash Plate Adjustment

These two torque equations can also be written as,

$$T = \bar{T}\, \xi \csc(\xi) \cos(\xi - \varphi_1) \quad \text{for} \quad 0 < \varphi_1 < 2\xi \ , \quad \textbf{(17)}$$

where,

$$\xi = \begin{cases} \frac{\pi}{2 N_o} & \text{odd number of pistons} \\ \frac{\pi}{N_e} & \text{even number of pistons} \end{cases} , \quad \textbf{(18)}$$

and,

$$\bar{T} = \frac{A_p\, r \tan(\alpha)\, P_d (1-\varsigma)}{\xi} . \quad \textbf{(19)}$$

It can be shown that for $\frac{\partial T}{\partial \varphi_1} = 0$,

$$\bar{T} = \bar{T}_s \cos(\xi) \sec(\xi - \varphi_1) \quad \textbf{(20)}$$

Where, $\bar{T}s$ is the starting torque when α is zero. From Equation (19) the average starting torque is,

$$\bar{T}_s = \frac{A_p\, r \tan(\alpha_s)\, P_d (1-\varsigma)}{\xi} \quad \textbf{(21)}$$

Equations (20) and (21) yield,

$$\bar{T} = \frac{1}{\xi} A_p\, r \tan(\alpha_s)\, P_d (1-\varsigma) \cos(\xi) \sec(\xi - \varphi_1) \quad \textbf{(22)}$$

Equating the Equations (19) and (22),

$$\tan(\alpha) = \tan(\alpha_s) \cos(\xi) \sec(\xi - \varphi_1) \quad \textbf{(23)}$$

If the swash plate is dithered according to the control law described in Equation (23), theoretically it should eliminate the ripple effect and generate a constant shaft torque. However, this may not be true when compressibility and leakage are considered. Regardless, this control law should reduce the magnitude of the torque ripple by a considerable amount. A swash plate control law is plotted in figure (3) for an initial swash plate angle of 18 deg.

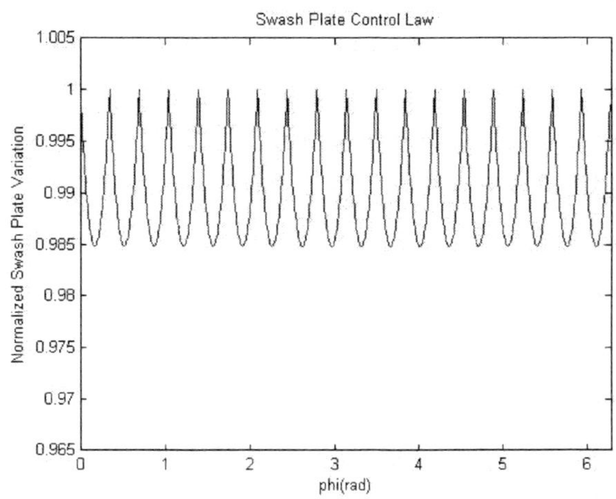

Figure 3 Swash plate control law

Piston Pump Model and Results

A more realistic piston pump model was made in Matlab® Simulink® software. In this model instantaneous pressure, fluid compressibility and leakage are considered unlike the closed-form solution where they were neglected. Manring [11] describes a numerical model of a piston pump with odd and even number of pistons. The piston kinematics, control-volume analysis and other relevant information and equations could be found in this reference. Two of the most important equations are pressure time rate of change (Equation (24)) and instantaneous piston chamber volume (Equation (25)),

$$\frac{dP}{dt} = \frac{\beta}{V_n}\left(Q - \frac{dV_n}{dt} - Q_l\right) \quad \textbf{(24)}$$

$$V_n = V_o - A_p\, r \tan(\alpha) \sin(\varphi) \quad \textbf{(25)}$$

The flow in and out of the piston chamber occurs at high velocity and thus at high Reynold's number and hence Q is modeled using the classical orifice equation. Table 1 lists the design parameter used in this model.

Table 1 Model Parameters

Design Parameter	Value
Piston pitch circle radius- r (m)	0.05982
Area of the piston- A_p (m)	7.144e-4
Piston chamber volume V_o when $\alpha = 0$, (m^3)	17.46e-6
Discharge Pressure- P_d (MPa)	25
Pump speed ω (rad/s)	235.6
Leakage coefficient- K	2e-11
Coefficient of discharge- C_d	0.62
Density of fluid- ρ (kg/m^3)	850
Bulk modulus- β (N/m^2)	14 GPa

Figure (4) shows a typical numerical result of Equation (24) as φ_1 varies from $\frac{\pi}{2}$ to $\frac{5\pi}{2}$ (one pump revolution). Here, both instantaneous piston pressure and area are plotted. As the piston passes over either the intake port or the discharge port of the valve plate, the port area remains at a maximum constant. Within these regions, the pressure within the nth piston chamber also appears to remain fairly constant. The two ports on the valve plate are bridged by transition regions where area goes from a maximum value to a minimum value, slowly grows within the transition slot and then quickly returns to the original maximum value. Pressure transition occurs almost linearly when piston passes over the transition regions. A closer look of this pressure profile shows the presence of a pressure spike at the transition region. This is due to a phenomenon called pressure undershoot and is a result of the volumetric compression and expansion of the fluid (explained in [11]). Figure (5) shows a numerical torque ripple. A very similar torque profile could be generated using the closed-form solution given in Equation (17) (not shown here). It is evident that the torque varies by amplitude of only 1.5 percent; however, even this small repeating

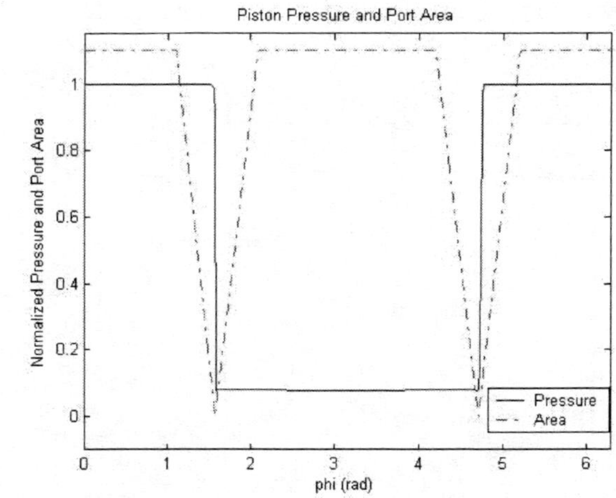

Figure 4 A numerical pressure profile

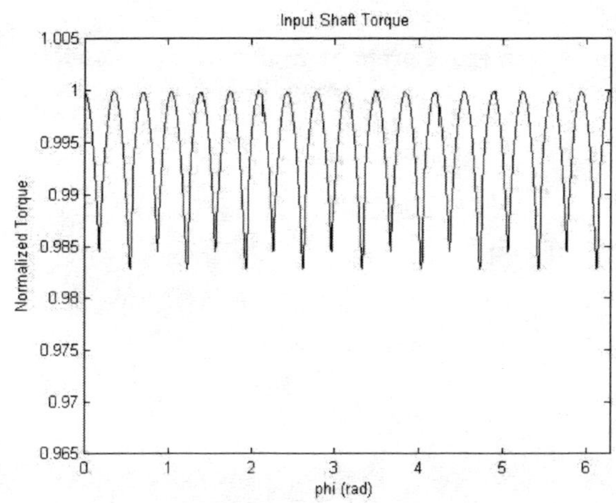

Figure 5 Normalized torque ripple

variation is capable of creating undesirable vibration effects [13].

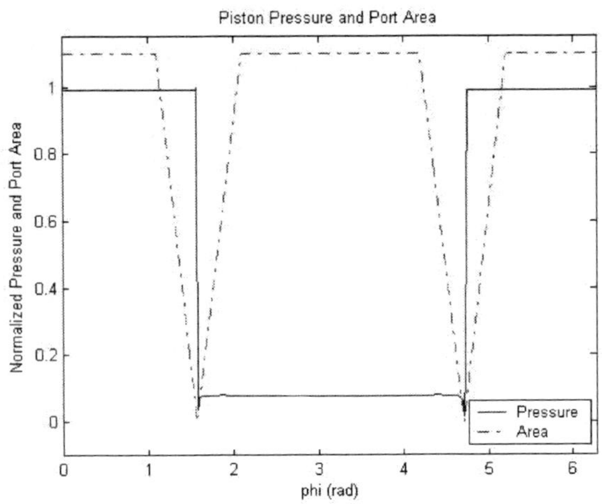

Figure 6 Numerical pressure profile with swash plate control law applied

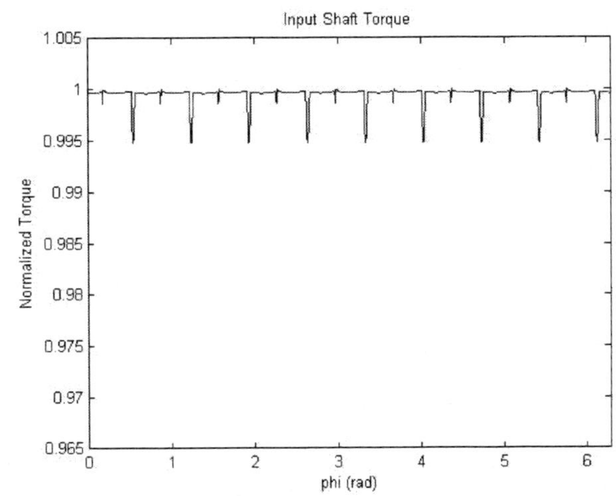

Figure 7 Normalized torque ripple with the swash plate control law applied

Figure (6) shows a numerical pressure profile when the swash plate control law governs the swash plate movement. As shown, the pressure profile does not change much from the earlier case but has a prominent undershoot at the pressure transition region. This is due to the fact that now the instantaneous fluid volume also changes in accordance with the control law and has major effect during the transition region. Figure (7) shows a numerical torque ripple when the swash plate variation is applied. As shown, the torque is mainly constant except spikes at regular intervals. The spikes in the torque ripple are due to the spikes in the pressure ripple. Even with spikes, the maximum variation in the torque ripple is 0.5 percent as against 1.5 percent earlier. If the pressure profile is made smoother, this instantaneous torque changes could be eliminated and a constant torque could be obtained with the gain of 1.5 percent.

Conclusion

The primary objective of this paper is to describe a new method to reduce the noise by varying the swash plate of the pump continuously to eliminate the torque ripple. The entire research is based on this hypothesis that the torque ripple is the main source of noise.

1. Against the commonly accepted theory that noise in an axial piston pump is generated by fluid pulsation, a case is produced supporting the shaft torque ripple as the lead source of the noise. Therefore, if the amplitude of the shaft torque is reduced, noise level should get reduced too.
2. A closed-form solution of the input shaft torque is produced as a function of the average shaft torque and total number of pistons.
3. A numerical torque ripple has been shown to vary by 1.5 percent of the maximum torque input.
4. A swash plate control law is presented according to which if the swash plate is adjusted continuously, the torque becomes constant. This may not hold entirely true for any real world application where the assumptions also don't hold true. However, application of this control law should reduce the amplitude of torque ripple by a good amount (1 percent) and should help lower the noise level and reduce the overall pump vibration considerably.

Acknowledgements

This project was proposed and sponsored by engineers at Caterpillar, Inc. who are devoted to the continuous improvement of off-highway equipment. The authors would like to give a note of special thanks to B.J. Landsberger who is now a research faculty member at the University of Nevada – Las Vegas for his initiation of this work and for his participation in discussions that have proven to be helpful.

References

1) Edge K. A., Darling J., "The Pumping Dynamics of Swash Plate Piston Pumps," *Journal of Dynamic Systems, Measurement, and Control,* Vol. 111 (1989), pp-307-312
2) Harrison A. M., Edge K. A., "Reduction of axial piston pump pressure ripple," *Proceedings of the Institution of Mechanical Engineers Part I- Journal of Systems and Control Engineering*, 214 (1), (2000), pp - 53-63
3) Kojima Eiichi, Yu Jianhong, Ichiyanagi Takayoshi; "Experimental Determining and Theoratical Predicting of Source Flow Ripple Generated by Fluid Power Piston Pumps," *SAE Technical Paper Series, Publication by SAE*, Warrendale, PA, USA, 2000-01-2617. pp
4) Ivantysynova Monika, Huange Changchun, Christiansen Sven-Kelana; "Computer Aided Valve Plate Design – An Effective Way to Reduce Noise," *SAE Technical Paper Series, Publication by SAE*, Warrendale, PA, USA, 2004-01-2621. pp
5) Pettersson, Maria E.; Weddfelt, Kenneth G.; Palmberg, Jan-Ove; "Methods of reducing flow ripple from fluid power piston pumps - an theoretical approach"' *SAE Technical Paper Series, Publication by SAE*, Warrendale, PA, USA, 911762. pp 1-10
6) Schutten; Herman P.; Wakefield; Danny M.; Zimmerer; Donald R.; Malaney; David W.; "Noise reduction at the second order frequency", *USPTO 5358388* October 25, 1994
7) Achten, P.A.J., Van den Brink, T.L., Paardenkooper, T., Platzer, T., Potma, H.W., Schellekens, M.P.A. and Vael, G.E.M., "Design and testing of an axial piston pump based on the floating cup principle", *The Eighth Scandinavian International Conference on Fluid Power- Proc. SICFP'03*, Vol. 2 (2004), pp 805-820, Tampere University of Technology.
8) Wilkes Roy G.; "Reduction of Noise in Hydraulic Systems," *SAE Technical Paper Series, Publication by SAE*, Warrendale, PA, USA, 952154. pp
9) Strunk Richard D.; "Silencer for Hydraulic Piston Pump Pressure Pulsations," *SAE Technical Paper Series, Publication by SAE*, Warrendale, PA, USA, 911759. pp
10) Landsberger B. J. "Using the ideal function concept for machine noise control". *146th Meeting of the Acoustic Society of America*. (2003) Austin,TX
11) Manring N. D., "The Torque on the Input Shaft of an Axial-Piston Swash-Plate Type Hydrostatic Pump," *Journal of Dynamic Systems, Measurement, and Control,* Vol. 120 (1998), pp-57-62
12) Ivantysyn J., Ivantysynova M., Hydrostatic Pumps and Motors: Principles, Design, Performance, Modeling, Analysis, Control and Testing. *Tech Book International, India 2001.*
13) Bolinger J. W., "Considerations for Hydrostatic System Noise Control," *Society of Automotive Engineers International Off-Highway Meeting and Exposition*, Milwaukee,WI.

Proceedings of IMECE2005
2005 ASME International Mechanical Engineering Congress and Exposition
November 5-11, 2005, Orlando, Florida USA

IMECE2005-80284

ON SELF-SENSING ACTUATORS FOR ELECTROHYDRAULIC VALVES: COMPARISONS BETWEEN BOXCAR WINDOW OBSERVER AND KALMAN FILTER

Qinghui Yuan
Eaton Corp. Innovation Center
Eden Prairie, MN 55344
Email: QinghuiYuan@eaton.com

Perry Y. Li
Department of Mechanical Engineering
University of Minnesota
Minneapolis, MN 55455
Email: pli@me.umn.edu

ABSTRACT

Self-sensing refers to extracting the position information from the electromagnetic signals instead of from a physical position sensor. Self-sensing actuator has the benefit of significantly reducing hardware effort. In this paper, we investigate two types of observers: Boxcar Window Observer and Kalman Filter. The performance of both observers has been compared in simulation and experimentally.

1 Introduction

Self-sensing refers to extracting the position information from the electromagnetic signals of an actuator, like currents and voltages, instead of from a physical position sensor such as LVDT. This technique has attracted the attention of researchers in some fields. One example is self-sensing magnetic bearing, which is useful for minimizing the number of wires in some applications, like heart pump [1]. Another example is self-sensing switched reluctance motor, which is the cost effective and robust solution for electric motors [2] [3].

Similarly, self-sensing actuators can be applied in fluid power control engineering. In the electrohydralic valves, spool position feedback, usually furnished by a linear variable differential transformer (LVDT), is needed for higher performance control. Incorporating LVDTs into electrohydraulic valves incurs the cost of the LVDTs themselves, the electric circuits for exciting and decoding the LVDTs, and the overhead of assembling the LVDTs. Therefore, the concept of self-sensing actuators can be very useful to reduce hardware requirement of electrohydraulic valves at the expense of the increased software complexity. Although software development is expensive, the overall cost can be significantly reduced when the product is produced in large quantities.

In this paper, we investigate self-sensing actuators applied to direct acting proportional valves. The study is based on the configuration of a dual-solenoid actuator where each solenoid provides an independent single direction force. An observer approach is proposed to estimate the position information. Unlike in [4] where both the electric and the mechanical ports of the actuator system are used to construct the unknown state vector, our solution only relies on the inductance characteristic of the two solenoids in a push-pull configuration. This results in lower order and more robust observers, since they circumvent the uncertainty due to mechanical load on the system. The main difficulty in the observer design is that the output equation of the observer is highly nonlinear. An exact linearizing approach based on multiple time samples and a Boxcar filter [5] and a Kalman filter [6] were proposed. However, experimental evaluation and comparisons have not been made. The objective of this paper is to compare the result of two different methods. It is shown that the Kalman filter offers the smoother estimate.

A simple solenoid model is used to model how the solenoid inductance varies with the actuator displacement, which is foundational to the self-sensing principle. The model neglects hysteresis, saturation, and a number of details of model which may be very significant for the solenoids in electrohydraulics. A self-calibration procedure has been proposed [7]. This approach is used in identifying the parameters experimentally in this paper.

Our approach that relies on simple solenoid models differ from other self-sensing approach [8] that uses much more complicated solenoid models.

The rest of the paper is organized as follows. In section 2, we formulate the model of a dual-solenoid actuator. Section 3 reviews the principle and the analysis of two types of observers: the Boxcar Window observer and the Kalman Filter in [5] and [6], respectively. In section 4, simulation is presented to verify the observer design. Some experimental results are presented in Section 5. Section 6 contains concluding remarks.

2 Modeling of a dual solenoid actuator

Consider two identical solenoid (with subscripts $i = 1, 2$) are set up in the push-pull configuration as in Fig. 1. Let the input voltages across the coils be u_1 and u_2, and the resulting currents through the coils be i_1 and i_2. The dynamics of the flux linkages λ_1, λ_2 are given by

$$\begin{aligned} \dot{\lambda}_1 &= -Ri_1 + u_1 \\ \dot{\lambda}_2 &= -Ri_2 + u_2 \end{aligned} \tag{1}$$

where R is the resistance of solenoid 1 and 2. By definition, the flux linkages are related to the currents via the inductances:

$$\begin{aligned} \lambda_1 &= L_1(x)i_1 \\ \lambda_2 &= L_2(x)i_2 \end{aligned} \tag{2}$$

where $L_1(x), L_2(x)$ are the inductances of solenoid 1 and 2 associated with the spool displacement x by

$$\begin{aligned} L_1(x) &= \frac{\beta}{d+x} \\ L_2(x) &= \frac{\beta}{d-x} \end{aligned} \tag{3}$$

where β, d are the parameters of the model, in which both solenoids are assumed to be identical. This meets most of case in ElectroHydraulics. Note that the pulling forces F_1 and F_2 (see Fig. 1) acting on the spool share the identical parameters to L_1, L_2

$$\begin{aligned} F_1 &= \frac{\beta}{2}\frac{i_1^2}{(d+x)^2} \\ F_2 &= \frac{\beta}{2}\frac{i_2^2}{(d-x)^2} \end{aligned} \tag{4}$$

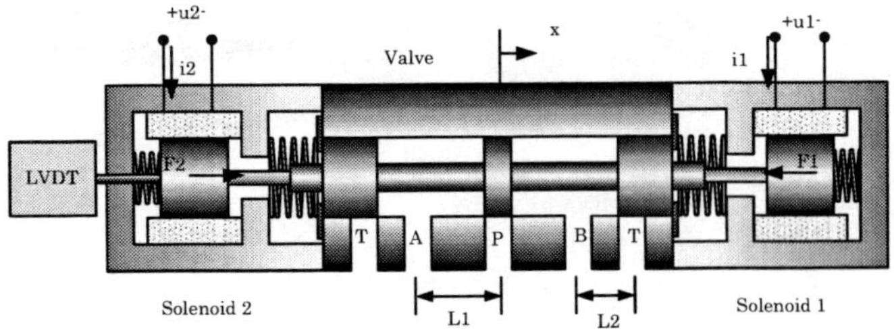

Figure 1. Configuration of a dual-solenoid actuator in a directional valve.

In the above analysis, saturation and hysteresis of the solenoids are not taken into account. Therefore, the inductance of the coil can be simplified as a function of the spool displacement. If saturation is considered, then the inductance would be a function of both the spool displacement and the current. If hysteresis is, then the inductance has to be modeled according to the $B-H$ curve, which is very complicated. In most of the high profile commercial valves, the solenoid is designed to work in the linear region with little hysteresis. In this case, the models are valid.

3 Flux linkage observer

We will construct an observer for the flux linkages in a dual-solenoid actuator. Once the flux linkages are obtained, the spool position can be easily estimated. The required signals are the voltages u_1, u_2 and the currents i_1, i_2. As we have mentioned, a full order observer with the states $(\lambda_1, \lambda_2, x, v)$ can be constructed, where the flux linkages λ_1, λ_2 are the electrical signals, and the displacement and velocity (x, v) are the mechanical signals. Since $\dot{v}$ is subject to unmodeled or poorly modeled forces, the estimated states obtained by an observer that estimate both the electrical (λ_1, λ_2) and the the mechanical (v, x) states would intuitively have poor precision and robustness performance. However, notice that if we are able to correctly estimate the flux linkages, position information can be obtained from the estimated flux linkages and the measured currents, as discussed next. Therefore, our task is to find the reduced-order observer that only involves the electrical dynamics $\dot{\lambda}_1, \dot{\lambda}_2$ in Eq. (1). The mechanical dynamics will not be important any more.

3.1 Principle

The principle of the observer is presented as follows. If the inductances L_1, L_2 are known, then from Eq. (3), the displacement x is given by

$$x(L_1, L_2) = \frac{\beta}{2}\left(\frac{1}{L_1} - \frac{1}{L_2}\right)$$

Unfortunately, the inductance values cannot be directly measured. We would like to find an alternative way to estimate inductances, and accordingly the displacement. Note that the inductance is associated with the corresponding flux linkage and the current. Therefore, our objective becomes to estimate the flux linkages, provided that the currents are measured. Then the position information can be captured via the following equation

$$x(\lambda_1,\lambda_2) = \frac{\beta}{2}\left(\frac{i_1}{\lambda_1} - \frac{i_2}{\lambda_2}\right) \tag{5}$$

where i_1, i_2 are the currents, and λ_1, λ_2 are the flux linkages.

In the section to follow, we will derive an observer to reconstruct the information of the flux linkages. In the observer design procedure, we use the following constraint between the inductances (derived from Eq. (3))

$$\frac{1}{L_1(x)} + \frac{1}{L_2(x)} = \frac{2d}{\beta} \tag{6}$$

or

$$\frac{i_1}{\lambda_1} + \frac{i_2}{\lambda_2} = \frac{2d}{\beta} \tag{7}$$

Eq. (7) serves as an output equation to the dynamics for the observer design. Notice that λ_1 and λ_2, which are to be estimated, appear nonlinearly.

3.2 Boxcar Window Observer

Due to nonlinearity of Eq. (7), we cannot easily design the observer. To overcome the difficulty, we define a moving Boxcar window in time domain, and solve the initial conditions of flux linkages in the wondow. Reader are referred to [5] for details of the Boxcar window observer.

From Eqs. (1), we can represent the flux linkages at $\forall t > t_0$ with respect to the initial conditions $\lambda_1(t_0), \lambda_2(t_0)$ as

$$\begin{aligned} \lambda_1(t) &= \lambda_1(t_0) + I_1(t_0,t) \\ \lambda_2(t) &= \lambda_2(t_0) + I_2(t_0,t) \end{aligned} \tag{8}$$

in which

$$\begin{aligned} I_1(a,b) &:= \int_a^b [u_1(\tau) - i_1(\tau)R]\, d\tau \\ I_2(a,b) &:= \int_a^b [u_2(\tau) - i_2(\tau)R]\, d\tau. \end{aligned} \tag{9}$$

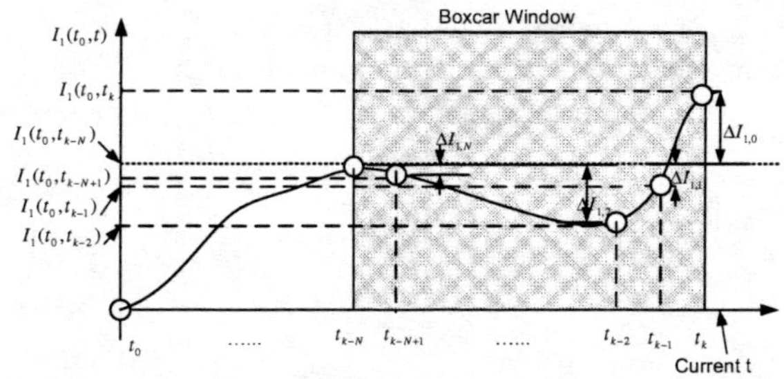

Figure 2. Least squares method is used to calculate the local initial condition $\lambda_{1,0}(k) = \lambda_1(t_{k-N}), \lambda_{2,0}(k) = \lambda_2(t_{k-N})$, for the horizon with the fixed width $N\Delta_T$. See text for details.

We assume that R is insensitive to the temperature variation, and can be measured precisely. Since t, $u_1(t), u_2(t), i_1(t), i_2(t)$ are known (via measurement) at all t, so are $I_1(t_0,t), I_2(t_0,t)$. The problem of finding λ_1, λ_2 can be transformed into one in which the initial conditions $\lambda_1(t_0), \lambda_2(t_0)$ are required to be solved.

Theoretically, we can evaluate Eq. (8) at $t = t_a$, and $t = t_b$ ($t_a \neq t_b$) respectively, and then combine them with (7)

$$\begin{aligned} \frac{i_1}{\lambda_1(t_0) + I_1(t_0,t_a)} + \frac{i_2}{\lambda_2(t_0) + I_2(t_0,t_a)} &= \frac{2d}{\beta} \\ \frac{i_1}{\lambda_1(t_0) + I_1(t_0,t_b)} + \frac{i_2}{\lambda_2(t_0) + I_2(t_0,t_b)} &= \frac{2d}{\beta} \end{aligned}$$

two unknown variables $\lambda_1(t_0), \lambda_2(t_0)$ can be obtained from the above equations. However, this method is not robust since the errors of $I_1(t_0,t), I_2(t_0,t)$, and hence λ_1, λ_2, would accumulate quickly due to the disturbance in the open loop dynamics.

In order to avoid the error accumulation, a Boxcar window is defined in the time history, and the initial conditions of the flux linkages are solved for the particular window. The window moves uniformly, therefore we will have the uniformly updated initial conditions. The details are explained as follows.

Define a update interval $\Delta_T > 0$, and $t_k = k\Delta_T$ for $k = 0, 1, 2, 3, \ldots$, as shown in Fig. 2. In our algorithm, a boxcar window is defined to cover a time horizon $[t_{k-N}, t_k]$ where N is a positive integer. In other words, the horizon has the width $N\Delta_T$. Given a horizon associated with t_k, we can define the local variables

$$\begin{aligned} \Delta I_{1,j}(k) = I_1(t_{k-N}, t_{k-j}) &= I_1(t_0, t_{k-j}) - I_1(t_0, t_{k-N}) \\ \lambda_{1,0}(k) &= \lambda_1(t_{k-N}) \\ i_{1,j}(k) &= i_1(t_{k-j}) \\ \Delta I_{2,j}(k) = I_2(t_{k-N}, t_{k-j}) &= I_2(t_0, t_{k-j}) - I_2(t_0, t_{k-N}) \\ \lambda_{2,0}(k) &= \lambda_2(t_{k-N}) \\ i_{2,j}(k) &= i_2(t_{k-j}) \end{aligned}$$

for $j = 0,1,\ldots,N-1$, where $I_1(\cdot,\cdot), I_2(\cdot,\cdot)$ refer to Eq. (9).

Combining the above variables at $j = 0,1,\ldots,N-1$ (or at $t = t_k, t_{k-1},\ldots,t_{k-N+1}$) with Eq. (7) yields N nonlinear equations

$$\frac{i_{1,j}(k)}{\lambda_{1,0}(k)+\Delta I_{1,j}(k)} + \frac{i_{2,j}(k)}{\lambda_{2,0}(k)+\Delta I_{2,j}(k)} - \frac{2d}{\beta} = 0 \quad for\ j = 0,1,\ldots,N-1 \tag{10}$$

Multiplying the above equation by $[\lambda_{1,0}(k)+\Delta I_{1,j}(k)][\lambda_{2,0}(k)+\Delta I_{2,j}(k)]$ gives

$$\begin{aligned} f(j) = {} & i_{1,j}(k)[\lambda_{2,0}(k)+\Delta I_{2,j}(k)] + i_{2,j}(k)[\lambda_{1,0}(k)+\Delta I_{1,j}(k)] \\ & - \frac{2d}{\beta}[\lambda_{1,0}(k)+\Delta I_{1,j}(k)][\lambda_{2,0}(k)+\Delta I_{2,j}(k)] = 0 \\ & for\ j = 0,1,\ldots,N-1 \end{aligned} \tag{11}$$

Subtracting $f(j-1)$ from $f(j)$ in Eq. (11), for $j = 1,\ldots,N-1$, can eliminate the nonlinear terms, and give $N-1$ equations

$$a_j(k)\lambda_{1,0}(k) + b_j(k)\lambda_{2,0}(k) = c_j(k) \quad for\ j = 1,\ldots,N-1 \tag{12}$$

where

$$\begin{aligned} a_j(k) :=& i_{2,j}(k) - i_{2,j-1}(k) - \frac{2d}{\beta}[\Delta I_{2,j}(k) - \Delta I_{2,j-1}(k)] \\ b_j(k) :=& i_{1,j}(k) - i_{1,j-1}(k) - \frac{2d}{\beta}[\Delta I_{1,j}(k) - \Delta I_{1,j-1}(k)] \\ c_j(k) :=& \frac{2d}{\beta}[\Delta I_{1,j}(k)\Delta I_{2,j}(k) - \Delta I_{1,j-1}(k)\Delta I_{2,j-1}(k)] + i_{1,j-1}(k) * \\ & \Delta I_{2,j-1}(k) + i_{2,j-1}(k)\Delta I_{1,j-1}(k) - i_{1,j}(k)\Delta I_{2,j}(k) - i_{2,j}(k)\Delta I_{1,j}(k) \end{aligned} \tag{13}$$

for $j = 1,\ldots,N-1$.

Notice that although Eq. (10) represent N nonlinear equations in the initial flux linkages $\lambda_{1,0}(k), \lambda_{2,0}(k)$, Eq. (12) represent $N-1$ equations that are linear. The least squares algorithm is then utilized to calculate the local initial conditions $\lambda_{1,0}(k)$ and $\lambda_{2,0}(k)$ associated with the boxcar window $[t_{k-N}, t_k]$

$$\begin{bmatrix} \lambda_{1,0}(k) \\ \lambda_{2,0}(k) \end{bmatrix} = [M(k)^T M(k)]^{-1} M(k)^T C(k) \tag{14}$$

where $M(k) = \begin{bmatrix} a_1(k) & b_1(k) \\ a_2(k) & b_2(k) \\ \vdots & \vdots \\ a_{N-1}(k) & b_{N-1}(k) \end{bmatrix}$, and $C(k) = \begin{bmatrix} c_1(k) \\ c_2(k) \\ \vdots \\ c_{N-1}(k) \end{bmatrix}$.

In short, we can summarize our observer as follows. For $t = t_k$, this is so called the update time point. All the local variables $i_{1,j}(k), i_{2,j}(k), \Delta I_{1,j-1}(k), \Delta I_{2,j-1}(k)$ for $j = 1,2,\ldots,N-1$, are refreshed. Eq. (14) are then utilized to obtain the local initial conditions $\lambda_{1,0}(k), \lambda_{2,0}(k)$. Next, for $\forall t \in (t_k, t_{k+1})$, we will use the following formulas to get the flux linkages.

$$\begin{aligned} \widehat{\lambda}_1(t) &= \lambda_{1,0}(k) + \Delta I_{1,0}(k) + \int_{t_k}^{t} (u_1 - i_1 R)\, d\tau \\ \widehat{\lambda}_2(t) &= \lambda_{2,0}(k) + \Delta I_{2,0}(k) + \int_{t_k}^{t} (u_2 - i_2 R)\, d\tau \end{aligned} \tag{15}$$

Notice that although the initial flux linkages are updated discretely, the flux linkage estimates are available at all times. The position estimate can then be obtained via

$$\begin{bmatrix} i_1/\widehat{\lambda}_1 \\ i_2/\widehat{\lambda}_2 \end{bmatrix} = \begin{bmatrix} d/\beta \\ d/\beta \end{bmatrix} + \begin{bmatrix} 1/\beta \\ -1/\beta \end{bmatrix} x \tag{16}$$

from Eq. (3) using the least squares method.

Before we conclude our algorithm in this section, it is worth mentioning the initial behavior of the observer starting from t_0. Because we cannot compute the local initial conditions before a complete horizon is formed, the Boxcar window observer cannot provide the correct flux linkage information in the period $[t_0, t_0 + N\Delta_T)$.

3.3 Kalman Filter

In the Boxcar Window observer design (section 3.2), we transform the flux linkage estimate problem into one of solving the initial conditions. However, from Eq. (1), it is also clear that we may use the Kalman Filter method if a term both measured and estimated can be found so that we may construct an output injection for the Kalman filter.

First, an output equation is developed. Again using Eqs. (8) (9), the inductance constraint in Eq. (7) can be rewritten as:

$$\begin{aligned} & [\beta i_2(t) - 2dI_2(t,t_0)]\lambda_1(t) + [\beta i_1(t) - 2dI_1(t,t_0)]\lambda_2(t) \\ & + 2dI_1(t,t_0)I_2(t,t_0) = 2d\lambda_1(t_0)\lambda_2(t_0) \end{aligned} \tag{17}$$

where Eq. (8) has been used repeatedly. The left side of Eq. (17) is linear in (λ_1, λ_2). The right hand side of Eq. (17), although nonlinear in the unknown initial states $(\lambda_1(t_0), \lambda_2(t_0))$, is not a function of time t. The unknown initial states can be eliminated via taking the difference of Eq. (17) evaluated at t and at t_0:

$$\begin{aligned} & [\beta i_2(t) - 2dI_2(t,t_0)]\lambda_1(t) + [\beta i_1(t) - 2dI_1(t,t_0)]\lambda_2(t) \\ & + 2dI_1(t,t_0)I_2(t,t_0) - \beta i_2(t_0)\lambda_1(t_0) - \beta i_1(t_0)\lambda_2(t_0) = 0 \end{aligned}$$

Then substitute $(\lambda_1(t),\lambda_2(t))$ for $(\lambda_1(t_0),\lambda_2(t_0))$ using Eq. (8):

$$\{\beta[i_2(t)-i_2(t_0)]-2dI_2(t,t_0)\}\lambda_1(t)+\{\beta[i_1(t)-i_1(t_0)]-2dI_1(t,t_0)\}\lambda_2(t) = -2dI_1(t,t_0)I_2(t,t_0)-\beta i_2(t_0)I_1(t,t_0)-\beta i_1(t_0)I_2(t,t_0) \quad (18)$$

Eq. (18) can be interpreted as the output equation for the flux linkages (λ_1,λ_2) in Eq. (1). The left side of Eq. (18) is the linear combination of the flux linkages, with the right side furnishing the "measurement".

Similar to the case in Section 3.2, the error may accumulate in $I_1(t,t_0),I_2(t,t_0)$. We may restrict the computation in a (finite) boxcar horizon right before the current time t, thus having an effect of forgetting the the past currents and voltages. We can take $t_0 = t-T$ where T is a fixed time interval. We denote the output "measurement" by

$$y(t) = -2dI_1(t,t-T)I_2(t,t-T)-\beta i_2(t-T)I_1(t,t-T) -\beta i_1(t-T)I_2(t,t-T) \quad (19)$$

The dynamics of the flux linkages Eq. (1), the linear output equation Eq. (18), and the measurement Eq. (19) are in a standard time varying linear state space form. Thus we can design a Kalman Filter [9] to reconstruct the unknown states $\lambda_1(t),\lambda_2(t)$:

$$\frac{d}{dt}\begin{bmatrix}\widehat{\lambda}_1\\ \widehat{\lambda}_2\end{bmatrix} = \begin{bmatrix}-Ri_1+u_1\\ -Ri_2+u_2\end{bmatrix} + G(t)[\hat{y}(t)-y(t)] \quad (20)$$

$$\hat{y}(t) = C(t)\begin{bmatrix}\widehat{\lambda}_1\\ \widehat{\lambda}_2\end{bmatrix}$$

where

$$C(t) = \begin{bmatrix}\beta[i_2(t)-i_2(t-T)]-2dI_2(t,t-T)\\ \beta[i_1(t)-i_1(t-T)]-2dI_1(t,t-T)\end{bmatrix}$$

And the Kalman Filter gain $G(t)$ is obtained by solving the Riccati Differential Equation (RDE):

$$\dot{P}(t) = -P(t)C^T(t)V^{-1}C(t)P(t)+W$$
$$G(t) = P(t)C(t)^T V^{-1}$$

where $P(t)\in R^{2\times 2}$ is covariance matrix of the states, $V\in R, W\in R^{2\times 2}$ are the spectrum density of the measurement noise and process noises.

The stability of the Kalman Filter ensures that in the absence of model uncertainty, process noises or measurement noise, we have $\widehat{\lambda}_1(t)\rightarrow\lambda_1(t)$, $\widehat{\lambda}_2(t)\rightarrow\lambda_2(t)$.

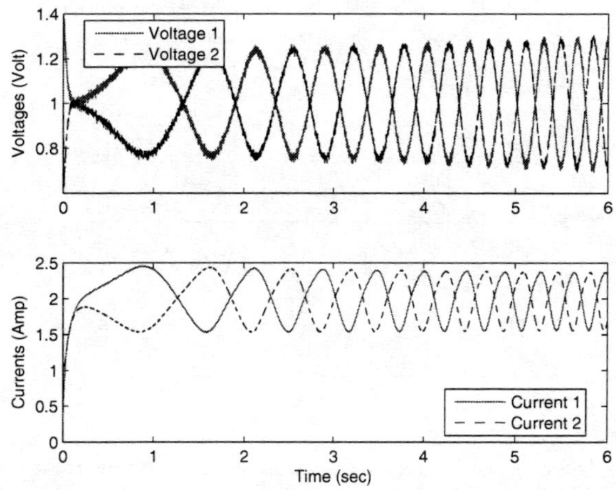

Figure 3. Simulation results: input voltages (with process noise) and output currents (with measurement noise) of solenoids .

4 Simulation study of self-sensing actuator

Both observers are tested by simulation in Matlab/Simulink. The parameters used in the simulation are $\beta = 2.64\times10^{-4}H-m$, $d = 7.76\times10^{-3}m$, and $R = 0.5\Omega$. The spool dynamics are given by a mass-spring-damper $(0.01K_g, 2029N/m, 6N-s/m)$ system with sufficiently large spring constant so that the valve is open loop stable. System nonlinearity is neglected in simulation. The spool is commanded to move according to a $0.1Hz$ to $3Hz$ chirp signal in a range of $[0,6]$ *sec*. The input voltages are corrupted by the processing noise, while the currents are disturbed by the measurement noise (Fig. 4). Both types of noises are implemented using "Band Limited White Noise" block in simulink. For the Boxcar Window observer, we choose $N = 4$, and $\Delta_T = 0.05sec$. For the Kalman Filter, we choose $T = 0.02sec$, $V = 2e5$, and $W = diag(2e5,2e5)$.

First, we assume the parameters d,β are well known. The simulation results for the Boxcar Window Observer can be seen in Figs. 4-5. Note that in the beginning, the estimation errors are significant. This is due to the Boxcar window can be only formed after a certain time. Then the estimated flux linkages approach the actual one very quickly. A property of this type of observer is that the estimates may be discontinuous because the local initial conditions are designed to be updating discretely.

Similarly, as can be seen in Figs 6- 7, the Kalman Filter is effective in estimating the unknown flux linkages and the spool displacement.

Next, we assume that the solenoid parameters are not precisely identified. In the simulation, both of the observers are implemented using the parameter $\hat{\beta} = 2.64\times10^{-4}H-m, \hat{d} = 6.9\times10^{-3}m$. In other words, $\hat{d}$ is 10% less than the real value d. As can be seen in Fig. 8, the displacement estimation errors,

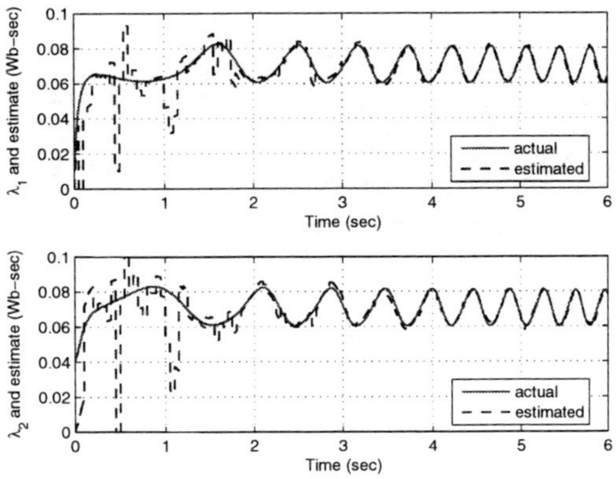

Figure 4. Simulation results for the Boxcar Window Observer: the actual and estimated flux linkages .

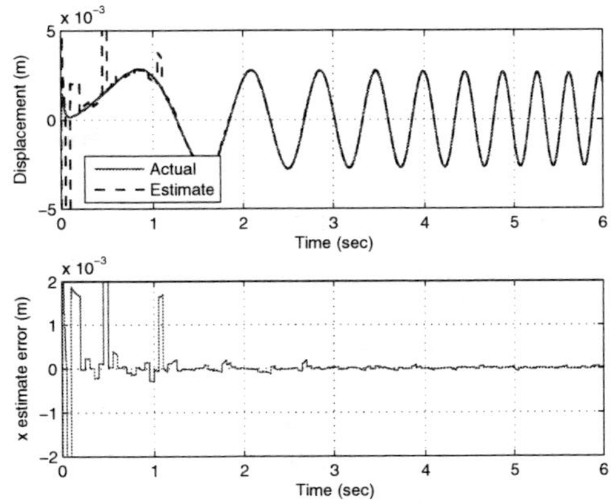

Figure 5. Simulation results for the Boxcar Window Observer. Top: the actual and estimated displacement; Bottom: the displacement estimation error.

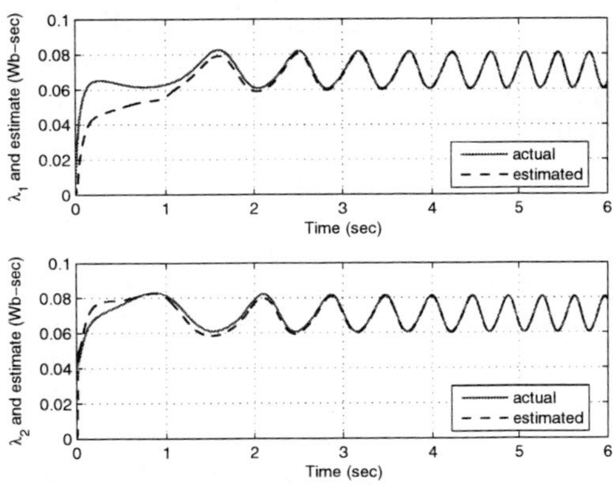

Figure 6. Simulation results for the Kalman Filter: the actual and estimated flux linkages .

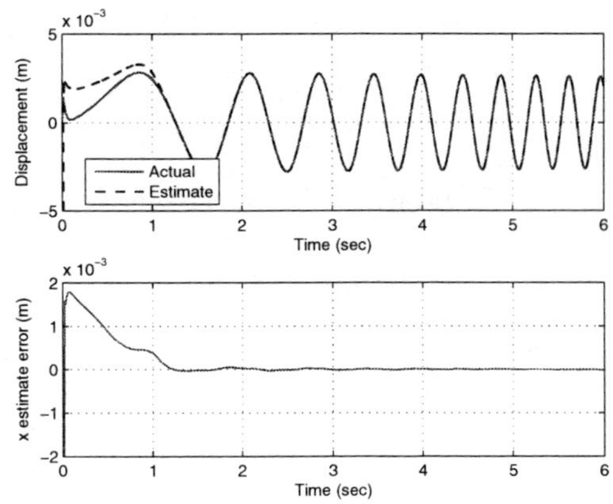

Figure 7. Simulation results for the Kalman Filter. Top: the actual and estimated displacement; Bottom: the displacement estimation error.

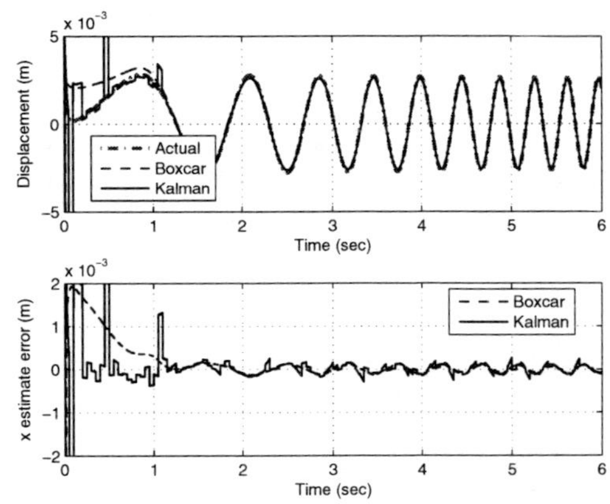

Figure 8. Simulation results for the case where the solenoid parameters are not well known ($\hat{d} = 90\% d$). Top: the actual and estimated displacement by both observers; Bottom: the estimation error for both observers.

either by the Boxcar Window Observer or by the Kalman Filter, are more significant than those in Figs. 5 and 7. However, the results are still good.

5 Experimental results

The experimental study has been conducted as well. We use xPC Target prototyping system that enables us to execute Matlab/Simulink models with physical systems in realtime. The diagram of the experimental setup can be seen in Fig. 9. The host computer is a Laptop (T40, IBM), while the target computer is a desktop (Dimension 2400, Dell). The TCP/IP communication is chosen between the host and the target system for the fast program-uploading and data-downloading. On the target computer are installed two I/O boards: CIO-DDA06 (Comput-

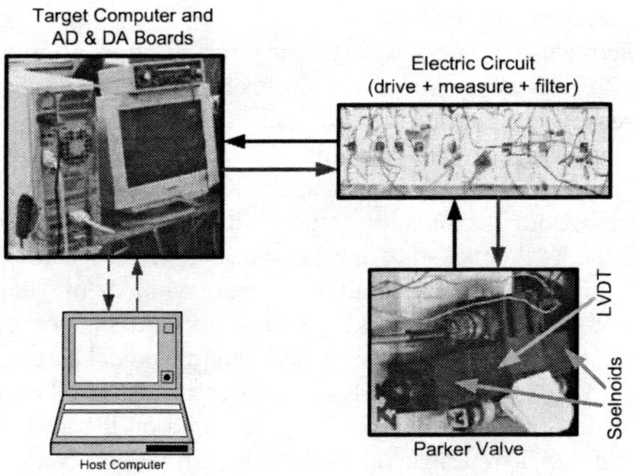

Figure 9. The diagram of the experimental apparatus.

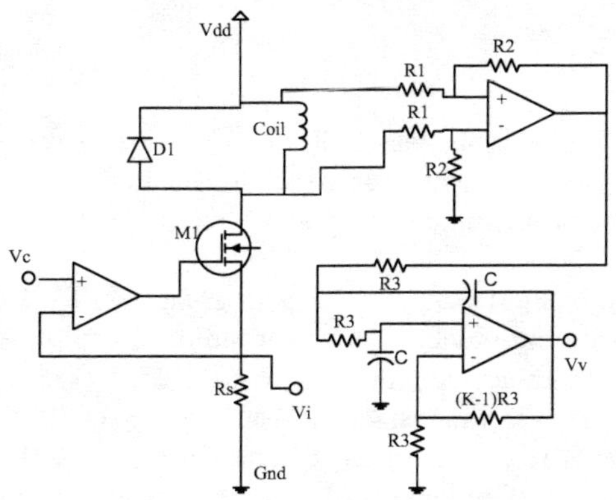

Figure 10. The schematic of electric circuit for driving the solenoids and measuring the voltage and current signals. The output signals V_v and V_i reflect the voltage across the solenoid and the current, respectively.

erboard, US) and AD512 (Humusoft, Czech). They total have 6 channels of DA and 8 channels of AD. We also build up a electric circuit on the breadboard to drive the solenoids and to measure the voltages and currents. The details of the electric circuit will be presented later. A commercial proportional valve (D1F, Parker Hannifin Corp.), with two solenoids at each end, is utilized in our experimental study. The LVDT in the valve provides us with the actual spool displacement that can be used to verify the observer estimates.

Next we will introduce the electric circuit, whose functionality consists of actuation and measurement. The left side of Fig. 10 (including an Op amp, a MOSFET M1, a diode D1, a coil/solenoid), represents a current driver circuit that is used to feed the desired current value over the solenoid (coil) according to control input V_c and the sensing resistor R_s. The actual current though the coil can be measured by the sensing resistor R_s (V_i). The right side of Fig. 10 is for measuring the voltage across the coil. First, the voltage across the coil is fed into a differential amplifier (upper right). The signal out of the differential amplifier is very noisy due to the high frequency switching of the MOSFET $M1$. Therefore, a voltage-controlled voltage-source (VCVS) Butterworth filter is placed afterwards (lower right). In Fig. 10, we set $V_{dd} = 10V$, $R_s = 1\Omega$, $R_{coil} = 6.2\Omega$, $R_1 = 54k\Omega$, $R_2 = 28k\Omega$, $R_3 = 54k\Omega$, $C = 2.2nF$, so the cutoff frequency of the low pass active filter is $f_c = \frac{1}{2\pi R_3 C} = 1.34K\ Hz$. The sample time of the xPC Target system is set to be $T_s = 1m\ sec$.

In Fig. 10, the output (V_v, V_i) are the measurements, i.e., V_i is for the current, while V_v is for the voltage. Since we have two solenoids, two independent circuits are built accordingly, with the outputs (V_{v1}, V_{i1}) and (V_{v2}, V_{i2}). In the experiment, we assume that the current measurements are precise. The output signals can be so calibrated that the actual voltages are

$$u_1 = 5.70V_{v1} - 4.44, i_1 = V_{i1}$$
$$u_2 = 6.17V_{v2} - 4.91, i_2 = V_{v2}$$

for solenoid 1 and 2, respectively.

The calibration of the solenoid parameters will be presented as follows. A self-calibration method in [7] is used to identify the model parameters. The basic idea is that the parameters of the solenoid could be retrieved from the limited knowledge of the displacement. For example, using two inexpensive switch sensors to detect the end stop displacement (please see [7] for more detail). Since our valve is equipped with a LVDT, the end-stroke information is obtained from the LVDT. Fig. 11 illustrates the iteration procedure of the Newton method. We arbitrarily choose the initial guess of the parameters to be $\hat{\beta}_0 = 1 \times 10^{-3} H - m, \hat{d}_0 = 0.03m$. The solution converges to $\hat{\beta} = 1.2 \times 10^{-6} H - m, \hat{d} = 1.3 \times 10^{-3} m$ at the 163rd iteration step. These values are used to to design the observers. Note that in reality, d, β do not have to be the same for both solenoids.

In our experiment, we design a PI controller to implement the spool displacement tracking. The controller output is connected to the V_c port in Fig. 10. The proportional gain and integral gain are tuned so that the system behaves mildly. We will compare two estimation methods, the Boxcar Window Observer and the Kalman Filter, with the actual spool displacement is measured by the LVDT. We set $N = 4, \Delta_T = 0.01sec$ for the Boxcar Window Observer, while $T = 0.02sec$ for the Kalman Filter. In Fig. 12, it can be seen that the prediction using self-sensing technique agrees well with the LVDT measurement, and the self-sensing concept is valid. It is also observed that there are some spikes in the the Boxcar window Observer estimates due to the discrete correction of the initial states. In addition, there is less steady state error (see the time period when the displacement is at a standstill in Fig. 12) for the Kalman filter than for the Boxcar Window observer. Larger errors in the transient procedure could be probably due to the simple solenoid model

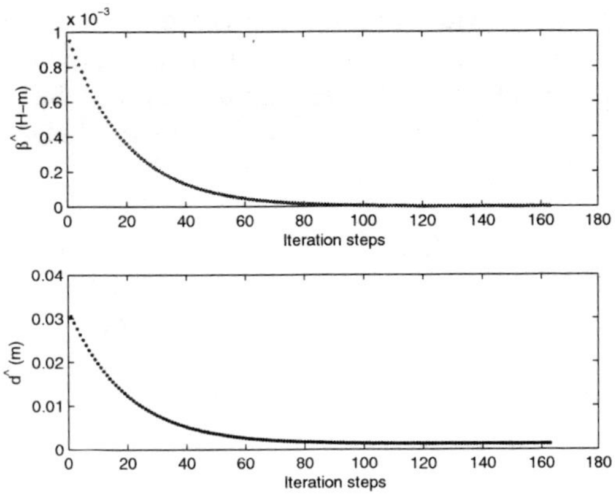

Figure 11. Self-calibration using experimental results: Estimated parameters $\hat{\beta}, \hat{d}$.

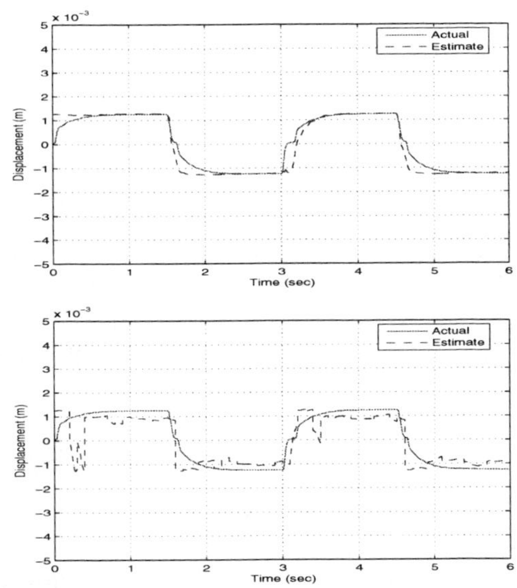

Figure 12. Experimental results for the Kalman Filter (upper) and the Boxcar Filter (lower).

we utilized, which does not match the actual solenoids installed in the Electrohydraulic valves. Therefore, the actual inductance-displacement curve passing through two points may be expressed other than $\frac{\beta}{d \pm x}$.

The self-sensing algorithm does not consider thermal effect. The experimental process lasts in the order of seconds, therefore, the temperature, and accordingly the resistance, can be assumed to be constant.

6 Conclusion

Self-sensing refers to extracting the position information from the electromagnetic signals instead of from a physical position sensor. Self-sensing actuator has the benefit of significantly reducing hardware effort. In this paper, two types of observers, the Boxcar window observer and the Kalman Filter, have been compared both in simulation and experimentally. It is shown that the Kalman Filter has smoother estimate and smaller steady error. Finally, it is also worth mentioning that due to lack of details of the simple solenoid model we used in algorithm, both observers may be sensitive to the model structure or the model parameters. In that case, more care needs to be taken to identify the model parameters precisely, or the observers based on the augmented model could be redesigned based on the current observers.

Acknowledgement: The paper is based on work supported by the National Science Foundation under grant ENG/CMS-0088964.

REFERENCES

[1] Noh, M. D., and Maslen, E. H., 1997. "Self-sensing magnetic bearings using parameter estimation". IEEE Transactions on Instrumentation and Measurement, **46** (1) Feb. , pp. 45–50.

[2] Zhan, Y. J., Chan, C. C., and Chau, K. T., 1999. "A novel sliding-mode observer for indirect position sensing of switched reluctance motor drives". IEEE Transactions on Industrial Electronics, **46** (2) April. , pp. 390–397.

[3] McCann, R. A., Islam, M. S., and Husain, I., 2001. "Application of a sliding mode observer for position and speed estimation in switched reluctance motor drives". IEEE Transactions on Industrial Electronics, **37** (1) Jan/Feb , pp. 51–58.

[4] Vischer, D., and bleuler, H., 1993. "Self-sensing active magnetic levitation". IEEE Transactions on Magnetics, **29** (2) march .

[5] Yuan, Q., and Li, P. Y., 2004. "Self-sensing actuators in electrohydraulic valves". In ASME Fluid System Technology Division Publication - 2004 International Mechanical Engineering Congress, Anaheim, CA, no. IMECE2004-62104.

[6] Li, P. Y., and Yuan, Q., 2005. "Flux observer for spool displacement sensing in self-sensing push-pull solenoids". In the Sixth International Conference on Fluid Power Transmission and Control (ICFP' 2005), Hangzhou, China,.

[7] Yuan, Q., and Li, P. Y., 2004. "Self-calibration of push pull solenoid actuators in electrohydraulic valves". In ASME Fluid System Technology Division Publication - 2004 International Mechanical Engineering Congress, Anaheim, CA, no. IMECE2004-62109.

[8] Eyabi, P., 2003. *Modeling and Sensorless Control of Solenoidal Actuators*. Phd thesis, The Ohio State University.

[9] Bryson, A. E., 2002. *Applied Linear Optimal Control: Exaples and Algorithms*. Cambridge university press.

Proceedings of IMECE2005
2005 ASME International Mechanical Engineering Congress and Exposition
November 5-11, 2005, Orlando, Florida USA

IMECE2005-80779

MODEL AND CONTROL FOR HYDRAULIC EXCAVATOR'S ARM

Daqing Zhang*, Qinghua He,P eng Hao, Xinhai Zhang
The College of Mechanical & Electrical Engineering,C entral South University,C hangsha,C hina

ABSTRACT

To control excavator's arm and realize autonomous excavation, firstly, full kinematic and dynamic models of the excavator arm, regarded as a planar manipulator with three degrees of freedom, were derived. Secondly, experiment excavator was retrofitted with electrohydraulic proportional valves, associated sensors, and a computer control system; then, the full nonlinear math model of electrohydraulic proportional system was achieved. Thirdly, this paper presents a discontinuous projection based on an adaptive robust controller to approximate the nonlinear gain coefficient of the valve and the nonlinear of the whole system, the error is deal with robust feedback and an adaptive robust controller was designed. Finally, the experiment of the boom motion controli s presented to illustrate the feasibility. These efforts had resulted in new control design methodologies that werea pplicable tos everal hydraulic proportional systems.

INTRODUCTION

A hydraulic excavator is a multi-functional construction machine. Workers in the construction industry use it for tasks such as excavating, dumping, finishing, lifting work, etc. However, operators who control hydraulic excavators must be trained for many years to do such work quickly and skillfully. A hydraulic excavator has three links: boom, dipper and bucket; and an operator have two arms. Thus, it is not easy for beginners to execute elaborate work that manipulates three links at the same time. Moreover, because the operators have to run work in various dangerous and dirty environments, the number of skillful operators is ever decreasing. For that reason, studying the automation of hydraulic excavators is necessary for improving productivity, efficiency, and safety [1, 2]. The automation of hydraulic excavators has been studied by several researchers in [1-6], and so on. Among the fundamental element used as a basis for developing more complicated tasks, the Controlo f excavator's arm is an importanta nd basic task in autonomous research of hydraulic excavator at present, and this work is difficult from the standpointo f the following problems: parameter variations in mechanical structures, various nonlinearities in hydraulic actuators, and disturbance due to the contact with the ground. For such reasonsd ifficultiese xist in controlling a robotic excavator. To solve these problems, several research works have been performed, which may be categorized as either studies based on rules (for instance, in [3, 5]) or model base studies (for example, in [1, 2, 4]). This paper presents a discontinuous projection based on an adaptive robust controller to approximate the nonlinear gain coefficient of the valve and the nonlinear of the whole system, the error is deal with robust feedback and an adaptive robust controller is designed. Finally, the experimento f the boom motion control is presented to illustrate the feasibility.

NOMENCLATURE

P_1 , P_2	The pressure of the boom cylinder at the head-side and rod-side y. MPa
A_1, A_2	The areas of pistons on two sides. m^2
I_2, I_3, I_4	The Moment of inertia of boom, stick and bucket kgm^2
m_2, m_3, m_4	The mass of boom, dipper,a nd bucket kg
m_L	The mass of unknown load in bucket. kg
β	The bulk modulus. N/m^2
Q_1 , Q_2	The supplieda nd returnf low to cylinder. $l/\min$
C_{ic} , C_{ec}	The internal and external leakage coefficient. $m^3/s/Pa$
β_e	The effective bulk modulus. N/m^2
V_1, V_2	The volume of fluid flow to and from the hydraulic cylinder. m^3
y	The displacement of piston. m
$\theta_2, \theta_3, \theta_4$	The boom,s tick ,bucket joint angles, rad
P_r	The pressure of return oil. MPa
ΔP	The pressure ofs pring ofl oad sense valve. MPa
C_d	The flow gain coefficient.
X_v	The displacement of spool. m
ρ	The density of oil. kg/m^3

* Corresponding author. (Postal address:4 10083,t elephone:1 3203133714 fax:0 731-4020612,E -mail:daqing_zh@163.c om).
This work was supported by the National "863" program of China under Grant 2003AA430200

1. OVERVIEW OF EXPERIMENT EXCAVATOR

In this paper, a robotic excavator as a testbed is retrofitted based on the Sunward SWE85 hydraulic excavator, and the hydraulic system retrofitted can be shown in Fig. 1.

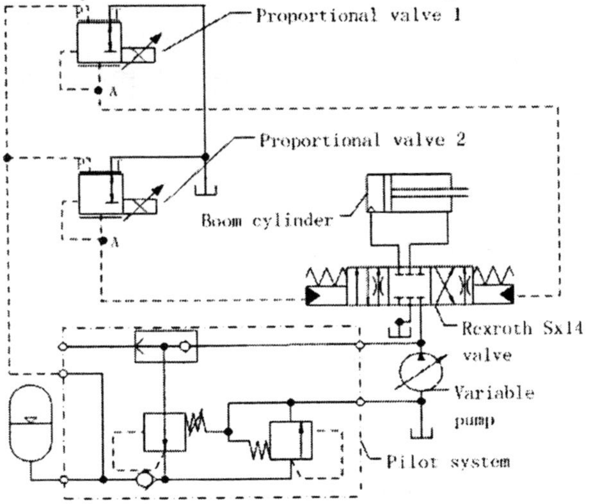

Fig. 1 The retrofitted electrohydraulic system of excavator (the boom cylinder)

On the base of original hydraulic system, electrohydraulic pilot control has replaced hydraulic pilot control by adding proportional relief valves and substituting hydraulic pilot handle with electrical handle. Using the button on the electrical handle can switch between manual control and computer control.

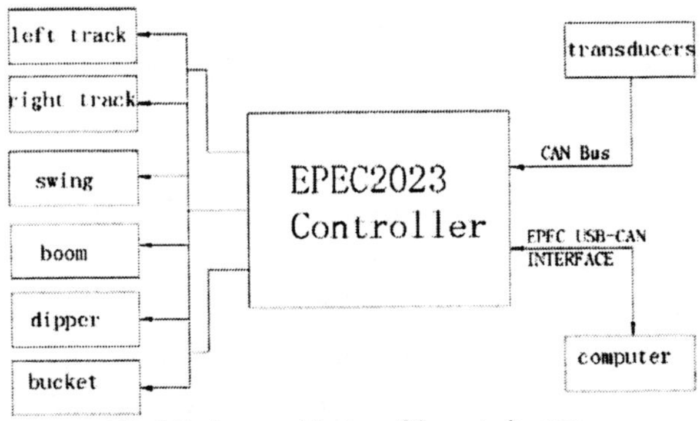

Fig.2 Hardware architecture of the controls ystem

The block diagram of hardware designed to control the robotic excavator is seen in Fig. 2. The control system consists of two levels; the high-level (PC) controller computes reference angles corresponding to reference trajectories and transfers orders to low-level controller; According to these orders, the low-level controller implements position control by measuring the joints position and pressure of cylinders. That is, controlling the pilot valve to drive the cylinders according to the output current.

2. MODEL OF ROBOTIC EXCAVATOR'S ARM

The robotic excavator's arm is made up of low level controller, hydraulic actuator system and excavator's manipulator. The manipulator contains boom, stick and bucket.

2.1 The kinematic model of manipulator

The aim of establishing the kinematic model of manipulator is to transfer bucket tip's reference trajectories to the corresponding and required reference angles sequence of each joint, and to get the motion sequences of hydraulic cylinders. The base coordinate $O_0\{x_0y_0z_0\}$ is defined in the Fig. 3 to descript the position of bucket tip, $O_1\{x_1y_1z_1\}$, $O_2\{x_2y_2z_2\}$, $O_3\{x_3y_3z_3\}$, $O_4\{x_4y_4z_4\}$ are put on each joint respectively. So the $D-H$ transform matrixes between each coordinate can be given as follows:

$$A_{i-1}^{i}=\begin{bmatrix}\cos\theta_i & -\cos\alpha_i\sin\theta_i & \sin\alpha_i\sin\theta_i & a_i\cos\theta_i\\ \sin\theta_i & \cos\alpha_i\cos\theta_i & -\sin\alpha_i\cos\theta_i & a_i\sin\theta_i\\ 0 & \sin\alpha_i & \cos\alpha_i & d_i\\ 0 & 0 & 0 & 1\end{bmatrix}$$

Where, $i=1,2,3,4$. The matrix denoting the position and orientation of bucket tip can be obtained as

$$A_0^4=\begin{bmatrix}c_1c_{234} & -c_1s_{234} & s_1 & c_1(a_4c_{234}+a_3c_{23}+a_2c_2+a_1)\\ s_1s_{234} & -s_1s_{234} & -c_1 & s_1(a_4s_{234}+a_3s_{23}+a_2s_2+a_1)\\ s_{234} & c_{234} & c_{234} & a_4s_{234}+a_3s_{23}+a_2s_2+d_1\\ 0 & 0 & 0 & 1\end{bmatrix} \quad (1)$$

Where $c_1=\cos(\theta_1), s_1=\sin(\theta_1) c_{234}=\cos(\theta_2+\theta_3+\theta_4)$, $s_{234}=\sin(\theta_2+\theta_3+\theta_4)$.

$f(t)$ is the reference trajectory of bucket, whose independent variable is t, and the reference position and orientation of bucket can be expressed by a rigid body transform in the base coordinate system. That is:

$$f(t)=\begin{bmatrix}n_x & s_x & a_x & p_x\\ n_y & s_y & a_y & p_y\\ n_z & s_z & a_z & p_z\\ 0 & 0 & 0 & 1\end{bmatrix} \quad (2)$$

Where n,s,a is orientation matrix, p is the position vector. The joint angles corresponding to the reference trajectory are obtained by resolving Eq. (3).

$$f(t)=g_{st}(\theta(t)) \quad (3)$$

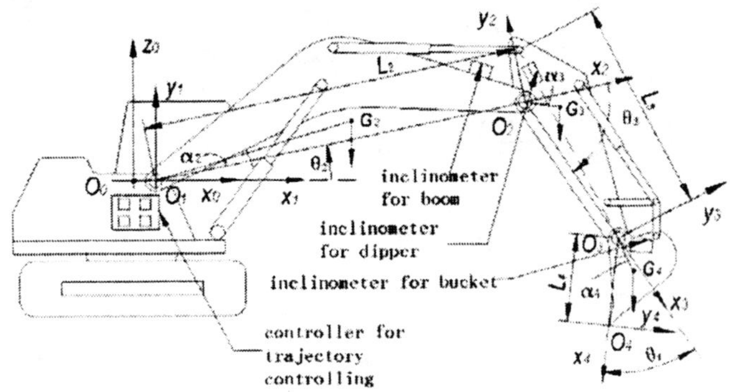

Fig.3 The Schematic modelo f excavator and the position of controller and transducers on the excavator

2.2 The dynamic model of manipulator

Figure 3 shows a schematic representation of the manipulator of hydraulic excavator with the assigned variables θ_2, θ_3 and θ_4 for the boom, stick and bucket joint angles, respectively. The cab swing angle (θ_1) is not shown in this figure, as it is not considered in this study. In Fig.3, G_i if the center of gravity for link i expressed in polar coordinates (r_i, α_i) relative to the link i. The link angles with respect to the horizontal plane are θ_2 for the boom, $\theta_{23} = \theta_2 + \theta_3$ for the stick, and for the bucket, $\theta_{234} = \theta_2 + \theta_3 + \theta_4$. The dynamic model can be developed according to Lagrange equation. This paper takes boom as an example, and gives out the dynamic equation [7-11] of boom as follows:

$$(D_C + m_L l_e^2)\ddot{\theta}_2 + G_c(\theta_2) + m_L g l_g(\theta_2) = \frac{\partial y}{\partial \theta_2}(P_1A_1 - P_2A_2) - T(t, \theta_2, \dot{\theta}_2) \quad (4)$$

$$D_c = I_2 + m_2 r_2^2 + (m_3 + m_4) l_2^2 + m_3 l_2 r_3 \cos(\theta_4 + \alpha_3) + I_3 + m_3 r_3^2 + m_4 l_3^2 + I_4 + m_4 r_4^2 + 2 m_4 l_2 l_3 c_3 + 2 m_4 l_3 r_4 \cos(\theta_4 + \alpha_4) + 2 m_4 l_2 r_4 \cos(\theta_{34} + \alpha_4)$$

$$G_c = m_2 g r_2 \cos(\theta_2 + \alpha_2) + (m_3 + m_4) g l_2 \cos\theta_2 + m_3 g r_3 \cos(\theta_{23} + \alpha_3) + m_4 g (l_3 \cos(\theta_{23}) + r_4 \cos(\theta_{234} + \alpha_4)), l_g = l_2 c_2 + l_3 c_{23} + l_4 c_{234},$$

$$l_e^2 = (l_2 \cos\theta_2 + l_3 \cos\theta_{23} + l_4 \cos\theta_{234})^2 + (l_2 \sin\theta_2 + l_3 \sin\theta_{23} + l_4 \sin\theta_{234})^2, i = 2,3,4$$

Where $T(t, \theta_2, \dot{\theta}_2)$ is the total disturbed moment including moment of load and friction.

2.3 The model of electro-hydraulic proportional system

2.3.1 The hydraulic cylinder equation of continuity

The driving cylinders of the manipulator of hydraulic excavator are control by Rexroth SX14 valve. The driving cylinders of the manipulator of hydraulic excavator are control by Rexroth SX14 valve. The orifice size of directional valve is decided by output pressure of proportional relief valve, which is decided by the output current of controller. Hydraulic system model can be seen in Fig. 4.

Following is a summary of the assumption that has been made in developing the model of a hydraulic cylinder:

1) The proportional valve is a symmetrical 3-way and 4-port valve. The dead band of valve is also symmetrical, and the flow in it is turbulent.

2) Possible dynamic behavior of the pressure in the transmission lines between valve and actuator is assumed to be negligible.

3) Pressure is equal everywhere in one volume of hydraulic cylinder, and the temperature and the bulk modulus are constants.

4) The leakage of flows is laminar.

The continuous equation [12] of compressible oil is given as follows:

$$\sum Q_{in} - \sum Q_{out} = \frac{dV}{dt} + \frac{V}{\beta} \bullet \frac{dP}{dt} \quad (5)$$

Here V is the initial volume of liquid subjected to compression. dV and dP are the changes in pressure and volume respectively. $\sum Q_{in}$ is the input flows of liquid, and $\sum Q_{out}$ is the output flows of liquid.

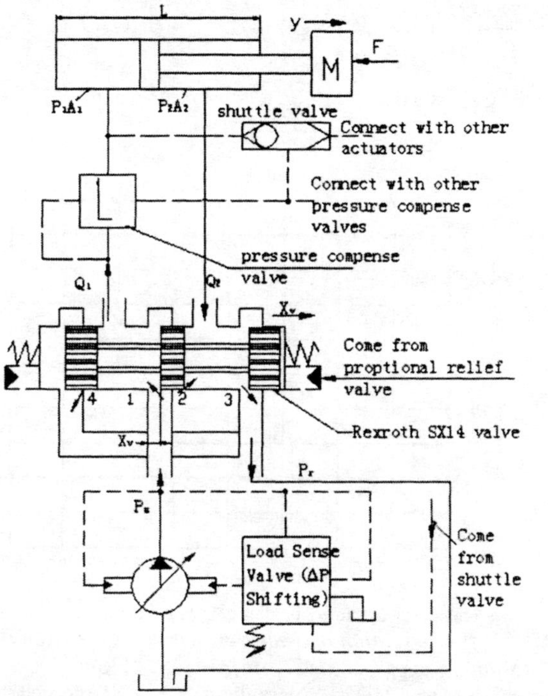

Fig.4 The diagram of LUDV hydraulic system after retrofitting

Considering the internal and external leakage of cylinder, the Eq. (6) and (7) are can be developed from Eq. (5) as follows:

$$Q_1 - C_{ic}(P_1 - P_2) - C_{ec}P_1 = \frac{dV_1}{dt} + \frac{V_1}{\beta_e} \bullet \frac{dP_1}{dt} \quad (6)$$

$$C_{ic}(P_1 - P_2) - Q_2 - C_{ec}P_2 = \frac{dV_2}{dt} + \frac{V_2}{\beta_e} \bullet \frac{dP_2}{dt} \quad (7)$$

Where Q_1 and Q_2 are the supplied and return flow to the cylinder respectively. β_e is effective bulk modulus(including liquid, the air in oil and so on). V_1, V_2 can be got as follows:

$$V_1 = V_{01} + A_1 y, \qquad V_2 = V_{02} - A_2 y \quad (8)$$

So the Eq. (9) can be given as:

$$\frac{dV_1}{dt} = A_1 \frac{dy}{dt}, \qquad -\frac{dV_2}{dt} = A_2 \frac{dy}{dt} \quad (9)$$

The external leakage is the leakage between the piston rod and external seals. Because external leakage on the construction machine is forbidden, and external leakage is disappear with the development of sealing technology. The influence of external leakage is neglected in this paper. y can be expressed as a function of θ_2.Then Eq. (6) and (7) can be rebuilt as:

$$\dot{P}_1 = \frac{\beta_e}{V_1}[Q_1 - C_{ic}(P_1 - P_2) - A_1\frac{\partial y}{\partial \theta_2}\dot{\theta}_2] \tag{10}$$

$$\dot{P}_2 = \frac{\beta_e}{V_2}[C_{ic}(P_1 - P_2) - Q_2 + A_2\frac{\partial y}{\partial \theta_2}\dot{\theta}_2] \tag{11}$$

Where C_{ic} is assumed to be a constant to simplify the system model.

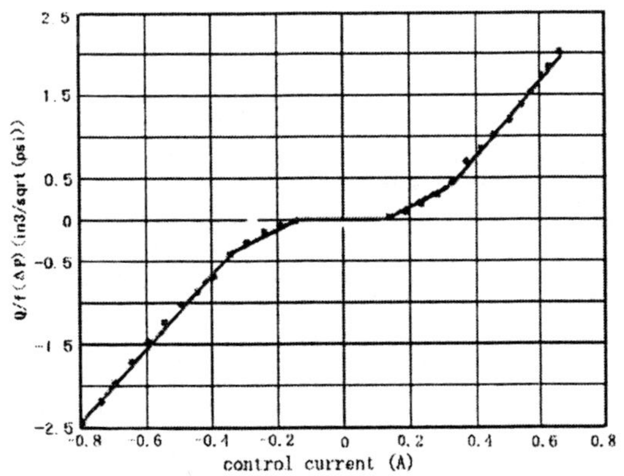

Fig.5 nonlinear characteristic flow gain of the proportional valve

2.3.2 The flow equation of electro-hydraulic proportional valve

In this paper, Load Independent Flow Distribution (abbreviated to LUDV) hydraulic system is adopted on the experimental robotic excavator. According to the theory of LUDV system [13], the flow equation can be gotten as follows:

$$Q_1 = C_d W X_v \sqrt{\frac{2}{\rho}\Delta P_1} = \begin{cases} C_d W K_I I(t)\sqrt{\frac{2}{\rho}\Delta P}, I(t) \ge 0 \\ -C_d W K_I I(t)\sqrt{\frac{2}{\rho}(P_1 - P_r)}, I(t) < 0 \end{cases} \tag{12}$$

$$Q_2 = C_d W X_v \sqrt{\frac{2}{\rho}\Delta P_2} = \begin{cases} -C_d W K_I I(t)\sqrt{\frac{2}{\rho}(P_2 - P_r)}, I(t) \ge 0 \\ C_d W K_I I(t)\sqrt{\frac{2}{\rho}\Delta P}, I(t) < 0 \end{cases} \tag{13}$$

Where $I(t)$ is the control current of proportional valve. K_I is the current gain of proportional valve.

In this paper, the proportional direction control valve has a quite large dead band, which is due to the close centered valve configuration. C_d is proved to be nonlinear by experiment [14]. The net actual valve opening x_{nav} can be defined as

$$x_{nav} = \begin{cases} x_v - x_d & , x_v > x_d \\ 0 & , -x_d \le x_v \le x_d \\ x_v + x_d & , x_v < -x_d \end{cases} \tag{14}$$

Where x_d is the dead band of valve.

The actual experimental test results reveal the nonlinear flow gain in the Fig. 5, and the flow gain coefficient of valve can be approximated by two lines to simplify the model. To be controlled conveniently, Q (the flow of valve) can be expressed as Eq. (15) and (16).

$$Q(x_v, \Delta P) = Q_M(x_v, \Delta P) + \tilde{Q}(x_v, \Delta P) \tag{15}$$

$$\begin{aligned} Q_{1M} &= C_{d1} W f_1(\Delta P_1) x_{nav} \\ Q_{2M} &= C_{d2} W f_2(\Delta P_2) x_{nav} \end{aligned} \tag{16}$$

Where Q_M represent certain simplified flow mapping function that will be used in the controller design. $\tilde{Q}$ is the model error of the flow projection, and will be treated by robust feed back . Generally speaking, C_{d1} and C_{d2} are constant for one working state.

3.THE ADAPTIVE ROBUST CONTROLLER

According to the model of excavator, the main difficulties to control the robotic excavator are that:(1)the dynamic of system is strongly nonlinear; (2) parameters of system are uncertain, for example, the uncertainty of m_L, β_e (valid bulk modulus) and T_n (the total disturb of system) which is brought by external disturb and un-modeled friction.

In this paper, the nonlinearity and uncertainty of parameters are treated by a discontinuous projection based on an adaptive robust controller [8]. In Fig. 5, the line is used to compensate for the dead band of vale, which is the symmetrical line of plus curve based on the axis that is through the coordinate origin and make an angle of 45• with the x axis. The flow projection error is compensated by robust feed back. m_L, β_e, T_n are mainly considered as

$$q_1 = \frac{1}{1 + (l_e / D_c)m_L}, \quad q_2 = \frac{T_n}{D_c + m_L l_e^2}, \quad q_3 = \beta_e \quad ,$$

and $q = [q_1, q_2, q_3]^T$ is defined as unknown parameter set. Then Eq. (4), (10) and (11) can be transform to Eq. (17).

$$\begin{aligned} \ddot{\theta}_2 &= \frac{q_1}{D_c}[\frac{\partial y}{\partial \theta_2}(P_1 A_1 - P_2 A_2) - G_c(\theta_2)] \\ &\quad + \frac{q_1}{l_e^2} g l_g - \frac{1}{l_e^2} g l_g + q_2 + \tilde{T}(t, \theta_2, \dot{\theta}_2), \end{aligned} \tag{17}$$

$$\dot{P}_1 = \frac{q_3}{V_1}(-C_{ic}(P_1 - P_2) - A_1\frac{\partial y}{\partial \theta_2}\dot{\theta}_2 + Q_{1M} + \tilde{Q}_1),$$

$$\dot{P}_2 = \frac{q_3}{V_2}(A_2 \frac{\partial y}{\partial \theta_2}\dot{\theta}_2 - Q_{1M} - \widetilde{Q}_2 + C_{ic}(P_1 - P_2))$$

Where, $\widetilde{T} = \frac{T(t,\theta_2,\dot{\theta}_2) - T_n}{D_c + m_L l_e^2}$, $q_{\min} \le q \le q_{\max}$, $|\widetilde{T}| \le \delta_T$, $|\widetilde{Q}_1| \le \delta_{Q1}$, $|\widetilde{Q}_2| \le \delta_{Q2}$, and $q_{\min}, q_{\max}$, δ_T, δ_{Q1} and δ_{Q2} are known number. It is assumed that $\hat{q}$ and $\widetilde{q}$ are the estimated value and estimated error of q respectively. Equation (18) is the adaptation law structure needed.

$$\dot{\hat{q}} = \Pr oj_{\hat{\theta}}(\Gamma \tau) \tag{18}$$

Where Γ is the diagonal matrix τ is the adaptive function. $\Pr oj_{\hat{\theta}}(\bullet)$ is the discontinuous projection [15].

According to the above work of this paper, the following task must be undertaken.

Step 1

Defining $P_L = P_1 A_1 - P_2 A_2$, P_L can be treated as the input of Eq. (17). The goal of this step is to get a virtual control law P_{Ld} for P_L so that output error $z_1 = \theta_2 - \theta_{2d}$ can converge to zero to get high trajectory tracking accuracy. Assuming that $\dot{\theta}_{2r} = \dot{\theta}_{2d} - k_1 z_1$, $\ddot{\theta}_{2r} = \ddot{\theta}_{2d} - k_1 \dot{z}_1$,

$$z_2 = \dot{\theta}_2 - \dot{\theta}_{2r}, \phi_2 = \left[\frac{1}{D_c}(\frac{\partial y}{\partial \theta_2}P_{Lda} - G_c) + g\frac{l_g}{l_e^2}, 1, 0\right]^T$$

and according to function of *Lyapunov*, the following equation can be gained.

$$P_{Ld} = P_{Lda} + P_{Lds}, \quad \tau_2 = \phi_2 z_2,$$

$$P_{Lda} = \frac{\partial \theta_2}{\partial y}[G_c(\theta_2) + \frac{D_c}{q_1}(-g\frac{\hat{q}_1}{l_e^2}l_g + g\frac{1}{l_e^2}l_g - \hat{q}_2 + \ddot{q}_{2r})]$$

$$P_{Lds} = P_{Lds1} + P_{Lds2}$$

$$P_{Lds1} = -\frac{D_c}{q_{1\min}}\frac{\partial \theta_2}{\partial y}(k_{2s1} + k_2)z_2, k_2 > 0 \tag{19}$$

Where k_{2s1} is a control gain to be synthesized later, P_{Lds2} is any function satisfying the following conditions

$$\text{Condition1} \quad z_2[\frac{1}{D_c}q_1\frac{\partial y}{\partial \theta_2}P_{Lds2} - \widetilde{q}^T\phi_2 + \widetilde{T}] \le \varepsilon_2$$

$$\text{Condition2} \quad z_2\frac{\partial y}{\partial \theta_2}P_{Lds2}, \varepsilon_2 > 0 \tag{20}$$

Step 2

$z_3 = P_L - P_{Ld}$ is assumed to be an error of input then the actual controlling principle will be developed to make z_3 approximate to zero. According to Eq. (17), (19), the Eq. (21) can be given as follows:

$$\dot{z}_3 = \dot{P}_L - \dot{P}_{Ld} = q_3[-(\frac{A_1^2}{V_1} + \frac{A_2^2}{V_2})\frac{\partial y}{\partial \theta_2}\dot{\theta}_2 + (\frac{A_1}{V_1}Q_{1M} + \frac{A_2}{V_2}Q_{2M}) + (\frac{A_1}{V_1}\widetilde{Q}_1 + \frac{A_2}{V_2}\widetilde{Q}_2)] - \dot{P}_{Ldc} - \dot{P}_{Ldu} \tag{21}$$

Where $\dot{P}_{Ldc} = \frac{\partial P_{Ld}}{\partial \theta_2}\dot{\theta}_2 + \frac{\partial P_{Ld}}{\partial \dot{\theta}_2}\hat{\ddot{\theta}}_2 + \frac{\partial P_{Ld}}{\partial t}$ is the calculable part of $\dot{P}_{Ld}$, and $\dot{P}_{Ldu} = \frac{\partial P_{Ld}}{\partial \dot{\theta}_2}[-\frac{1}{D_c}(\frac{\partial y}{\partial \theta_2}P_L - G_c(\theta_2))\widetilde{q}_1 - \frac{1}{l_e^2}gl_g\widetilde{q}_1 - \widetilde{q}_2 + \widetilde{T}] + \frac{\partial P_{Ld}}{\partial q}\dot{\hat{q}}$ is the error part of $\dot{P}_{Ld}$ .this error part is treated by robust feedback method. $\frac{A_1}{V_1}Q_{1M} + \frac{A_2}{V_2}Q_{2M} = Q_L$ is considered as a valid input in Eq. (21). In this process, Q_{Ld} which is the valid input of Q_L must be developed to make P_L track P_{Ld} which is developed in step (1). So Q_{Ld} (controlling function) and τ_3 (adaptive function) can be given respectively as follows:

$$Q_{Ld}(\theta_2,\dot{\theta}_2,P_1,P_2,\hat{q},t) = Q_{Lda} + Q_{Lds}, \quad \tau_3 = \tau_2 + \phi_3 z_3$$

$$Q_{Lda} = -\frac{1}{q_3}Q_{Lde}, \quad Q_{Lds} = Q_{Lds1} + Q_{Lds2},$$

$$Q_{Lds1} = -\frac{1}{q_{3\min}}(k_3 + k_{3s1})z_3, \quad k_3 > 0 \tag{22}$$

$$Q_{Lde} = \frac{1}{D_c}\frac{\partial y}{\partial \theta_2}z_2\hat{q}_1 - \hat{q}_3(\frac{A_1^2}{V_1} + \frac{A_2^2}{V_2})\frac{\partial y}{\partial \theta_2}\dot{\theta}_2 - \dot{P}_{Ldc},$$

$$\phi_3 = [\frac{1}{D_c}\frac{\partial y}{\partial \theta_2}z_2 - \frac{\partial P_{Ld}}{\partial \dot{\theta}_2}[\frac{1}{D_c}(\frac{\partial y}{\partial \theta_2}P_L - G_c(\theta_2)) + \frac{1}{l_e^2}gl_g], \frac{-\partial P_{Ld}}{\partial \dot{\theta}_2}, -(\frac{A_1^2}{V_1} + \frac{A_2^2}{V_2})\frac{\partial y}{\partial \theta_2}\dot{\theta}_2 + Q_{Lda}]^T$$

Where k_{3s1} is a control gain to be synthesized later, P_{Lds2} is any function satisfying the following conditions:

$$\text{Condition 1} \quad z_3[q_3 P_{Lds2} + q_3\widetilde{Q}_L - \widetilde{q}^T\phi_3 - \frac{\partial P_{Ld}}{\partial \theta_2}\widetilde{T}] \le \varepsilon_3$$

$$\text{Condition 2} \quad z_3 Q_{Lds2}, \varepsilon_3 > 0 \tag{23}$$

After Q_{Ld} being gained, the valid displacement of spool can be computed according to Eq. (12), (13) and (16) as follows:

$$x_{nav} = \frac{Q_{Ld}(\theta_2, \dot{\theta}_2, P_1, P_2, \hat{q}, t)}{\frac{A_1}{V_1} C_{d1} f_1(\Delta P_1) + \frac{A_2}{V_2} C_{d2} f_2(\Delta P_2)} \quad (24)$$

x_v is assumed to be proportionable to $I(t)$, that is, $I(t) = Kx_v$, then

$$I(t) = Kx_v = \begin{cases} K(x_{nav} + x_d), & x_{nav} > 0 \\ 0, & x_{nav} = 0 \\ K(x_{nav} - x_d), & x_{nav} < 0 \end{cases} \quad (25)$$

Assuming $\tau = \tau_3$ in Eq. (18), k_{2s1} and k_{3s1} are selected to make

$$k_{2s1} \geq \frac{1}{2}\|\Gamma\phi_2\|^2, k_{3s1} \geq \|\partial P_{Ld}/\partial\hat{q}_2\|^2 + \frac{1}{2}\|\Gamma\phi_2\|^2 \quad (26)$$

As a result, high accuracy can be achieved by using Eq. (25) as a controlling principle.

4. EXPERIMENT AND CONCLUSION

The motion controlling experiment of boom is done on the robotic excavator, which is based on the Sunward SWE85 excavator. It is assumed that the reference angle is a sine curve, whose amplitude is 40 degrees, and whose center line is -2 degree. The experimental result is showed in Fig. 6. The result exhibits good tracking performance for boom cylinder under the controller developed in this paper. The peak error is less than 4 degrees.

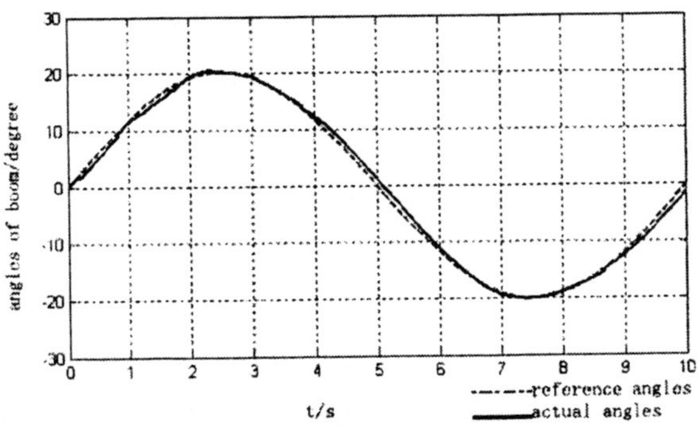

Fig. 6 Experiment of the boom's angle control

The motion control of hydraulic excavator is the base to achieve automatic excavation. The motion control principle is developed in this paper, under which good tracking performance can be gained. So base on the work done in this paper, the telecontrol of excavator can be realized if communication system is developed; and automatic excavation can be achieved if the trajectory programming is completed.

ACKNOWLEDGMENTS

The author would like to thank all those students, research associates who have contributed to the development of the robotic excavator in this paper. In addition they would like to thank the assistance given by the company: Sunward Intelligent Machine Corporation.

REFERENCES

[1] Sung-Uk Lee, Pyung Hun Chang,C ontrolo f a heavy-duty robotic excavator using time delay Controlw ith switching action with integral sliding surface, Proceeding of the 2001 IEEE international conference on Robotics & Automation,S eoul, Korea, May 21-26,2 001,p p.3955-3960

[2] Sung-Uk Lee, Pyung Hun Chang,C ontrolo f a heavy-duty robotic excavator using time delay Controlw ith integrals liding surface,C ontrol Engineering Practice 10(2002),p p.697-711

[3] Bradley D, Seward D, the development,c ontrola nd operation of an autonomous robotic excavator [J], Journalo f Intelligent and Robotic Systems, 1998,2 1:73-97.

[4] Nguyen Hong Quang,R obust low levelc ontrolo f robotic excavation, Ph.D. dissertation, University of Sydney. March 2000

[5] Patrick Sean Rowe, Adaptive motion planning for autonomous mass excavation,P h.D.d issertation,C arnegie Mellon University, January 1999

[6] Masakazu Haga, Watanabe Hiroshi,K azuo Fujishima, Digging control system for hydraulic excavator, Mechatronics 11 (2001),p p.665-676

[7] A. J. Koivo, M. Thoma,E .Kocaoglan, and J. Andrade-Cetto,M odeling and controlo f excavator dynamics during digging operation, Journal of aerospace engineering,J anuary 1996, pp.10-18

[8] FanpingBu and BinYao, Nonlinear Adaptive Robust Controlo f Hydraulic Actuators Regulated by Proportional Directional ControlV alves with Dead band and Nonlinear Flow Gains, Proceedings of the American ControlC onference Chicago, Illinois June 2000,pp.4129-4133

[9] Shahram Tafazoli,P eter D. Lawrence, and S.E.S alcudean,I dentification of inertial and friction parameters for excavator arms, IEEE transactions on robotics and automation, VOL.15 NO.5 October 1999, pp.966-971

[10] Wang Jian Zhi,R esearch on the working condition monitoring and automatic operation for hydraulic excavator,M aster. Dissertation, Zhe Jiang University, January 1997

[11] FanpingBu and BinYao, Observer based coordinated adaptive robust controlo f robot manipulator driven by single-rod hydraulic actuators, Proceeding of the 2000 IEEE internationalc onference on Robotics & Automation San Francisco,C A• April 2000

[12] Li HongRen, hydraulic controlling system, Defense industry publishing company, June 1981,p p.59-60

[13] Chen Qian Gen,J i Yun Feng, Wu Wang Rong, Load independent flow distribution controls ystem, Hydraulic and Pneumatic,1 0(2003),p p.10-11

[14] Yao B. High performance adaptive robust controlo f nonlinear systems:a generalf ramework and new schemes [A]. In: Proceeding of IEEE Conference on Decision and Control[C],1 997,pp.2489–2494

[15] B.Yao and M.T omizuka, Smooth robust adaptive sliding mode controlo f robot manipulators with guaranteed transient performance,i n Proc.o f American ControlC onference, pp. 1176-1180

APPENDIX:

Fig.7 The experimental robotic excavator based on Sunward SWE85 excavator

Proceedings of IMECE2005
2005 ASME International Mechanical Engineering Congress and Exposition
November 5-11, 2005, Orlando, Florida USA

IMECE2005-80842

REAL-TIME SIMULATED PRESSURE RELIEF VALVES IN PHYSICAL HYDRAULIC CIRCUIT

Tero Eskola **Heikki Handroos** **Takao Nishiumi**[+]

Department of Mechanical Engineering
Lappeenranta University of Technology
P.O. Box 20, FIN-53851 Lappeenranta
Finland

[+]Department of Mechanical System Engineering
National Defense Academy
1-10-20 Hashirimizu, Yokosuka 239-8686
Japan

ABSTRACT

The present paper deals with hardware-in-the-loop (HIL) simulation of hydraulic components and systems. The main idea is to develop hydraulics by using a simulation model of it as a part of the machine. The interface between the real and simulated parts is defined by means of pressures and flows. The proposed idea makes it possible to test a variety of design parameters of a hydraulic system or single component on-line while running the practical experimental machine. The method is demonstrated in its application to a single-stage pressure relief valve in a simple hydraulic circuit. The real valve is replaced in the circuit by a HIL simulator mimicking the dynamic behavior of the valve. Finally the dynamic responses of original pressure relief valve and the HIL simulator are compared.

NOMENCLATURE

A	spool surface area, m^2
$C_1 \ldots C_4$	empirical constants
C_D	discharge coefficient
D	spool diameter, m
F_j	jet flow force, N
F_f	stationary flow force, N
F_p	pressure force, N
F_0	preload setting force of poppet spring
K, K_c	variable including discharge coefficient, orifice geometry, poppet displacement and square root of density of oil, m^3/s$\sqrt{\text{Pa}}$
k_m	mechanical spring constant, N/m
m	poppet mass, kg
p_1, p_2	pressure, Pa
Q_i	flow, m^3/s
x	poppet displacement, m
α	flow angle, rad

INTRODUCTION

HIL simulator consists of a real-time simulator and experimental hardware. HIL simulation has been used in most cases for testing existing control hardware with a virtual process model. The method has rarely been proposed in the form in which part of the process is simulated while rest of the process is real physical. Some examples in the fields of high speed trains [1], vehicle suspensions [2] and spacecraft dockings [3] are documented. In this research the hydromechanical system is interfaced by means of pressures and flows instead of forces and motions.

The method used is based on a real-time simulation model of hydraulics, which is simulated while running the physical prototype. The simulation model inputs are the measured pressures and it calculates the virtual flows that are transformed into real flows by means of flow feedback control systems that use high bandwidth flow transducers, PID-controllers and real high performance servo solenoid control valves.

In the previous studies the proposed R&D method has been studied mostly for testing different kind of directional valves for controlling physical mechanical prototypes [4], [5], [6] and [6]. In this paper the method is utilized for a study of pressure relief valves (PRV) in existing hydraulic circuit. The method is valid since the modeling accuracy and bandwidth of the interface is sufficient compared with the dynamics of simulated component. The method can be used for designing and testing of fluid power driven machines. Developed simulator makes it possible to study alternative hydraulic components and systems in real machine environment.

SIMULATION MODEL

Fig. 1 shows the schematic construction of a single-stage PRV [7], [9], [10]. If the flow angle is assumed to be equal α (see Fig. 1) the force equation of the poppet is

$$m\ddot{x} + k_m x = F_p - F_f + F_j - F_0 \tag{1}$$

where

$$F_p = p_2 A - p_1 \pi D x \sin\alpha\cos\alpha$$
$$F_f = 2C_D \pi D x p_1 \sin\alpha\cos\alpha$$
$$F_j = 2C_D \pi D x p_1 \sin\alpha\cos\alpha$$

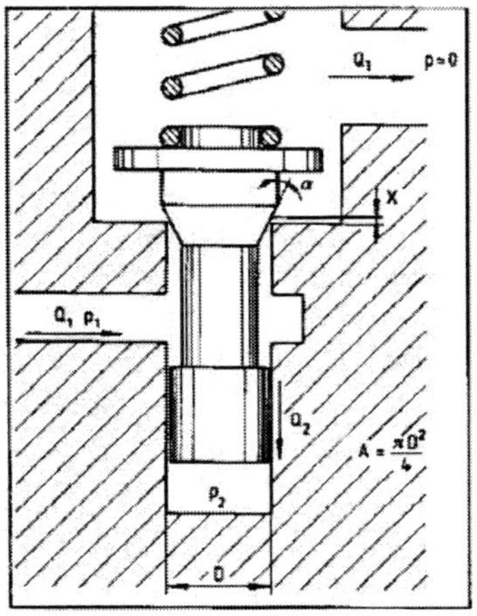

Fig. 1 Schematic representation of a commonly used single-stage pressure relief valve type.

Allowing the flow through the valve as a nonlinear function of the poppet displacement x and the system pressure p_1, the continuity equation is

$$Q_1 = C_d \pi D x \sin\alpha\sqrt{\frac{2p_1}{\rho}} = K\sqrt{p_1} \tag{2}$$

The laminar flow to the damping chamber is

$$Q_2 = K_c(p_1 - p_2) = A\dot{x} \tag{3}$$

Equations (1), (2) and (3) can be brought into semi-empirical form as follows

$$\left.\begin{aligned} &\ddot{K} + 2C_3C_4\dot{K} + (C_1 + C_2p_1)C_3^2K = C_3^2(p_1 - p_{ref}) \\ &Q_1 = K\sqrt{p_1} \end{aligned}\right\} \tag{4}$$

where

$$C_1 = \frac{\mathrm{k_m}}{\mathrm{AC_d}\pi D\sin\alpha\sqrt{\frac{2}{\rho}}}, \quad C_2 = \frac{\cos\alpha}{\mathrm{AC_d}\sqrt{\frac{2}{\rho}}}$$

$$C_3 = \sqrt{\frac{\mathrm{AC_d}\pi D\sin\alpha\sqrt{\frac{2}{\rho}}}{m}}, \quad C_4 = \frac{\mathrm{A}^2}{2\mathrm{K_c m C_3}}$$

Instead of using the analytical formulas for the parameters $C_1,\ldots, C_4$ in equation (4) in the following, a method which is independent of the constructional differences of the valves which conform to the same model as equation (4), to identify the values of the parameters is presented.

Under steady-state condition equation (4) can be simplified as follows

$$Q_1 = \frac{p_1 - p_{ref}}{\frac{C_1}{\sqrt{p_1}} + C_2\sqrt{p_1}} \tag{5}$$

By using equation (5) and selecting two points from measured steady-state characteristic curves it is possible to define the semi-empirical parameters C_1 and C_2. To identify the values of the parameters C_3 and C_4 a single measured dynamic response is needed.

If the effects of the inertia of the poppet are assumed to be negligible the equation (4) can be reduced into following form

$$\left.\begin{aligned} &\frac{2C_4}{C_3}\dot{K} + (C_1 + C_2p_1)K = (p_1 - p_{ref}) \\ &Q_1 = K\sqrt{p_1} \end{aligned}\right\} \tag{6}$$

HIL SIMULATION AND EXPERIMENTAL WORK

Fig. 2 shows the schematic construction of the experimental rig used. In the present system the cylinder is controlled by using the directional valve *v1*. Valve *v4* is used to select the active valve: either the PRV *v3* or the HIL simulator with high bandwidth servo solenoid valve *v2*. Released flow is measured by flow transducer *a1* and pressure by pressure sensor *a2*. It is important to notice that the dynamical properties like bulk modulus and volumes of pipes *p1* and *p2* must be equal. The physical structure of the experimental rig is presented in Fig. 3.

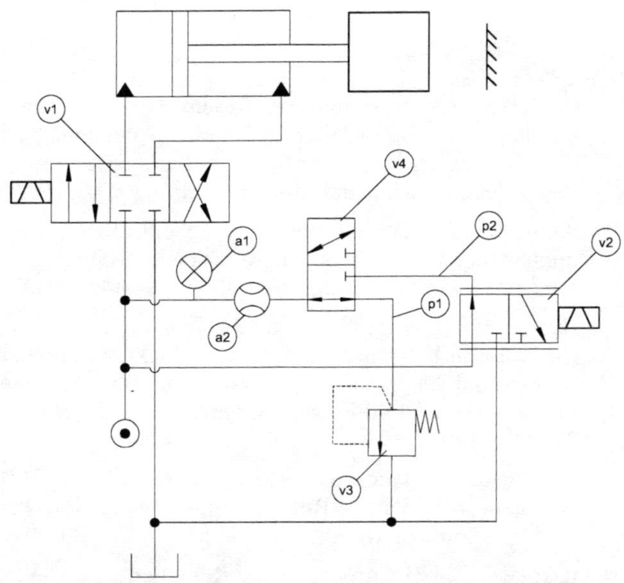

Fig. 2 Schematic representation of the used test rig.

Fig. 3 Picture of the test rig used.

The natural behavior of PRV can be seen when the oil pressure is suddenly increased. In this study this was done by driving the cylinder at the end of its stroke. At first this was done by using the real PRV in action. During the first work cycle two points of measured pressure and flow were taken and used to solve the empirical constants C_1 and C_2 by using equation (5). Constants C_3 and C_4 were defined based on the dynamic response of the flow. After parameter settings similar work cycles were done by using the HIL simulator with the defined models.

The defined simulation models of the present study were modeled first in Matlab/Simulink environment and then converted to c code by using Simulink/RTW. After this the code is compiled and loaded into a dSPACE processor card. The bandwidth of servo solenoid valve used is 35 Hz, the reaction time of the flow transducer is less than 2 ms, the bandwidth of the pressure sensor is 2 kHz and the used sample-time is 1 ms.

RESULTS

Two different simulation models of PRV are used. The models are based on the equations (4) and (6). The functionality of the HIL simulator can be reviewed by comparing the measured flows and pressures of both cases etc. by using the original PRV and by using the HIL simulator. In Fig. 4 and Fig. 6 three oil flows are compared. Q_{Model} is the output flow of the simulation model, Q_{HIL} is the output flow of the HIL simulator's interface and Q_{Real} is the output flow of the real PRV. Measured pressures are compared in Fig. 5 and Fig. 7.

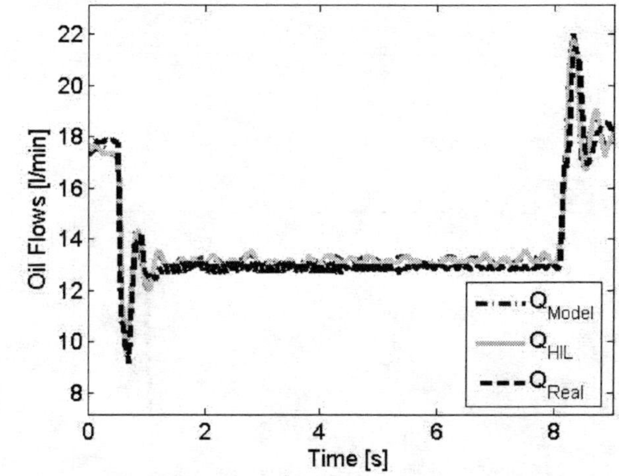

Fig. 4 Flows by using 1st order model.

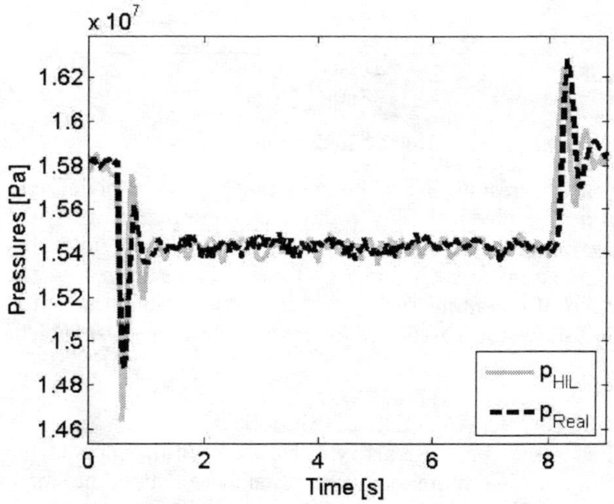

Fig. 5 Pressures by using 1st order model.

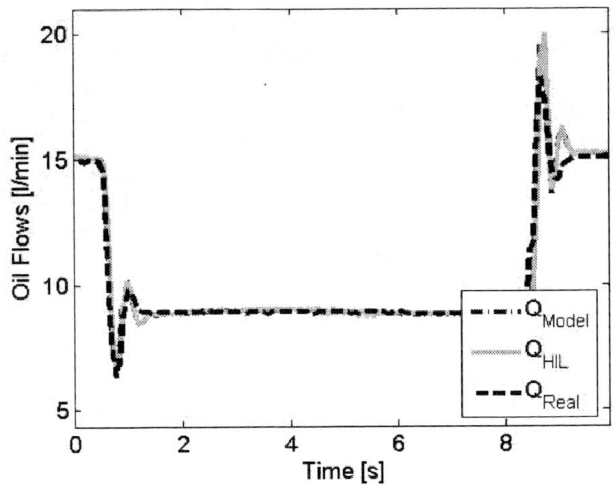

Fig. 6 Flows by using 2nd order model.

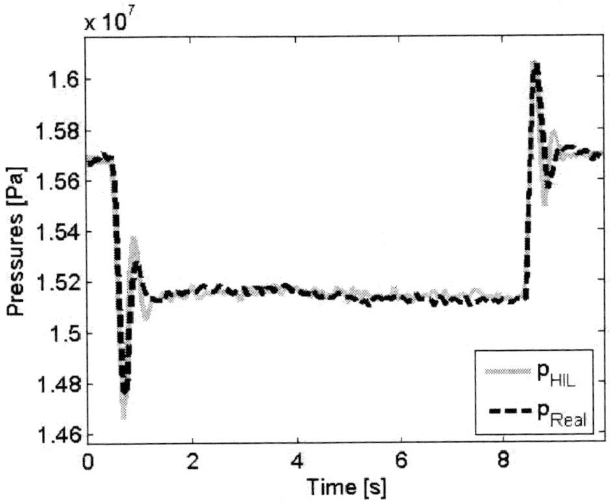

Fig. 7 Pressures by using 2nd order model.

From the results it can be seen that the both models are simulating the physical PRV pretty well. Improvement for the accuracy could be reached by using more effort for defining the empirical parameters C_3 and C_4. It can also be seen that the bandwidth of the simulator's interface is high enough and it is able to follow the calculated reference flows with sufficient accuracy.

CONCLUSIONS AND EXTENSIONS

The present paper deals with hardware-in-the-loop (HIL) simulation of hydraulic components and systems. The main idea is to develop hydraulic systems or individual components by using their simulation model or models as a part of the physical machine. The interface between the real and simulated parts is defined by means of pressures and flows instead of forces and motions. The paper describes the modeling of PRV in detail and proposes two alternative simulation models that are made by employing the semi-empirical approach.

A simple test rig was built and promising results achieved. From the results it can be concluded that the studied HIL simulation based method is valid for R&D of hydraulically driven machines. The method needs further investigations to determine the constraints in applicability. The method may appear to be a novel powerful R&D tool for hydraulic machines.

REFERENCES

[1] Helduser, S., Stuewing, M., Liebig, S., Dronka, S., Development of Electro-hydraulic Actuators Using Linked Simulation and Hardware-in-the-loop Technology. Power Transmission and Motion Control, pp. 49-56, 2001.

[2] Plummer, A., A Servohydraulic Control System for Implementing Components in Mechanical Systems. Power Transmission and Motion Control, pp. 131-145, 2003.

[3] Huang, Q., Zhang, S., Han, J., Jiang, H., Ye, Z., A Hardware-in-the-Loop Spacecraft Docking Simulation System and Its Stability Analysis. The 6th International Conference on Fluid Power Transmission and Control, pp. 799-802, 2005.

[4] Eskola, T., Handroos, H., Rouvinen, A., Virtual Hydraulics as a Part of Real Machine construction. Power Transmission and Motion Control, pp. 17-23, 2002.

[5] Eskola, T., Handroos, H., Rouvinen, A., Virtual Hydraulics in Hardware-in-the-loop Simulation of Mechatronic Machines. The fifth Japan Fluid Power Society's International Symposium on Fluid Power, pp. 895 – 898. 2002.

[6] Eskola, T., Pedersen, P., Handroos, H. Control of Interface of HIL Simulator for Hydromechanical Machines. The 6th International Conference on Fluid Power Transmission and Control, pp. 18-22. 2005.

[7] Eskola, T., Pedersen, P., Handroos, H. Experimental Analysis of HIL Simulator with Flow/Pressure Interface. The 6th International Conference on Fluid Power Transmission and Control, pp. 14-17. 2005.

[8] Handroos, H., Methods for Combining a Theoretical and an Empirical Approach in Modeling Pressure and Flow Control Valves for CAE-programs for Fluid Power Circuits. Acta Polytechnica Skandinavica, Mechanical Engineering Series No.96, Helsinki 1990, 52pp. Published by Finnish Academy of Technology.

[9] Handroos, H., Vilenius, M. The Utilization of Experimental Data in Modelling Hydraulic Single Stage Pressure Control Valves. Journal of Dynamic Systems, Measurements, and Control, Vol. 112. pp. 482-488. 1990

[10] Handroos, H., Vilenius, M. The Influence of Semi-Empirical Control Valve Model Parameter Accuracy on Simulated Hydraulic System Responses. Journal of Dynamic Systems, Measurements, and Control, Vol. 114, 1992.

Proceedings of IMECE2005
2005 ASME International Mechanical Engineering Congress and Exposition
November 5-11, 2005, Orlando, Florida USA

IMECE2005-81743

EXPERIMENTAL OPERATION AND CHARACTERIZATION OF A FREE PISTON COMPRESSOR

Jose Riofrio and Eric J. Barth
Department of Mechanical Engineering
Vanderbilt University
Nashville, TN 37235-1592

ABSTRACT

The ongoing design evolution of a free piston compressor (FPC) is presented in this paper. The FPC is a proposed device that utilizes combustion of a hydrocarbon fuel to compress air into a high-pressure supply tank. This device is designed to extract chemically stored energy from the fuel and convert it to potential energy of compressed air, while achieving high conversion efficiency relative to other small-scale portable power supply systems. The chemically stored energy of the hydrocarbon fuel is first converted into kinetic energy of the free piston by the end of the combustion phase. Subsequently, the moving piston acts as a pump and air compressor during a compression phase. The proposed technology is intended to provide a compact and efficient pneumatic power supply source appropriate for human-scale robots. The design and implementation of this version of the FPC is shown, and experimental results relating all phases (combustion, expansion and pumping) are discussed. The total efficiency of the system is experimentally measured and compared to its theoretical prediction.

1.0 INTRODUCTION

The need for an effective portable power supply for untethered human-scale robots has increasingly become a matter of interest in robotics research. Current prototypes of humanoid robots, such as the Honda P3, Honda ASIMO and the Sony QRIO, show significant limitations in the duration of their power sources in between charges (the operation time of the Honda P3, for instance, is only 25 minutes). This becomes a strong motivation for the implementation of a more adequate source of power. The motivation details are discussed more thoroughly in [1].

This paper presents the design of a free piston compressor (FPC) as a power supply for pneumatically actuated systems. The FPC serves the function of converting chemically stored energy of a hydrocarbon fuel into pneumatic potential energy of compressed air. More specifically, it extracts the energy by producing combustion of a stoichiometric mixture of propane and air, and the combustion-driven free piston acts as an air pump to produce the compressed air. The FPC, coupled with pneumatic actuators, is intended as an alternative to electrical batteries coupled with electrical motors. The main objective of this idea is to exploit the high mass specific energy density of hydrocarbon fuels and the high mass specific power density of linear pneumatic actuators, in order to provide at least an order of magnitude greater combined energy and power density (power supply and actuation) than state of the art electrical power supply and actuation systems.

Given their inherent penalization for carrying their own mass, the total energetic merit of an untethered power supply and actuation system is a combined measure of the source energy density of the energetic substance being carried, the efficiency of conversion to controlled mechanical work, the energy converter mass, and the power density of the actuators. With regard to a battery powered electric motor actuated system, the efficiency of conversion from stored electrochemical energy to shaft work after a gear head can be high (~ 50% to 80%) with very little converter mass (e.g. PWM amplifiers); however, the energy density of batteries is relatively low (about 180 kJ/kg for NiMH batteries), and the power density of electrical motors is not very high (on the order of 50 W/kg) rendering them heavy in relation to the mechanical work that they can output. With regard to the hydrocarbon-based pneumatic power supply and actuation approach presented here, the converter mass is high relative to a battery/motor system, and the total conversion efficiency is shown in [2] to be low in relative terms. However, the energy density of hydrocarbon fuels is in the neighborhood of 45 MJ/kg (where the oxidizer is obtained from the environment and therefore has no associated mass penalty), which is more than 200 times greater than the energy density of conventional electrical batteries. This implies that even with poor conversion efficiency (< 10%), and with only mild expectations of miniaturizing the energy converter, the available energy to the actuator per unit mass of the energy source (mass of fuel plus mass of energy converter) is still at least one order of magnitude greater than the battery/motor system. Additionally, linear pneumatic actuators have roughly one order of magnitude greater power density than traditional electrical motors. Therefore, the proposed technology offers the potential of greatly increased energetic characteristics over state-of-the-art electrical power supply and actuation systems.

The idea of using a free piston combustion-based device as a pump has been around since the original free-piston patent by Pescara in 1928 [3]. The automotive industry conducted a large amount of research in the 1950s. Ford Motor Company considered the use of a free piston device as a gasifier in 1954 [4]. General Motors presented the "Hyprex" engine in 1957 [5]. Such endeavours were aimed at an automotive scale engine and were largely unsuccessful. In more recent times, the free piston engine concept has been considered for small-scale power generation. Aichlmayr, et. al. [6, 7] have considered the use of a free piston device as an electrical power source on the 10 W scale meant to compete with batteries. Beachley and Fronczak [8], among others, have considered the design of a free-piston hydraulic pump. McGee, et. al. have considered the use of a monopropellant-based catalytic reaction as an alternative to combustion, as applied to a free piston hydraulic pump [9].

The FPC presented here is intended as a power supply for a mobile pneumatic robotic system of human comparable power, mass and size. It is shown analytically in [2] that the use of a free piston engine as a direct air compressor offers nearly ideal loading characteristics necessary for high efficiency (relative to similar scale combustion based devices), in a simple and small package.

A first design of the FPC was presented in [10]. In this previous design, particular emphasis was placed into the combustion portion of the device, and it outlined the main features of the FPC concept and design considerations of that particular prototype. The main features should be re-stated here, since they constitute the essence of the FPC idea:

Inertial Loading – The free piston is not rigidly attached to a crankshaft or any timing linkage alike, so it offers purely inertial loading to the expanding combustion gases. This allows the free piston to load up with kinetic energy resulting from the work done by the ideal adiabatic expansion [10] of the combustion gases. The combustion gases are allowed to expand until they reach atmospheric pressure, all while still contributing to the inertial loading. This full expansion contributes to a higher efficiency than if full expansion were not allowed, as is the case with most small-scale IC engines. As an additional consequence, the FPC has a quiet exhaust, since no high-pressure gases will be exhausted into the atmosphere.

Breathe-in Mechanism – When the combustion gases reach atmospheric pressure, the free piston will still be traveling (with maximum kinetic energy), and thus will induce a drop of pressure in the combustion chamber as the motion continues. This pressure drop will cause an intake check valve to open and allow fresh atmospheric air to enter the chamber. This fresh air will both cool down and dilute the combustion products. The breathe-in mechanism ensures a low temperature operation of the device.

Start on Demand – Since the intake valves and spark plug are electrically actuated, and since high-pressure injection of air and fuel eliminate the need for a conventional intake and compression stroke, the FPC does not require the implementation of a starter. This allows the engine to start on demand, without the need for a separate starting cycle. This feature highlights the compatibility between the FPC and a pneumatic robotic system, since they can be tied together by implementing a simple control loop to maintain a particular pressure in a supply reservoir. The FPC would receive a signal and start operating as soon as the actuation pressure supply drops, and likewise turn off once reaching the desired pressure.

The features outlined above, among others, are discussed more in detail in [10], and become the platform for the work presented in this paper. What follows is an introduction of the new FPC prototype and some new design considerations, a comprehensive analysis of the design evolution, analysis of new experimental data showing combustion and pumping, and a calculation of the total efficiency of the system.

2.0 NEW VERSION OF FPC

The previous prototype of the FPC [10] consisted of one combustion cylinder and two pumping cylinders. The three cylinders were aligned in parallel, with the combustion cylinder in the center. The pistons were rigidly attached to each other, and their main functions (combustion and pumping) occurred at the right of the pistons. The combustion would drive the piston assembly to the left, and a set of springs reversed the motion back to the right, thus initiating the pumping phase. Figure 1 shows a schematic of this previous prototype.

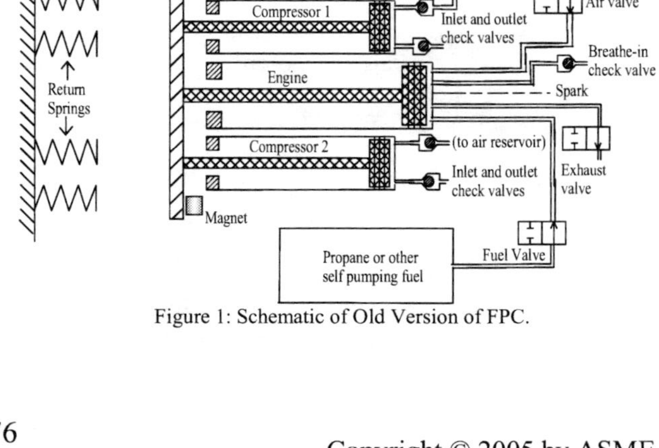

Figure 1: Schematic of Old Version of FPC.

The most appealing aspect of this configuration was that combustion and pumping occurred on separate strokes, so the inertial loading during combustion was not resisted by the compressed air. While this prototype provided some useful and insightful data, it also showed some limitations. These were mostly due to energy losses associated with collisions with the springs, the non-ideality of the springs in storing and returning energy, and the high friction in the cylinders. The rigid linking of the pistons also provided a very slight misalignment, which accounted for even greater friction. Additionally, the exhaust valve, breathe-in check valve, and inlet and outlet check valves were offering severe flow restrictions, which suggested the use of a smaller cylinder bore, or alternatively, larger valves.

Figure 2 shows a schematic of the new version of the FPC. This prototype highlights combustion and pumping within the same stroke. The device consists of two cylinders in-line and opposing one another. Both cylinders have a combustion side (back of the piston) and a pumping side (rod side of the piston).

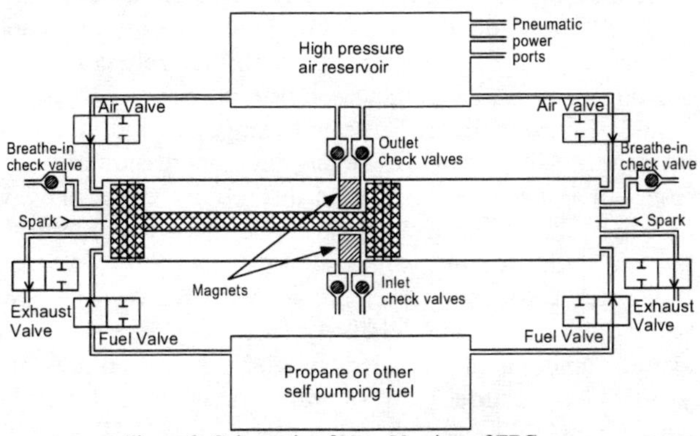

Figure 2: Schematic of New Version of FPC.

The device is completely symmetrical, so its starting position can be on either side. Assuming the piston's initial position is at the left, the piston is held in place by the magnets while injecting a mixture of pressurized air and propane into the combustion chamber. Once the proper amount of mixture has entered the chamber, the air and propane valves close and a sparkplug initiates combustion. The piston will then travel to the right while serving four functions: (1) pump fresh air into the air reservoir; (2) exhaust the diluted combustion products from previous combustion out of the right cylinder's combustion chamber; (3) breathe in fresh air into the right cylinder's pumping chamber; and (4) breathe in fresh air into the combustion chamber after the pressure has dropped below atmospheric, thus cooling down and diluting the combustion products. At the end of the stroke the piston will be held in place by the magnets on the right side, and the cycle can occur on the opposite side in the same fashion (from right to left). The work required to break the magnetic holding force after combustion is retrieved at the end of pumping. The force-distance profile of the magnets also allows dominantly inertial loading presented to the combustion pressure after a very short distance after break-away has occurred.

3.0 DESIGN

Figure 3 shows a picture of the new FPC prototype. Most of the hardware was inherited from the first prototype (valves, magnets, sensors, spark plugs). The cylinders were replaced with two 4-inch stroke, ¾-inch bore BIMBA® standard air cylinders, and ported appropriately. Figure 4 shows a close-up of the new cylinder configuration, with the piston fully retracted (and the opposite piston disconnected).

By examining Figure 3, it should be noted that the hardware implementation of this prototype mainly differs from the generalized schematic (Figure 2) in that the piston rods are connected to each other through a moving mass, which is needed for inertial loading purposes. Another important difference to note is that the moving mass carries the magnets (2 on each side), which snap onto the ferrous plates at the end of each stroke. Figure 4 shows the magnets pressed against the ferrous plate.

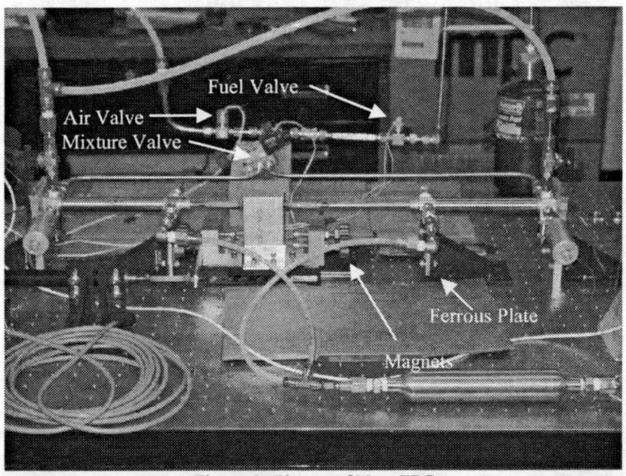

Figure 3: Picture of New FPC

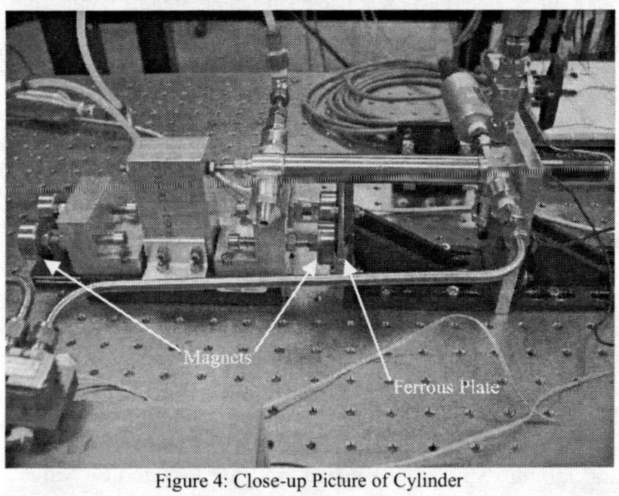

Figure 4: Close-up Picture of Cylinder

This new configuration addresses some of the limitations found in the previous version of the FPC. The cylinder rods were connected to the moving mass with small pieces of plastic tubing in order to avoid a purely rigid connection, which would yield friction due to misalignments. Added to that, the cylinders chosen are much smoother than the previous ones, so the frictional losses in this system are reduced. The lack of return springs makes the system quieter and a bit more energy efficient. Also, with these cylinders' bore, flow restrictions through valves are no longer contributing noticeable energetic losses.

Since the main objective of this paper is to show the total efficiency of the system from stored chemical energy of propane to stored pneumatic potential energy of compressed air, it is imperative to be as precise as is reasonably possible with all measurements. The most difficult, yet one of the most important measurements to determine is the mass of propane used for combustion. Knowing this mass and the energy density of propane, the total amount of initial energy can be determined and compared to the potential energy of compressed air in the reservoir. This potential energy is equal to $(P_s - P_{atm})V_s$, where P_s and V_s are the pressure and volume in the reservoir, and P_{atm} is atmospheric pressure.

3.1 Measuring the Mass of Propane

As shown in Figure 3, the air/fuel mixture enters the chamber by controlled opening and closing of the air, fuel and mixture valves. The air and fuel valves are timed such that a stoichiometric mixture exists within a common mixing line at the exit of both valves. This mixture is then admitted to one of the two combustion chambers via a three-way mixture valve. The air and fuel valves are Parker® Series-9 valves, and operate as two-way on/off valves. The nominal response time of these valves is 12 ms, while their commanded opening times for this application will range between 8-12 ms for propane and 50-80 ms for air. A mass flow meter could not be used to determine the mass of propane flowing because these are small pulses, and not steady flows. Also, since the propane valve operates at opening times close to the valve response time, the small pulse flow dynamics would not be similar to the steady flow dynamics. The mass was instead calculated by injecting propane into an inverted graduated cylinder of a beaker with water, and observing the water displacement in the cylinder for various durations of the commanded opening pulse sent to the fuel valve. This was done for 8, 9, and 10 ms pulses, and each one was performed several times to obtain an average. The standard deviation of measured displaced water for each pulse duration was very close to the readability of the scale. The same was done for the air, in order to find stoichiometric values based on air and fuel valve opening times. It should be noted that this match of opening times yields accurate stoichiometric mixtures only if flowing into atmospheric pressure.

For precise data acquisition purposes, the engine was run only in single fire shots. Before each fire, the exhaust port and mixture valve was opened, and the fuel and air valves opened for specific durations previously determined to yield a stoichiometric mixture. This was done several times such that the entire mixing line would contain a precise air-fuel ratio. The exhaust port was then closed and some of the stoichiometric air/fuel mixture remaining in the mixture line was pushed into the combustion chamber, by opening the mixture valve, until reaching the desired injection pressure. This pressurized mixture can be taken as an ideal gas, so the mass in the chamber can be calculated as $m = \frac{PV}{RT}$, where P and V are the pressure and volume in the chamber, T is the temperature of the mixture (approx. room temperature), and R is the gas constant of the stoichiometric propane/air mixture. Then, with knowledge of a stoichiometric mass ratio, the mass of propane subject to combustion can be determined.

3.2 Adjusting Mixture while Pumping

Ideally, every firing should be strong enough for the piston to just barely make it to the other side. If combustion is stronger than it needs to be, the piston will carry extra energy that will not be used for pumping, but dissipated in colliding with the other side. This is quite wasteful and brings down the total efficiency of the system. Conversely, if the combustion is weak and the piston carries less kinetic energy than needed to pump a full stroke, it will not make it to the other side, and some energy would need to be re-invested (by utilizing the air injection valve and energy stored in the reservoir) to push the piston to the other side (note that the area of the rod side being less that the piston side will allow this to occur when both sides are ultimately presented with the reservoir pressure). Precisely controlling the strength of the combustion for continuous operation would require more sophisticated equipment and is beyond the scope of this paper. However, full compression of the air reservoir is achieved here with discrete increases of valve opening times as needed. It should be noted that due to the low resolution of the valves at these extremely low mass flows, the combustion pressure increase between any two valve-opening times is relatively large. Figure 5 shows the average amounts of propane mass and combustion pressures yielded by all three sets of valve timings used for this experiment.

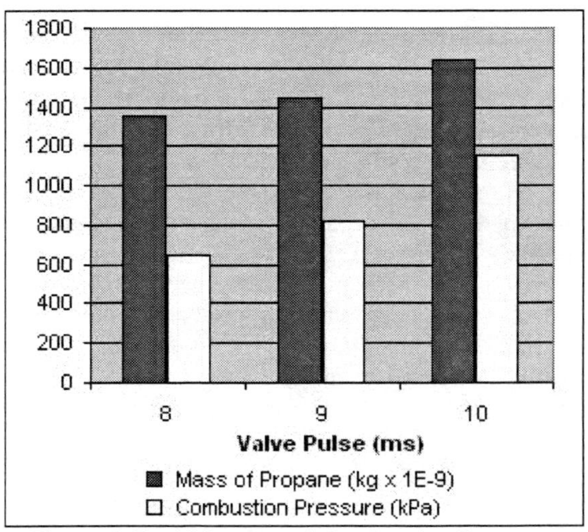

Figure 5: Fuel Masses and Combustion Pressures

For data gathering, the FPC is fired first with the lowest set of valve opening times. After a certain number of firings, the pressure in the air reservoir will reach a certain threshold, at which point the combustion pressure will not be adequate for the free piston to complete the stroke, and the injection is re-adjusted to the next set of opening times.

It should be intuitive that the first few shots after a threshold are the least efficient, while the last few shots before a threshold are the most efficient.

4.0 EXPERIMENTAL RESULTS AND EVALUATION

Experimentation consisted of a series of single fire shots. Combustion took place in a 6.4 mL volume, with a moving mass of 1.82 kg. Three different opening times of the fuel valve were used for combustion (8, 9, and 10 ms), depending on the combustion pressure desired. The data shown in Figures 6-13 are taken from two single shots of the device. The first set (Figures 6-9) show a less than ideal shot and the second set (Figures 10-13) show a more ideal shot. This variation is a result of the lack of adequate valve opening resolution resulting from the hardware used with this experimental setup. Figures 6, 7, 8, and 9 show combustion pressure, air reservoir pressure, position of the piston and velocity of the piston, respectively. These were taken from a typical firing at a low pumping pressure and exhibit the inadequacy of resolution of the air and fuel valves – note the collision with the opposing side apparent from the sharp decrease in velocity shown in Figure 9. The spark occurs at 0.4 seconds. Figure 6 also shows the injection pressure in the combustion chamber before ignition.

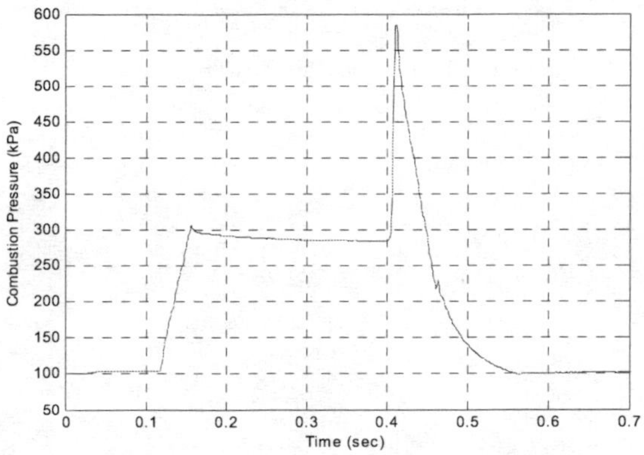

Figure 6: Pressure in the Combustion Chamber

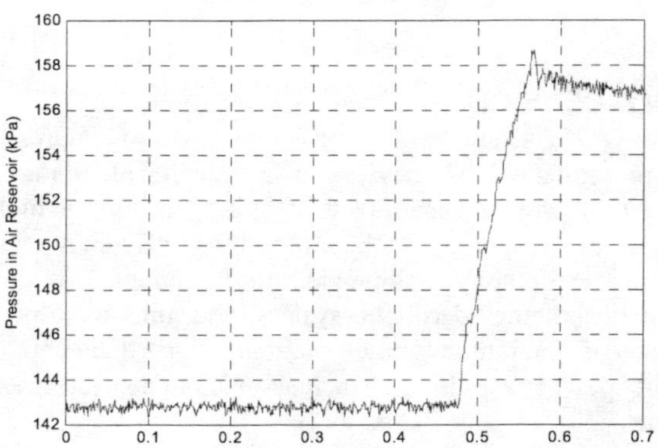

Figure 7: Pressure in the Air Reservoir

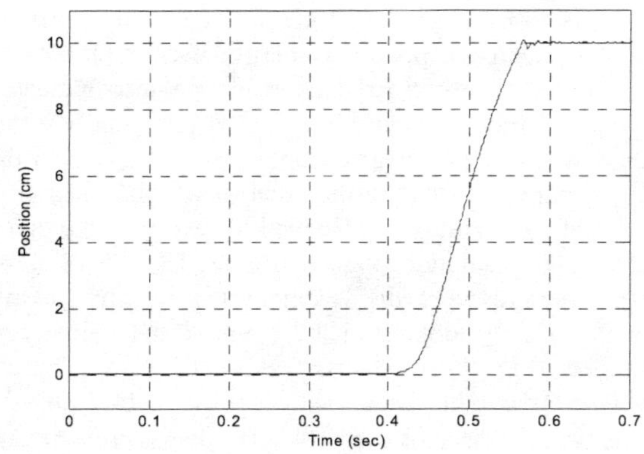

Figure 8: Position of the Free Piston

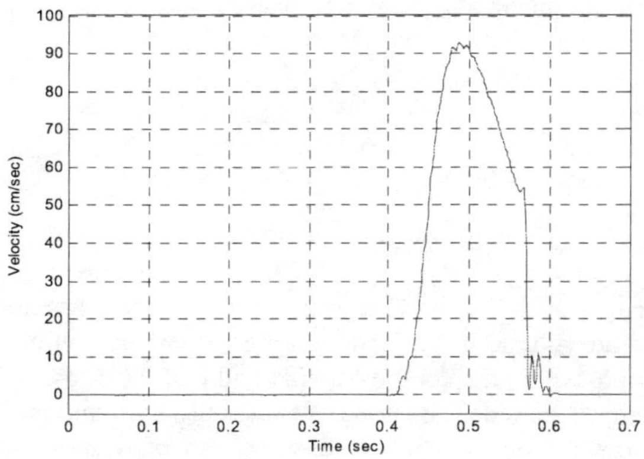

Figure 9: Velocity of the Free Piston

Despite the inefficient operation shown for Figures 6, 7, 8 and 9, it should be noted that these plots still exhibit the main features of the PFC outlined in [10], such as the inertial loading (nearly adiabatic expansion in Figure 6) and the breathe-in mechanism (combustion pressure decreases to atmospheric pressure before the end of the stroke). Also, the frequency for continuous operation can be maximized by reducing the time between signals. With basic knowledge of the injection, combustion and expansion timings, the duration of each stroke can be reduced to just under 0.4 seconds. This yields a total operational frequency of 2.5 strokes per second.

Through single fire shots, the air reservoir was successfully compressed from atmospheric pressure up to 310 kPa. It took 52 strokes to achieve this pressure, which can be achieved in 20 seconds of continuous operation (based on the previously described operational frequency). This compression pressure is limited by the ratio of stroke volume (engine displacement) over dead volume in the rod (compressing) side of the chosen cylinders. This dead volume was quite large (about 4.1 mL), and is a nonideality (as noted in [2]) that should be reduced in future design considerations. This nonideality presents the main drawback in this version of the FPC.

The total energetic merit of this system is represented by the efficiency of conversion from chemically stored energy in propane to pneumatic potential energy of compressed air. As shown analytically in [2], this efficiency also increases for larger combustion pressures. Experimentally, this efficiency was calculated for a single fire shot in which the piston just barely made it to the other side – the ideal case which could conceivably be achieved for every stroke given adequate valve resolution. Figures 10, 11, 12, and 13 show the respective pressures, displacement and velocity for this particular shot. By looking at the position and velocity in Figures 12 and 13, it can be observed that the piston "barely made it" to the other side, with a little help from the magnetic force. This shot used 1.42×10^{-6} kg of propane and resulted in an increase of the pressure in the air reservoir (a volume of 1.866×10^{-4} m^3) from 222.75 kPa to 235 kPa, yielding a pneumatic potential energy increase of 2.29 Joules. The total efficiency of this shot is calculated as,

$$\eta_{total} = \frac{0.002286\,\text{kJ}}{\frac{46350\,\text{kJ}}{\text{kg fuel}}\times 1.42*10^{-6}\ \text{kg fuel}}\times 100 = 3.47\%$$

Based a simple thermodynamic analysis presented in [2], the total efficiency of the system with these parameters is calculated to be 7.62%. However, this calculation assumes zero dead volume in the compressing side of the cylinder, no heat losses in the combustor side and high heat losses in the compressor side. The issue of excess dead volume in the compressor side present in this experimental prototype would be difficult to fully address with standard pneumatic equipment (such as those used here), and would require custom made parts in order to reduce the minimum rod-side volume. As also shown in [2], the efficiency of the device increases with increased combustion pressure (up to 20% efficiency). With these points in mind, the experimental efficiency met reasonable expectations, while leaving room for improvement in future designs.

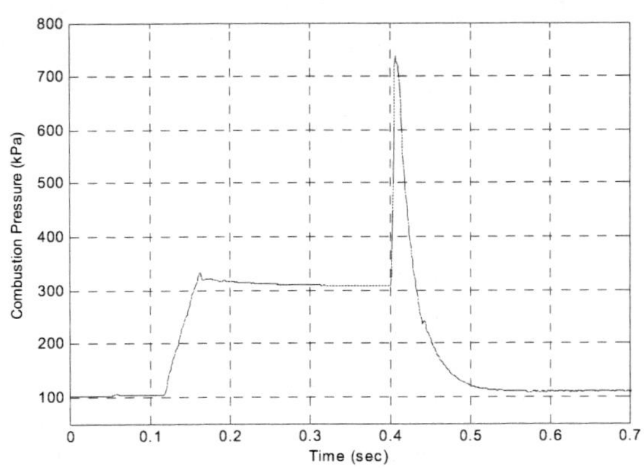

Figure 10: Pressure in the Combustion Chamber

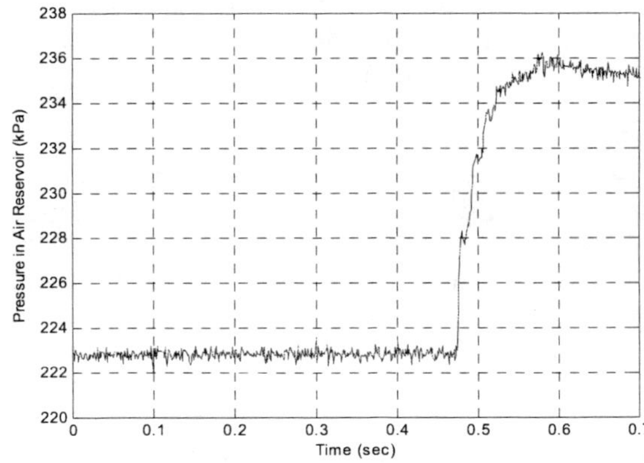

Figure 11: Pressure in the Air Reservoir

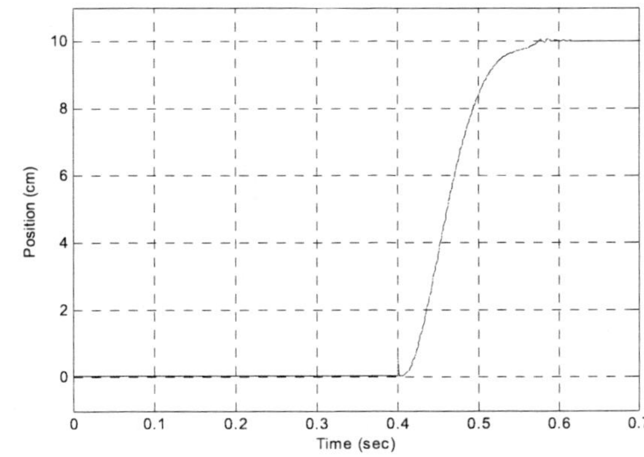

Figure 12: Position of the Free Piston

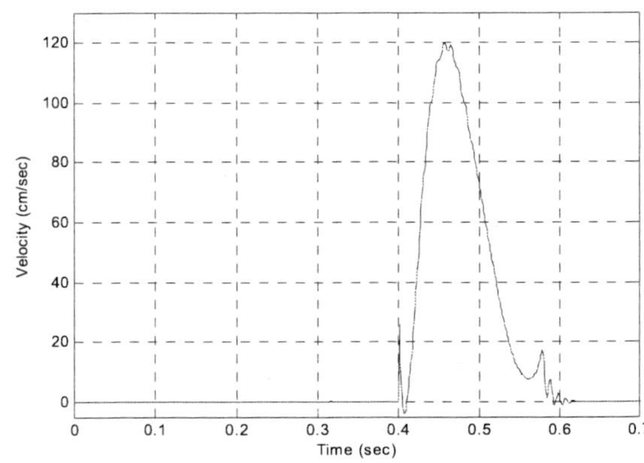

Figure 13: Velocity of the Free Piston

5.0 CONCLUSIONS

The design of a new version of a Free Piston Compressor was presented. This device is a small-scale internal combustion engine capable of pumping air into a high-pressure air reservoir, with satisfactory efficiency. The proposed technology is intended to be coupled with a pneumatic untethered robotic system, and aims to provide an order of magnitude greater energetic merit than state of the art power supply and actuation systems (electrical batteries coupled with DC motors).

The experimental prototype FPC exhibited predicted desirable design features such as: reliable and consistent firing, inertial loading and nearly adiabatic expansion of combustion gases, combustion pressures that decrease to atmospheric pressure before the end of the stroke thereby employing the breathe-in mechanism for cooling and low noise, and a predicted frequency of continuous operation of 2.5 strokes per second. It was shown that further design work regarding the resolution of the air and fuel valves is needed to secure decent efficiencies across different reservoir pressures, and further work regarding the dead volume in the rod-side of the piston is needed to increase efficiency.

A measured efficiency of 3.47% was achieved in converting stored chemical energy into stored pneumatic potential energy with a datum at atmospheric pressure: $(P_s - P_{atm})V_s$. In comparing this power supply system with state-of-the-art rechargeable batteries, an energy density of 180 kJ/kg for NiMH batteries and a conversion efficiency of 50% for an electromagnetic motor and mated gearhead would yield 90 kJ of delivered controlled mechanical work per kilogram of source energetic material. For the FPC under consideration, an energy density of 46350 kJ/kg for propane, a measured conversion efficiency of 3.47% from stored chemical to pneumatic energy, a 75% efficiency associated with the computed mass reinvestment from the air reservoir for the next combustion event (see [2]), and an assumed conversion efficiency of 30% from stored pneumatic energy to delivered controlled mechanical work of an associated pneumatic actuator, would yield 365 kJ of controlled work per kilogram of source energetic material. This increase, coupled with the equally important high power density of pneumatic actuators over electromagnetic motors (approximately 450 W/kg versus 50 W/kg), should contribute to a power supply and actuation system more appropriate for untethered human scale and power comparable robots and actuated devices. It is expected that design changes to the next generation of this device will result in further improvements.

REFERENCES

[1] Goldfarb, M., Barth, E. J., Gogola, M. A., Wehrmeyer, J. A., "Design and Energetic Characterization of a Liquid-Propellant-Powered Actuator for Self-Powered Robots". *IEEE/ASME Transactions on Mechatronics*, vol. 8, no. 2, pp. 254-262, June 2003.

[2] Barth, E. J., and Riofrio, J., "Dynamic Characteristics of a Free Piston Compressor," *2004 ASME International Mechanical Engineering Congress and Exposition (IMECE)*, IMECE2004-59594, November 13-19, 2004, Anaheim, CA.

[3] Pescara, R. P., "Motor Compressor Apparatus," U.S. Patent No. 1,657,641, Jan. 31, 1928.

[4] Klotsch, P., "Ford Free-Piston Engine Development," *SAE Technical Paper Series*, 590045, vol. 67, pp. 373-378, 1959.

[5] Underwood, A. F., "The GMR 4-4 'Hyprex' Engine: A Concept of the Free-Piston Engine for Automotive Use," *SAE Technical Paper Series*, 570032, vol. 65, pp. 377-391, 1957.

[6] Aichlmayr, H. T., Kittelson, D. B., and Zachariah, M. R., "Miniature free-piston homogenous charge compression ignition engine-compressor concept – Part I: performance estimation and design considerations unique to small dimensions," *Chemical Engineering Science*, 57, pp. 4161-4171, 2002.

[7] Aichlmayr, H. T., Kittelson, D. B., and Zachariah, M. R., "Miniature free-piston homogenous charge compression ignition engine-compressor concept – Part II: modeling HCCI combustion in small scales with detailed homogeneous gas phase chemical kinetics," *Chemical Engineering Science*, 57, pp. 4173-4186, 2002.

[8] Beachley, N. H. and Fronczak, F. J., "Design of a Free-Piston Engine-Pump," *SAE Technical Paper Series*, 921740, pp. 1-8, 1992.

[9] McGee, T. G., Raade, J. W., and Kazerooni, H., "Monopropellant-Driven Free Piston Hydraulic Pump for Mobile Robotic Systems," *ASME Journal of Dynamic Systems, Measurement, and Control*, vol. 126, pp. 75-81, March 2004.

[10] Riofrio, J., and Barth, E. J., "Design of a Free Piston Pneumatic Compressor as a Mobile robot Power Supply," *Proceedings of the 2005 IEEE International Conference on Robotics and Automation (ICRA)*, pp. 236-241, Barcelona, Spain, April 2005.

[11] Black, William Z., and James G. Hartley, *Thermodynamics, Third Edition*, New York: Harper Collins, 1996.

Proceedings of IMECE2005
2005 ASME International Mechanical Engineering Congress and Exposition
November 5-11, 2005, Orlando, Florida USA

IMECE2005-82002

INTEGRATED PRESSURE-FLOW-TEMPERATURE SENSOR FOR HYDRAULIC SYSTEMS

Charles Groepper
University of Minnesota
1100 Mechanical Engineering
111 Church St SE
Minneapolis Minnesota 55455
Groep004@umn.edu
(612) 845-8690

Dr. Tianhong Cui
University of Minnesota
1100 Mechanical Engineering
111 Church St SE
Minneapolis Minnesota 55455
Tcui@me.umn.edu
(612) 626-1636

Dr. Perry Li
University of Minnesota
1100 Mechanical Engineering
111 Church St SE
Minneapolis Minnesota 55455
Pli@me.umn.edu
(612) 626-7815

Dr. Kim Stelson
University of Minnesota
1100 Mechanical Engineering
111 Church St SE
Minneapolis Minnesota 55455
Kstelson@me.umn.edu
(612) 625-6528

ABSTRACT

The design of a miniature low-cost MEMS (Micro-Electronic Mechanical Systems) based sensor for sensing pressure, flow and temperature in fluid power systems is presented. The sensor is small enough to be incorporated into hydraulic components such as pumps, motor and cylinders. Considerable literature exists on the development of MEMS-based pressure and temperature sensors; therefore the primary challenge addressed in this paper is a low cost flow sensing method that does not impose extra energy dissipation into the system.

The flow sensing principle adopted for use with the sensor utilizes the small pressure differences that develop as fluid flows through any geometry. As an example to prove the validity of this principle, pressure-flow relationships are determined for an elbow geometry. Specifically, the pressure differences arising from the momentum change of the fluid as it passes through the elbow are correlated to the flow rate of the fluid through the bend. The pressure-flow correlations derived from theoretical modeling, computational fluid dynamic (CFD) simulation, and experimental measurement using the flow bend geometry are compared and a calibration scheme determined.

The first generation MEMS sensor incorporates pressure, flow and temperature measurement capability on a single sensor die. Pressure sensing is accomplished with piezoresistive strain elements, and temperature sensing is done with a thermister.

KEYWORDS:
MEMS, Flow meter, Pressure sensor, Fluid power

INTRODUCTION

Presently, the flow rate, pressure, or temperature of a fluid power system can only be obtained by adding a stand-alone measurement component to the system. The cost of each component separately is highly dependant of the type and quality of the measurement one wishes to make. However, it is not uncommon for the cost of pressure, flow and temperature measurement combined to be on the order of a thousand dollars or more. In addition to high cost, adding separate measurement components to the system also has the disadvantages of increasing the size, complexity and weight of the system.

Many pressure, flow and temperature sensing components currently on the market have been designed for laboratory use. The durability of these sensors for use in a rigorous industrial service is often questionable. Since energy loss is usually not a concern in a laboratory environment, the use of these sensors often imposes significant energy dissipation into the system, which is not desirable in an industrial setting. In this paper, an alternative method for obtaining the pressure, flow and temperature in a fluid power system is presented, in which additional components will not need to be added to the system. The method entails the integration of sensors directly into existing system components to detect preexisting pressure and temperature phenomena. This approach results in a durable, economical sensing method that is extremely well suited for industrial applications.

The integration of pressure flow and temperature sensors directly into fluid power components has several advantages compared to stand alone components. First, since the sensor is integrated into existing components, the size and weight of the system is not increased. For sensors to be integrated into existing components they will need to be of a very small size. The size requirement will necessitate the use of MEMS manufacturing methods for production, which also has many inherent advantages, including economical production costs. One major challenge with MEMS based sensors is usually packaging; however, if the sensor is to be directly integrated in to fluid power components, the packaging is no longer a major concern. Since the sensor is to be placed inside the fluid power components, the sensor will be as durable as the component it resides within.

Another advantage of the proposed sensing method is the sensor will not impose extra energy dissipation on the system. No extra energy will be dissipated because the sensor is to be designed to utilize pressure drops that occur due to geometry that already exists in the system. Many fluid power components contain sudden expansion, sudden contraction, elbow, etc geometry, which can be shown to exhibit repeatable pressure drop-flow characteristics. If these pressure-flow characteristics are harnessed, the flow can easily be inferred by simply measuring these pressure drops.

Implementation of the flow sensing method requires the use of MEMS pressure sensors. Considerable research effort has already been devoted to the development of piezoresitive strain type pressure sensors. The result of this research is extensive documentation regarding the design and characterization of such sensors. Many of the designs have been commercialized, and are available with a variety of pressure ranges, spanning from a very low scale of 0-5000 Pa [1] to a very high scale of 0-50 MPa.[2] With such a varied range of pressure scales, a sensor to measure pressure drops due to losses though geometry and infer a flow rate can easily be developed.

Similarly, work done by Steinhart and Hart [3] characterizes the behavior of semiconductor resistors as the operating temperature is altered. This work simplifies the use of common resistors as temperature measurement devices with a simple logarithmic polynomial curve fit calibration equation. Reproduction of this work will allow a temperature sensing method to be easily integrated into the sensor design.

In contrast, little research has been done on the development of MEMS based flow rate sensors for the macro-domain. Currently, the primary thrust of MEMS flow measurement research is based on microfluidic systems, and has little application to the measurement of a macro scale flow rate in a fluid power system. Some work has been done on the development of thermal and magnetic type sensors, however, these sensors cannot be implemented in fluid power systems due to working fluid restrictions of each. [4] . Our approach of inferring a flow rate from pressure drops shows promise in the development of MEMS based flow rate sensors.

Since the design will infer a flow rate through the measurement of small pressure drops, a calibration scheme for the pressure-flow relationship will need to be developed. The accuracy in determining this relationship will determine the accuracy of the flow measurement. As such, the determination of the pressure-flow calibration is a very important aspect of the sensor design.

In order to prove the validity of the flow sensing principle, an elbow geometry will be used as an example. The pressure flow relationships for this geometry will be studied in detail using numerical, CFD and experimental procedures in order maximize understanding of the fluid flow phenomena, so a sensor with the highest possible accuracy can be developed. Once the flow phenomenon has been characterized, the hardware aspects of the sensor will be designed.

NOMENCLATURE

ρ	Fluid density, Resitivity
p	Pressure
n	Normal vector
V	Velocity
S	Radius of curvature
r	Radius of flow area
Q	Volumetric flow rate
π	Pi
A	Flow Area
D	Flow Diameter
μ	Dynamic Fluid Viscosity
w	Deflection
a	Distance from center to plate edge X direction
b	Distance from center to plate edge Y direction
x	Arbitrary distance from center of plate X direction
y	Arbitrary distance from center of plate Y direction
υ	Poisons ratio
R	Resistance
e	Strain
G	Gauge Factor
w,wr	Resistor width
l	Resistor length
t	Resistor thickness
V	Voltage
n	Number of resistor turns
η	Normal vector
ws	Resistor spacing
ls	Length of resistor on substrate
lp	Length of resistor onto diaphragm

ELBOW FLOW MEASUREMENT

1.1 Numerical Results

The use of a typical elbow geometry as a flow meter has been studied, and is commerciality available for use in water flow measurement.[5-11] However, no such flow meter exists for use in fluid power systems, in the laminar flow regime. The primary difference between the water and oil as the working fluid is the viscosity. The difference between the dynamic (absolute) viscosity of water and typical hydraulic oil at room temperature is roughly a factor of 100, and at an elevated temperature of 100°C, the difference drops to only a factor of 10. The difference in factors is attributed to a large viscosity temperature dependence of hydraulic oil compared to water. In order for the elbow to be used as a precision flow meter for hydraulic systems, the high viscosity low Reynolds number flow through the elbow will need to be understood and characterized. Other parameters including the entrance length preceding the elbow, pressure tap location, fluid density variation, and the effects of secondary flow fields will also need to be studied in order to maximize the sensitivity of the meter.

To begin, a rough mathematical estimate of the pressure difference between the inner and outer radii of the elbow will be developed. Consider the diagram of an elbow presented in Fig. 1.

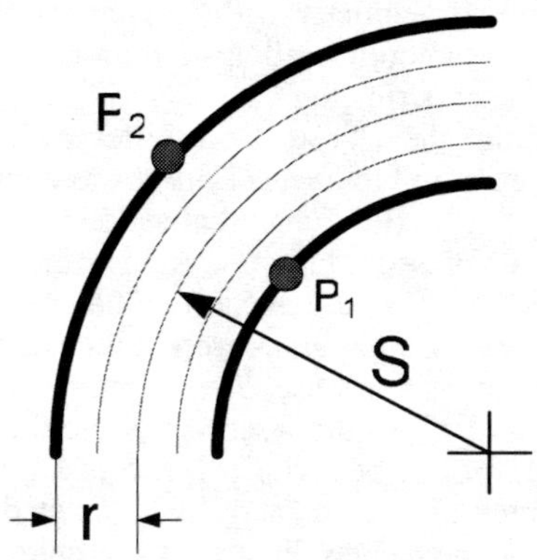

Fig. 1 Elbow Diagram

If the flow is assumed invisid, Euler's equation can be applied in streamline coordinates. Equation (1) summarizes this relationship in a direction normal to the flow:[12]

$$\frac{1}{\rho}\frac{\partial p}{\partial \eta} = \frac{V^2}{S} \tag{1}$$

Equation (1) can be integrated with respect to the elbow dimensions. If the general definition of the volumetric flow rate is applied to the result, the expression presented as Eq. (2) is obtained.

$$Q = \pi r^2 \sqrt{\frac{p_2 - p_1}{\rho \ln\left[\frac{S+r}{S-r}\right]}} \tag{2}$$

The primary shortcomings with this relationship are the assumption of 2-dimensional invisid flow which suggests a constant velocity profile across the section during the integration. With these shortcomings aside, the relationship can be effectively used to estimate the pressure difference between the inner and outer radii of a bend if the elbow geometry is known.

1.2 CFD Results

The results obtained with the Euler estimate shown as Eq. 2 predict a very small pressure difference between the inner and outer radii of the elbow, when compared to the usual pressures found in a fluid power system. The Euler relationships predicts a maximum pressure difference for a given flow rate will occur with the sharpest bend. In order to verify this result, and further study the suitability of this pressure difference for use to infer a flow rate, CFD methods are applied.

The geometry selected for study using CFD was that of standard cast iron off-the-shelf elbows. Measurement of these elbows determined they had a very sharp bend, and their broad availability and economical price further enhanced their suitability for the project.

The software used for the CFD study was CFX. This software is commercially available through ANSYS Ltd. Using CFX, many of the variables of the elbow could be quickly evaluated. These variables include: viscosity effects, entrance length, and pipe velocity.

In order to effectively capture the fluid behavior in the elbow, the flow regions including the elbow, sudden expansions that would occur before and after the elbow itself in an experimental apparatus, and substantial 10 diameter length entrance and exit regions were modeled.

To obtain accurate CFD predictions, careful consideration needs to be given during the meshing process. It was expected the largest gradients would occur in the boundary layers, and as such, the mesh was created with refinements in these areas. To study the effect of the mesh density on the solution, a sample mesh was created and a solution obtained. This solution was then compared to a similar mesh that had been automatically refined in the areas of the highest gradients. The heavily refined mesh demonstrated a 2.5% lower pressure difference between P_1 and P_2, and contained approximately 50% more nodes. For the purpose of this design, this small variation is not significant to warrant using the heavily refined mesh, and also provides reassurance the problem is not mesh dependant.

The geometry, flow rates, and viscosity of the problem presented itself as laminar in nature. For each of the simulations, a laminar model and also the κ-ε turbulence model was compared. This was done in order to better understand the flow behavior, especially because of the possibility of laminar to turbulent transitions occurring as a result of the presence of the elbow. For each of the tests in which the κ-ε turbulence model was used, the intensity parameter at the inlet was set to medium, or 5% intensity. Please refer to ANSYS Ltd for more information regarding the turbulence models in the CFX software.[13]

Once the effect of the mesh density on the solution was understood, the effect of the entrance length on the solution was examined. This was done by creating geometries with various entrance lengths, and subjecting each to the same set of boundary conditions. The result of the entrance length study is shown in Fig. 2. If the longest length is assumed to be the most accurate data point, the total variation between all of the entrance lengths is only 1.2%. For the purpose of modeling the elbow, Fig. 2 shows an entrance length of 10 pipe diameters is adequate. Longer entrance lengths are not reported because the large size of the geometry and CFX license restrictions would not allow for an adequate mesh density to be used. The CFD entrance length study was also compared to experimental data, and the result showed good correlation.

The elbow geometries typically found in fluid power system components such as pumps typically have little or no development length preceding them. The results of the entrance length study show these geometries can be used to infer flow rate with success, and the resulting observed pressure difference per given flow rate will only be reduced a very small amount due to the short development length.

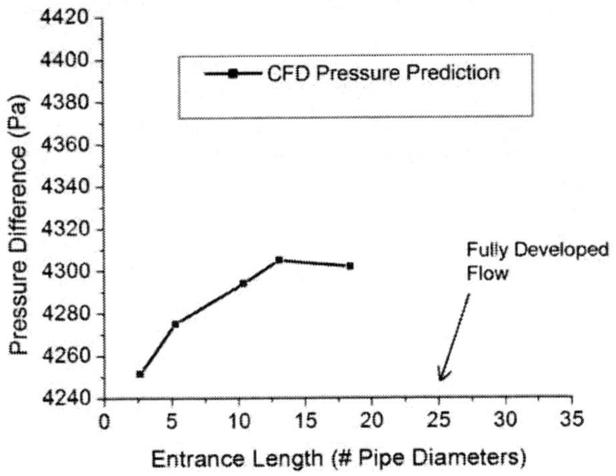

Fig. 2 Entrance Length Study Results

One advantage to using CFD simulations was the ability to characterize the effect of fluid viscosity on the observed pressure difference across the elbow. The viscosity of standard hydraulic oil roughly varies from 0.04 to 0.06 N s/m^2 for the standard range of operating temperatures, which is quite high compared with water and other such fluids. Since the Euler prediction given as Eq. 2 assumes invisid flow, the results of this study can only be compared to data obtained experimentally. During this investigation, the results given by the laminar and turbulent models differed. The laminar model correctly predicted the viscosity-pressure correlation seen experimentally, while the turbulent model predicted an inverse relationship. CFX predicted roughly a 2% difference in observed pressure difference due to the range of viscosity change.

One example of the phenomena that was examined in detail for the elbow geometry is the resulting secondary cross flow in the boundary layers due to the fluid viscosity. Investigation of the secondary flow phenomena showed the magnitude of the cross flow is viscosity and velocity dependant. Investigation with CFX shows that although this phenomenon is occurring within the bend with varying magnitude, its effect on measured pressure difference is minimal, suggesting the pressure difference is primarily momentum driven.

Compilation of the CFD data reveals a simple underlying relationship; for a given r/S geometry, pressure non-dimensionalized with the Euler number is simply a function of the Reynolds number.

1.3 Experimental Verification

In order to fully characterize the flow in the elbow, CFD predictions were compared to pressure-flow data obtained experimentally.

In order to minimize the amount of uncertainty in the experimental measurements, a two piece precision machined billet steel elbow was used for the experimental data collection. The primary motivation for machining the elbow was the parameters r and S could be controlled to within 0.013mm. The elbow contained two independent elbow channels, one of 9.525mm diameter, and the other of 6.35mm diameter. The r/S ratio for both channels was ½, which is representative of the sharp bends observed in off–the-shelf cast iron elbows, and also in many other typical fluid power elbow fittings. The flow channels of the elbow were machined with smooth walls with no evidence of machining marks, scratches, etc in the passageways. This allowed for the experimental apparatus to directly compare to the CFD predications, as the CFD models were done assuming geometry with a smooth wall. Machining the elbow with a smooth wall also eliminated unnecessary variables from the experiments.

Machining the elbows from billet steel also allowed for very small pressure tap areas (1.5mm^2) to be used such that there was very little flow disturbance due to the pressure measurement. The pressure tap dimensions were selected based on the work of Shaw [14] to have a less than 1% measurement error. Smaller pressure tap areas were attempted, and resulted in very slow pressure response. This effect was magnified at lower temperatures, when the viscosity of the fluid was higher. One explanation for this behavior was the possibility of trapped air within the pressure sensor and the associated plumbing. If this were the case, changes in pressure would require small flow rates of fluid through the pressure tap channels as the volume of the trapped air adjusted to a new equilibrium pressure.

The elbow was connected to the fluid power test bench using standard weight pipe nipples. The internal diameter of the nipples was machined such that it matched the channel diameters of the billet elbow, and was also controlled to within 0.013mm. The advantage to using standard nipples was the entrance length could be easily varied; however, the tradeoff was an introduction of a sudden expansion/ contraction in the threaded area where the nipples attached to the billet elbow block. CFD simulations and experimental measurement showed the sudden expansion/contraction would have a minimal effect on the resulting pressure gradient across the elbow.

Considerable experiments were performed using the machined elbow at various flow rates, entrance lengths, and at varying temperatures conditions. Figure 3 demonstrates typical experimental results when standard weight hydraulic oil at 41°C is used as the working fluid. The experimental results show good correlation with the CFX predication at higher flow rates, where the viscosity dependence of the turbine meter is less pronounced. The Euler estimation under predicts the observed pressure difference at all flow rates. Similar results were noted when the smaller 6.35 mm diameter channel was evaluated. In all cases, an alteration to the Euler predication with a simple constant multiplier improved the correlation, but did not result in a relationship that could be used as a calibration curve for the device. This suggests the viscosity effects are more complex, and a more in depth calibration procedure is necessary.

The pipe velocity is higher for the smaller diameter channel for a given flow rate. For the flow rates used in the experimental investigation, this resulted in higher observed pressure difference. Since the errors associated with pressure measurement were only slightly dependant on the observed pressure differential, the use of the smaller diameter channel results in data with a smaller percentage of pressure measurement uncertainty. Therefore, the smaller diameter geometry channel was calibrated for use as the flow measurement device.

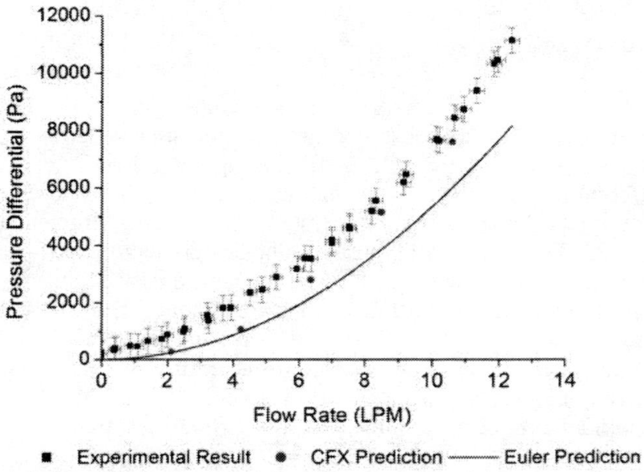

Fig. 3 Typical experimental results obtained with 9.525mm diameter elbow channel

The design of the flow sensing aspect of the device is dependant on the calibration of the pressure-flow relationship In order to determine the calibration of this relationship, the effects of the fluid viscosity and density on the measured pressure difference needs to be known experimentally. To accomplish this task, the system pressure was held constant while the temperature and flow rate was varied using the 6.35 mm diameter channel. Since the density and the viscosity of the fluid can easily be determined as a function of temperature and pressure, this allows for the effect of both density and viscosity to be more easily investigated as a function of temperature alone. The experimental data derived from this procedure is shown in Fig. 4.

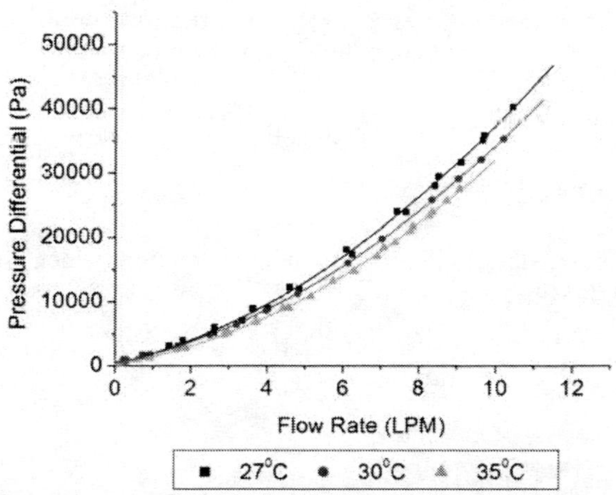

Fig. 4 Experimental results with constant system pressure

The error bars are not shown in Fig. 4. in order to improve the clarity. The magnitude of the error bars in Fig. 4 is similar to the bars shown in Fig. 3. Figure 4 demonstrates that the measured differential pressure is strongly dependant on the fluid temperature. The same behavior was also noted for the larger 9.25 mm diameter channel. When the temperature density effects are compensated, the pressure dependence on viscosity change are on the order of 13%. This does not correlate well with the CFX predication, which was approximately 2%. Some reasons for this deviation include error introduced in the experimental data by the turbine flow meter and small errors associated with the actual viscosity of the oil. In light of this fact, it was understood that CFX could not be used to develop an exact calibration procedure for the sensor that could be substituted for laboratory work. Instead, CFX was used to gain understanding of the flow phenomena, and evaluate the suitability of a given geometry.

In order to characterize the effect of both the temperature and the measured flow rate on the resulting differential pressure measurement, a standard Buckingham Pi analysis of the pertinent variables was performed.[15] For this analysis, the variables of importance were considered to be V, D, ΔP, μ, and ρ. The results of the analysis was two pi terms, one velocity or flow term and one unit-less pressure term. The velocity or flow type term worked out to be the Reynolds number, and the pressure term was simplified to be the ratio of the measured differential pressure to that of the Euler predicted pressures. The pi terms are summarized as Eq. 3.

$$\pi_1 = R_e = \frac{\rho V D}{\mu}$$

$$\pi_2 = \frac{\Delta p_{measured}}{\Delta p_{Euler}}$$

(3)

The reciprocal of the Reynolds number can be plotted versus pi 2. This plot is presented as Fig. 5, and it is quite apparent the result is a simple linear calibration curve for Reynolds numbers greater than 50.

It is expected the case of the infinite Reynolds number (zero viscosity case) would have a pressure ratio term of unity. The least square fit linear interpolation of the data presented in Fig. 5 results in the relationship presented as Eq. 4.

$$\pi_2 = 279.52\left[\frac{1}{R_e}\right] + 0.732$$

(4)

It is quite apparent from the relationship in Eq. 4 the zero viscosity case does not result in a pressure ratio of unity. The difference is most likely attributed to the assumption of a 2d geometry, and a constant velocity profile around the bend of the elbow (invisid flow). However, measurements were taken using hydraulic oil, a fluid which is quite viscous, and then the result was extrapolated to the zero viscosity case. This extrapolation is another likely reason for the difference, as the complex secondary flow phenomena is not considered except in the form of experimental measurement.

An alternative calibration procedure could be done using the temperature as the sole variable responsible for the pressure differences between the runs. However, it is a well established fact that the viscosity and density of hydraulic fluid also vary as a function of pressure [16]. Viscosity can vary by as much as a factor of 2 in the pressure ranges found in typical fluid power systems. If the density and viscosity variables are separated independently as shown, the calibration procedure can easily be used to account for the changes due to pressure. This procedure results in a flow-meter that can accurately measure flow rates at all process pressures and temperatures. The calibration procedure included fitting the data to a known flow rate. While gathering the experimental data, a turbine type flow meter was used to measure the flow rate. It is well established that turbine meters do exhibit some viscosity dependence on their resulting flow rates. Due to lack of documentation for the meter, the amount of variation in the measured flow due to viscosity only was unknown. However, the overall uncertainty of the meter was known, and the associated error bars are presented in Fig. 3. The calibration procedure described in the preceding paragraphs works to calibrate the elbow to this turbine meter. In practice it would be best to develop the calibration curve from a most exacting device, such as a positive displacement, or proving type meter.

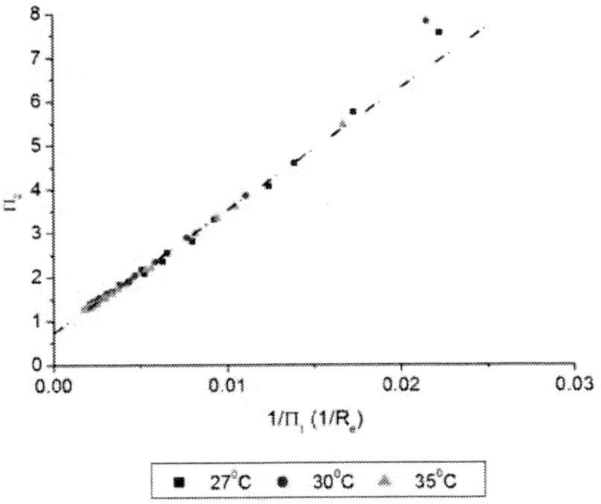

Fig 5. Calibration Curve For 6.35 mm Diameter Elbow

MEMS SENSOR DESIGN AND FABRICATION

The design of the hardware aspect of the sensor is based on the fluid phenomena that was both numerically simulated and experimentally observed. The sensor is to have the capacity to infer a flow rate, and extract the absolute system pressure and temperature.

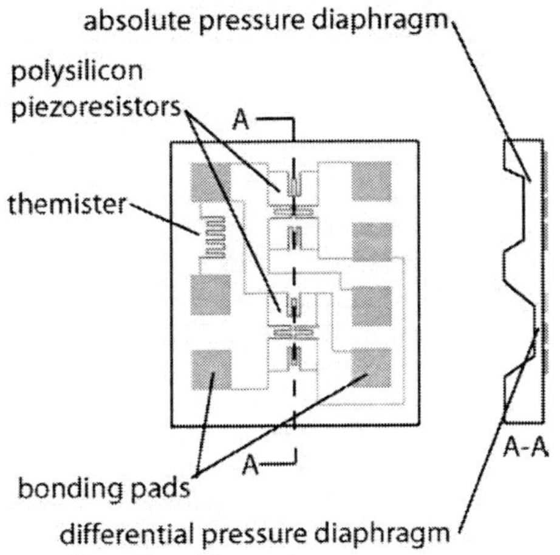

Fig 6. Sensor Functional Layout

Therefore, the hardware of the sensor is to measure two pressures, one small differential pressure to infer the flow rate, and another higher absolute (system) pressure. The design also requires a themister to be used on the die so temperature information can be extracted. The basic layout of the sensor can be seen in Fig. 6.

2.1 Diaphragm Design

In the laboratory investigations, the maximum differential pressure observed across the small elbow was 35kPa. This pressure will be the design maximum for the differential pressure sensing element. Also in the laboratory experiments, the maximum system pressure is limited to 3.5MPa, This value will be used as the design maximum for the absolute pressure sensing element. For both pressure sensors, a design factor of safety of 2 will be used. This factor of safety was selected for the design in case of accidental over pressurization and for uncertainty in material properties. Since the diaphragms are to be made of silicon dioxide, there is a certain amount of uncertainty regarding impurities, pinholes etc in the film.

The design of the pressure sensing diaphragms was done by using conventional numerical means, and also by using commercially available Finite Element Analysis (FEA) software.

In order for conventional numerical means to be used to design the diaphragms, the diaphragms were assumed to be plates of fixed periphery. Using this assumption, simple plate theory can be applied to determine the stress and strain relationships present at all locations on the plate. Development of the plate theory is lengthy, and therefore only the final result will be presented below. Please refer to Szilard and Boresi et al. [17,18] for a complete derivation of this underlying theory.

The general plate equation has no solution for the case of a fixed periphery plate, so alternative methods of solution were employed. One common method to indirectly solve the plate equation is the Ritz Energy method. For this method, all displacements which satisfy the boundary conditions and make the total potential energy of the structure a minimum are sought solutions.

The series given as Eq. (5) shows the solution of the Ritz Energy Method applied to the plate of fixed periphery.

$$\omega(x,y) = \sum_{m}^{\infty}\sum_{n}^{\infty}\frac{W_{mn}}{4}\left[1-(-1)^m\cos\frac{m\pi x}{a}\right]\left[1-(-1)^n\cos\frac{n\pi y}{b}\right]$$

$$n,m = 1,3,5,\ldots$$

(5)

This series satisfies the clamped boundary conditions which are shown as Eq (6).

$$\omega_{x=+-a} = 0$$

$$\omega_{y=+-b} = 0$$

$$\left(\frac{\partial\omega}{\partial x}\right)_{x=+-a} = 0$$

$$\left(\frac{\partial\omega}{\partial y}\right)_{y=+-b} = 0$$

(6)

If only the first term of the series is considered, the expression for the deflection of the plate reduces to the relationship shown in Eq (7):

$$\omega = \frac{W_{11}}{4}\left(1+\cos\frac{\pi x}{a}\right)\left(1+\cos\frac{\pi y}{b}\right)$$

(7)

Minimization of the potential energy allows W_{11} to be determined as shown in Eq (8).

$$W_{11} = \frac{16p_0a^4}{\pi^4 D}\frac{1}{3+3\frac{a^4}{b^4}+2\frac{a^2}{b^2}}$$

(8)

With the deflection of the plate determined, the stress at any arbitrary location can be equated from the relationship shown in Eq. (9).

$$\sigma_x = E\varepsilon_x + \upsilon\sigma_y$$
$$\sigma_y = E\varepsilon_y + \upsilon\sigma_x$$

(9)

Equation (10) shows the relationship between the second derivative of the deflection and the strain of the plate.

$$\varepsilon_x = -z\frac{\partial^2\omega}{\partial y^2}$$

$$\varepsilon_y = -z\frac{\partial^2\omega}{\partial x^2}$$

(10)

The expressions presented as the result of the Ritz method were coded in Matlab in order to produce 3 dimensional renderings of the results, and to more easily compare the results to that of commercial FEA software.

Figure 7 shows the deflection of the top surface of the pressure sensor diaphragm when the cavity is exposed to the maximum design pressure. To generate this figure, conditions matching that of the differential pressure cavity were used. The stress and strain in each of the coordinate directions could also be easily determined.

To simplify the design and manufacture of the sensor, both the differential and absolute pressure sensing diaphragms were designed as squares. With a square design, the stress and stain is symmetric in the respective directions. The strain in the x direction of the plate can be seen in Fig. 8. The stress in the x direction of the plate has a similar appearance to the strain map presented in Fig. 8.

The results shown in Fig. 7 and Fig. 8 were compared to an ANSYS FEA simulation with the same boundary conditions. The Ritz method with only 1 term in the series resulted in a 12% underestimation of the stress along the edges of the plate. This deviation can be attributed to using only 1 term in the series expansion, and also due to the stress concentration that occurs at the edge of the plate in the ANSYS simulation. To circumvent this problem, an additional 12% safety factor was introduced into the code in order to account for the variation.

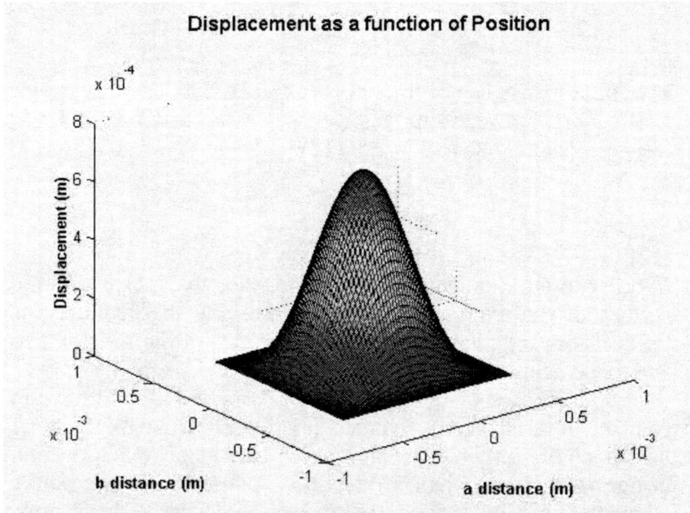

Fig. 7. Diaphragm displacement as a function of position at maximum pressure for the differential pressure measurement cavity

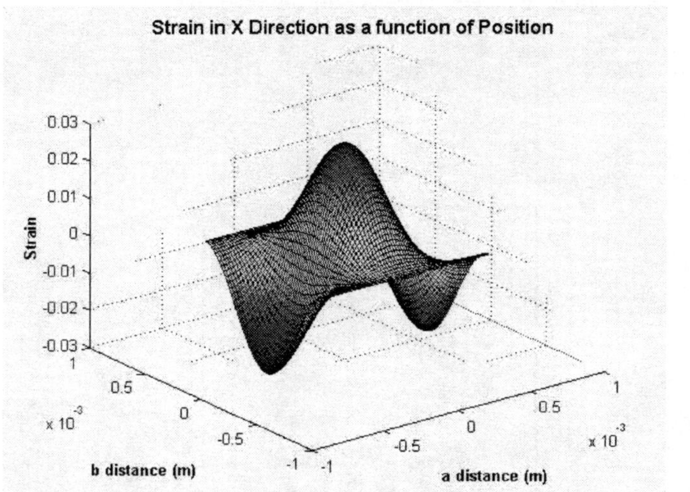

Fig. 8. Strain in the x direction as a function of position at maximum pressure for the differential pressure measurement cavity

2.2 Resistor Placement Optimization

With the thickness and the size of the pressure sensing diaphragms established, the optimum location for the piezoresitive strain elements could be determined. The procedure followed to optimize the placement of the piezoresitive strain elements is similar to that described by Bae et. al. In this procedure, a computer code is written to systematically place the resistors, and then simulate the output of the resulting design. In this way, the placement of the resistors can be iterated such that the maximum sensitivity of the sensor is quickly obtained. [19] Following are the fundamental governing equations that are used to write this code. The general relationship between strain and change in resistance is given by Eq. (11)[20].

$$\frac{\Delta R}{R} = G\varepsilon$$

(11)

The initial resistance of the polysilicon resistors is given by the relationship between the resistivity of the material and the resistor geometry shown in Eq. (12).

$$R = \frac{\rho l}{wt}$$

(12)

In order to programmatically size and place the resistors on the surface of the diaphragm, variables defining the geometry of the resistors are established. The definitions of the resistor size and placement variables are shown in Fig. 9.

Piezoresistive strain sensors operate by converting changes in resistance to changes in voltage with a Wheatstone bridge configuration. The maximum output of a Wheatstone bridge is realized when all resistors in the bridge are active elements, i.e. a full bridge design. For the design of this sensor, the goal is to maximize the sensitivity, so a full bridge design is used. Another advantage to this design is the sensor is less temperature dependant, as all elements of the sensor are in an isotheral environment. Any changes in resistance due to temperature will be reflected on all elements of the bridge, resulting in only small output changes.

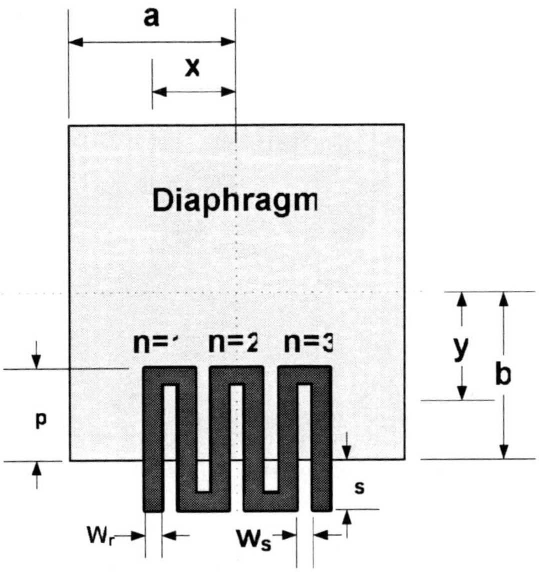

Fig. 9. Resistor Layout Diagram

To implement the full bridge design, two of the piezoresistive elements need to be placed in regions of positive (tensile) strain, while the opposing two elements need to placed in negative (compressive) strain regions. The effect can be understood by studying the general Wheatstone bridge configuration described by Eq. (13) and Fig 10. [21]

$$\frac{V_{out}}{V_{in}} = \frac{R_1R_4 - R_3R_2}{(R_1 + R_3)(R_4 + R_2)}$$

(13)

Equation (11), Eq. (12), and Eq. (13) were combined with the strain map presented in Fig. 8 to programmatically produce a method to predict the output voltage of the sensor. For simplicity, the resistor width values, wr and ws were set to 50um, the smallest feature size that can be cost effectively prototyped.

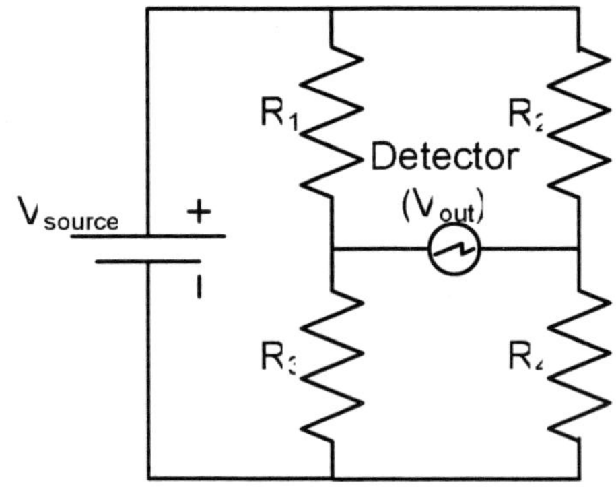

Fig. 10. General Wheatstone Bridge Layout

The length ls of the resistor away from the diaphragm edge was set to 100um for all simulations in order for the connections to the resistors to be made in areas of lower strain. The length puts the connection point far from the areas of maximum strain which occurs at the edge of the plate. In the simulations, this length worked to reduce the overall sensitivity of the sensor, as there was strain simulated only on the deflected diaphragm, and not on the surrounding substrate. In actuality, the strain does not suddenly stop at the edge of the diaphragm, but extends a certain distance beyond. Therefore, it is expected the optimization simulation will under predict the sensitivity of the sensor.

The variable lp, the length of the resistor extending onto the diaphragm, and n, the number of resistor turns were iterated such that the optimum output voltage was obtained. Figure 11 shows the result of the iterations.

It is apparent the maximum output of the sensor is realized when the number of turns of the resistor is 1 and the length ls of the resistor onto the diaphragm area is approximately 200um. From the work of Bae et all [19], it was assumed the overall length of a resistor with this configuration would be too short and result in noisy output signal. Instead, the number of turns of resistor that was selected was 2, and the a length ls of 250 um was selected. From Fig 11 it is apparent the tradeoff to a longer resistor configuration results in very little reduction in sensor sensitivity.

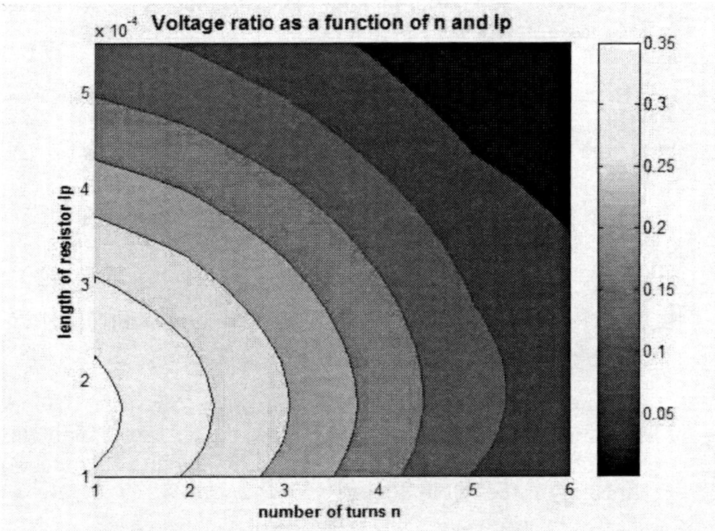

Fig. 11. Resistor placement optimization results

The same simulation program was also used to design the second pressure sensing cavity. This cavity is designed to sense the absolute (system) pressure, and will be subject to pressures of much larger magnitudes. The only difference between the design of the absolute cavity and the differential pressure sensing cavity is the diaphragm material. For the absolute pressure cavity, the material is silicon, whereas with the differential cavity, the diaphragm material was silicon dioxide. Attempts were made to use the thin dioxide material for the absolute sensor, however, the resulting cavity size was too small, and could not be cheaply fabricated

2.3 Temperature Design Considerations

The temperature sensing aspect of the design consisted of placing a resistor of substantial length on the device. The resistance was selected to be a very high value as to minimize self heating errors. The resistance value of the resistor will be calibrated to measure temperature using the Stienhart-Hart polynomial log fit method [3].

The change in resistance of the thermister will be measured with a simple quarter Wheatstone bridge design, with the 3 fixed resistance elements placed in an isothermal environment. A quarter bridge design was selected in lieu of a more sensitive half bridge design in order to keep the number of contacts from the die to the package to a minimum.

2.5 MEMS Fabrication Process

The fabrication sequence for the sensor is shown graphically in Fig. 12 The sequence starts by growing silicon dioxide on both sides of a double polished wafer with a wet thermal oxidation process. After oxidation, LPCVD is used to apply a layer of in situ heavily doped polysilicon, which is used as the piesoresitive strain material. The doped polysilicon is then patterned using a dry etching process to form resistors. A layer of Cr to promote adhesion is evaporated next, followed by the evaporation of a gold layer which is used as the contact medium. The Cr and gold layers are then patterned to form wire traces and contact pads. Next, the attention focuses on the bottom of the wafers. The topside is protected using a combination of a special mechanical holding apparatus and chemical means. The opening for the differential pressure cavity is then etched in the silicon dioxide, and the silicon substrate beneath is KOH etched to a depth of approximately 75 um. The cavity for the absolute pressure sensor is then opened in the silicon dioxide, and then both cavities are simultaneously KOH etched. The cavities are finish etched using DRIE to eliminate the sharp edge stress concentrations that resulted from KOH etching. The advantages of the DRIE process is slightly rounded corners in the cavities, and the ability to accurately control the etch depth and diaphragm thickness. After etching the finished silicon component is then anondicly bonded to glass, which has provisions to allow for the differential pressure measurement.

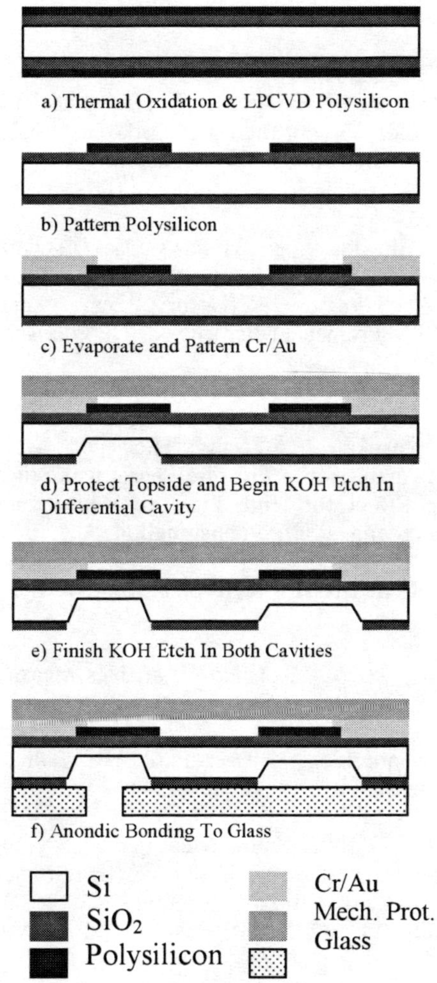

Fig.12 Sensor Fabrication Diagram

Conclusion

The design of a novel pressure, flow and temperature sensing scheme for use in fluid power systems was presented. The scheme required the flow rate in the system to be inferred from small pressure drops that result from flow through preexisting system geometry.

As a basis to prove the validity of the flow sensing concept, standard elbow geometry was evaluated. The geometry was evaluated using numerical, CFD, and experimental methods. It was determined the approach could be easily calibrated and used as an effective flow measurement method in fluid power systems.

After the validity of the flow measurement concept was established, sensor hardware was designed to extract pressure and temperature information from the system. The design of the hardware required coding of several optimization routines in order to design a sensor with the maximum sensitivity. The fabrication process to build the optimized prototype sensors was then presented.

Future work will entail building a prototype sensor device, and evaluating its performance. Based on the performance of the device, some of the hardware design aspects including the overall device size may need to be reconsidered.

ACKNOWLEDGMENTS

The help offered by PhD candidates Wei Xue and Yi Liu has been indispensable. Without their aid, work on the fabrication process would not have gone nearly as smoothly.

REFERENCES

[1] Bistue G., Elizalde J.G., Garcia-Alonso I., Olaizola S., Castano E., Gracia F.J., and Garcia-Alonso A., 1997, "A Micromachined Pressure Sensor for Biomedical Applications," J Micromech. Microeng. 7. IOP Publishing Ltd., UK., pp 244-46

[2] Matsuoka Y., Yamamoto Y., Yamada K., Shimada S., Tanabe M., Yasukawa A., and Hidetoshi M., 1994., "Characteristic Analysis of a Pressure Sensor using the Silicon Piesoresistance Effect for High Pressure Measurements." J. Micromech. Microeng. 5. IOP Publishing Ltd., UK., pp 25-31

[3] Steinhart I.S., and Hart S.R., 1968., Deep Sea Res. 15. pp 497

[4] Webster J.G., 2000 "Mechanical Variables Measurement" CRC Press LLC. Boca Raton FL. pp 9.63-9.69, 9.99-9.115

[5] Lambda Square Inc., 71 Deer Park Ave. Babylon, NY 11702 USA. http://www.lambdasquare.com

[6] Einhellig R. F., 2002, "Flow Measurement Opportunities using Irrigation Pipe Elbows." Hydraulic Measurements and Experimental Methods: 2002 Proceedings of the Specialty Conference. July 28-Aug 1 2002, Estes Park, Colorado. pp. 1111-1118

[7] Sanchez Silva, F., Toledo M. V., Hernandez Ruiz J., 1997. "Experimental Study for the Use of Elbows as Flowmeters" 1997 ASME Fluids Engineering Division Summer Meeting FEDSM97 July 22-26

[8] Replogle, J.A., Myers L.E., F.ASCE, Brust, K.J., 1966 "Evaluation of Pipe Elbows as Flow Meters" J. Irrigation and Drainage. September. IR3 pp. 17-34

[9] Taylor, D.C., McPherson, M.B., 1954. "Elbow Meter Performance." Jour. AWWA. Nov. pp. 1087-1095.

[10] Anderson, A.G., Straub, L.G.1948 "Hydraulics of Conduit Bends" St. Anthony Falls Hydraulic Laboratory University of Minnesota Bulletin No. 1.

[11] Lansford, W.M. 1936. "The Use of an Elbow in a Pipe Line for Determining the Rate of Flow in the Pipe." Bulletin :(University of Illinois(Urbana-Champaign campus) Engineering Experiment Station) No. 289. Dec. 1936

[12] Fox R.W., and McDonald A. T., 1998. *"Introduction to Fluid Mechanics 5th ed."* John Wiley and Sons, Inc. New York, NY. pp 242-45, 345-49

[13] ANSYS Inc. Southpointe. 275 Technology Drive. Canonsburg, PA. 15317. http://www.ansys.com/products/cfx

[14] Shaw, R. 1960. "The Influence of Hole Dimensions on Static Pressure Measurements" J. Fluid Mech. Vol 7 pp. 550-564

[15] Munson B.R., Young D.F., Okiishi T.H., 2002 *"Fundamentals of Fluid Mechanics 4th ed"* John Wiley and Sons, Inc. New York, NY. pp. 388-411

[16] Radhakrishnan M., 2003., "Hydraulic Fluids: A Guide to Selection, Test Methods, and Use." ASME Press. New York, NY pp. 20-22

[17] Szilard, R., 2004 *"Theories and Applications of Plate Analysis: Classical, Numerical and Engineering Methods."* John Wiley and Sons, New York, NY. pp.26-31, 187-196.

[18] Boresi, A., and Sidebottom, O., 1985 *"Advanced Mechanics of Materials."* John Wiley and Sons. New York, NY. pp. 442-452,457-471

[19] Bae B., Flachsbart B.R., Park. K., Shannon M.A., 2004 "Design optimization of a piezoresistive pressure sensor considering the output signal to noise ratio" J. Micromech. Microeng. 14 pp. 1597-1607.

[20] Madou M..J., 2002 *"Fundamentals of Microfabrication: The Science of Miniturization."* CRC Press LLC. Boca Raton, FL. pp 294

[21] Hambley, A.R., 1997 *"Electrical Engineering: Principles and Applications."* Prentice Hall, pp.104

AUTHOR INDEX

FPST-Vol. 12
Proceedings of the ASME
Fluid Power Systems and Technology -- 2005

Alirand, M. 93
Amundson, Kurt R. 27
Barth, Eric J. 175
Book, Wayne J. 99, 111
Boyaghchi, F. A. 139
Burton, Richard 9, 35, 121
Cao, Jianwei 85
Chase, Thomas R. 63
Cui, Tianhong 183
Driscoll, Scott 111
Ehghaghi, M. B. 139
Ellman, A. 131
Eskola, Tero 171
Fales, Roger 51
Gao, Yingjie 73
Groepper, Charlie 183
Gu, Linyi 85
Habibi, Saeid 35, 121
Handroos, Heikki 1, 171
Hao, Peng 165
He, Qinghua 165
Huggins, James D. 111
Ichiyanagi, Takayoshi 45
Jirawattana, P. 17
Kajaste, J. 131
Kauranne, H. 131
Kazerooni, H. 27
Kong, Xiangdong 73
Lebrun, M. 93
Li, Cassie Y. 63
Li, Perry Y. 63, 157, 183
Li, Zeliang 9
Manring, Noah D. 147
Mehta, Viral S. 147
Nikiforuk, Peter 9
Nishiumi, Takao 45, 171
Opdenbosch, Patrick 57
Orand, N. 93
Pietola, M. 131
Qiu, Minxiu 85
Raade, Justin W. 27
Radpukdee, T. 17
Riofrio, Jose 175
Sadegh, Nader 57
Sampson, E. 121
Shenouda, Amir 99
Shojaee Fard, M. H. 139
Stammen, C. 79
Stelson, Kim 183
Wang, Feng 85
Wang, Shu 35
Yousefi, Hassan 1
Yuan, Qinghui 157
Zhang, Daqing 165
Zhang, Qin 73
Zhang, Xinhai 165

Book Number: G01255